PHYCODIVERSITY
Aspects and Prospects

PHYCODIVERSITY
Aspects and Prospects

A Festschrift
to
Professor B.N. Verma

— Editor —
Prem Kumar Prasad
Principal
B.M. College
Rahika
Madhubani, Bihar

2011
DAYA PUBLISHING HOUSE
Delhi - 110 002

ISBN 81-7035-678-4
ISBN 978-81-7035-678-3

Published by : **Daya Publishing House**
4760-61/23, Ansari Road, Darya Ganj,
New Delhi - 110 002
Phone: 23245578, 23244987
Fax: (011) 23260116
e-mail : dayabooks@vsnl.com
website : www.dayabooks.com

Laser Typesetting : **Classic Computer Services**
Delhi - 110 035

Printed at : **Chawla Offset Printers**
Delhi - 110 052

PRINTED IN INDIA

Dr. B.N. Verma

Brief Resume of Dr. B.N. Verma

Dr Bhola Nath Verma was born on 8th June 1943 and was superannuated from the active service as University Professor, University Department of Botany, B. R. A. Bihar University, Muzaffarpur, Bihar (India). He is the Pioneer Phycologist who initiated and propagated Phycological Research in present Bihar. During thirty-three years of active service, he successfully conducted a number of major research projects sponsored by U.G.C., C.S.I.R., D.S.T., Ministry of Environment and Forests under various ecological as well as Man and Biosphere Programme and Ministry of Non-Conventional Source of Energy, Govt. of India, New Delhi. At least twenty research students were awarded on various aspects of basic as well as applied Phycology and limnology. He supervised C.S.I.R. sponsored post-doctoral research and made a basic contribution on the phenolics of fresh water algae in collaboration with Indian Toxicological Research Institute, Lucknow under the C.S.I.R. scheme for developing Countries granted to one of his research student. He conducted a maiden exploration of edible mushrooms in Arunachal Pradesh. He has published more than hundred research papers including establishment of many new taxa of Green Algae in scholarly journals of National and International repute and authored three reference books. In addition to teaching, research and writings, Prof. Verma actively participated in academic deliberations and chaired a number of technical sessions of National as well as International levels Symposia/Conferences besides delivering invited lectures. He has been associated as Life Member of seven learned Societies, and as executive of several of them. He is fellow of four Learned Societies. He was conferred Scientist of the year 2001 award by National Environmental Science Academy of India and was felicited by Tribhuvan University, Kathmandu, Nepal in an International Seminar held at its Birganj Campus in 2002 on the recognition of his outstanding contributions in the field of Plant Science. He was elected member of Indo-Nepal Plant Science Forum under the banner of Tribhuvan University. He was honored by Indian Botanical Society in its 26th Annual Conference held in the year 2003 at Zamia Hamdard, New Delhi and by Phycological Society of India in a National Symposium on Biology and Biodiversity of fresh water algae held in the year 2004 at University of Madras, Chennai.

Publications

1. Verma, B.N., 1966. Cytological studies on four species of *Rhizoclonium* Kuetz. *J.R.U.*, 3: 57–63.
2. Verma, B.N., 1967–68. Chromosome number in four Indian species of *Cladophora* Kuetz. *J.R.U.*, 4–5: 61–64.
3. Verma, B.N., 1968. Artificial induction of polyploidy in *Rhizoclonium hieroglyphicum* (Ag) Kuetz. *var. macromeres. Indian Jour. Sc., India,* 2(2): 99–102.
4. Sinha, J.P. and Verma, B.N., 1968. On the occurrence of amitotic nuclear division in *Pithophora cleveana* Wittr. *Phycos,* 7(1 and 2): 139–143.
5. Sinha, J.P. and Verma, B.N., 1969–70. Effect of growth promoting substances on the rate of nuclear division in *Rhizoclonium hieroglyphicum* (Ag) Kuetz. *J.R.U.*, 6–7.
6. Sinha, J.P. and Verma, B.N., 1970. Cytological analysis of the Charophytes of Bihar. *Phycos*, 9(2): 92–99.
7. Verma B.N. and Verma, S.K., 1976. A note on the cytology and life history of *Chaetomorpha gracilis* Kuetz. *Phycos,* 15(1 and 2): 89–94.
8. Verma, B.N. and Verma, M.P., 1977. A new record of *Oedocladium prescottii* Islam from Bihar. *Phycos,* 16(1 and 2): 51–53.
9. Verma, B.N., 1978. New record of *Dichotomosiphon tuberosus* (A. Br.) Ernst *var. indicum var. nov* from Bihar. *Phycos,* 17(1 and 2): 59–62.
10. Verma, B.N., 1979. Cytological studies in four species of *Pithophora* Wittr. *Cytologia,* 44: 29–38.
11. Verma, B.N., 1980. Karyotype analysis in three species of *Rhizoclonium* Kuetz. *Cytologia,* 45: 433–440.
12. Verma, B.N., 1980. Analysis of induced nuclear division in *Rhizoclonium hieroglyphicum* (Ag) Kuetz. *var. macromeres. Acta Botanica Indica,* 8: 215–218.
13. Verma, B.N., 1981. New record of *Cladophora* Kuetz. From Bihar. *Phycos,* 20(1 and 2): 44–48.
14. Verma, B.N., 1982. Karyotype analysis in three species of *Cladophora* Kuetz. *Cytologia,* 47: 137–145.
15. Dogra, A. and Verma, B.N., 1983. First report of *Gongrosira debaryana* Rabenhorst from India. *Bibliotheca Phycologica, J. Cramer,* 66: 371–372.
16. Verma, B.N. and Kumar, A., 1983. First report of chromosome number in *Vaucheria hamata* Wal. *Phycos,* 22(1 and 2): 1–3.
17. Labh, L. and Verma, B.N., 1983. *T. jwalai*, a new species of *Tolypella* from India. *Bibliotheca Phycologica, J. Cramer,* 66: 379–388.

18. Sinha, B.P. and Verma, B.N., 1983. A fresh water Florideae new to India. *Bibliotheca Phycologica, J. Cramer*, 66: 373–377.

19. Labh, L. and Verma, B.N., 1984. Cytological studies in three species of *Tolypella*. *Cytologia*, 49: 717–723.

20. Verma, B.N., 1985. Karyomorphological studies on three species of *Cladophora* Kuetz. *Cytologia*, 50(1): 25–30.

21. Verma, B.N., 1985. Cytology of *Cladophora liniformis* Kuetz. and *C. fracta* (Dillw.). *Cytologia*, 50 (1): 49–53.

22. Labh, L. and Verma, B.N., 1985. New record of chromosome numbers for the Charophytes. *Cytologia*, 50(1): 55–58.

23. Prasad, Prem K., Sinha, R.S.P. and Verma B.N., 1985. *Some Rare Green Algae from Nepal*. National Academy of Science Letters, India, 8(2): 33–34.

24. Labh, L. and Verma, B.N., 1985. First report of chromosome number in four taxa belonging to the genus *Chara* Linn. *Cytologia*, 50(1): 21–24.

25. Prasad, Prem K. and Verma, B.N., 1985. Aneuploid count for *Chara setosa* Klein ex Willd. *Cytologia*, 50(2): 241–245.

26. Verma, B.N., 1986. Cytotaxonomical studies in the genus *Rhizoclonium* Kuetz. *Cytologia*, 51: 177–183.

27. Verma, B.N., 1986. New counts for *Rhizoclonium riparium* Complex. *Cytologia*, 51: 507–512.

28. Labh, L. and Verma, B.N., 1986. New counts for *Chara fibrosa* Complex. *Cytologia*, 51: 185–191.

29. Verma, B.N. and Labh, L., 1986. Karyomorphological studies in the genus *Chara* Linn. *Cytologia*, 51: 501–506.

30. Verma, B.N. and Jha, C. N., 1988. Effect of diethyl sulphate on *Rhizoclonium* Kuetz. *Cytologia*, 53: 283–286.

31. Verma, B.N. and Prasad, Prem K., 1989. On the cytology of *Microthamnion strictissimum* Rabenh. from Muzaffarpur, Bihar. *Phycos*, 28(1 and 2): 251–253.

32. Jha, C.N. and Verma, B.N., 1989. Non-nitrogen fixing cyanobacteria as biofertilizer. *Advances in Applied Phycology*, 11: 153–157.

33. Shukla, C.P. and Verma B.N., 1989. Estimation of total intracellular phenolics and their leachates in culture filtrate during different growth phases of *Oscillatoria subbrevis*. *Environment and Ecology*, 7 (3): 730–732.

34. Shukla, C.P. and Verma, B.N., 1989. Estimation of total intracellular phenolics during different growth phases of *Chroococcus minor* (Kuetz.) Nag. *Biojournal*, 1(2): 111–112.

35. Sahu, K.K. and Verma, B.N., 1990. New records of *Pithophora* Wittr. from Bihar. *Environ. and Ecol.*, 8(3).

36. Kumar, A. and Verma, B.N., 1990. Cellular and extracellular nitrogen content of two N_2 fixing cyanobacteria. *Proc. Nat. Symp. Cyanobacterial Nitrogen Fixation*, p. 163–167.

37. Shukla, C.P. and Verma, B.N., 1990. Quantitative screening of intracellular phenolics during different growth phases of *Synechocystis pevalekii* Ercegovic. *Environ. and Ecol.*, 8(3).

38. Sahu, K.K. and Verma, B.N., 1990. Comparative studies on the effect of chloramphenicol on a nitrogen fixing and non-nitrogen fixing cyanobacteria. In: *Proc. Nat. Symp. Cyanobacterial Nitrogen Fixation*, p. 175–182.

39. Shukla, C. P. and Verma, B.N., 1990. Cellular N_2 content and leaching with reference to some non heterocystous cyanobactria. In: *Proc. Nat. Symp. Cyanobacterial Nitrogen Fixation*, p. 169–173.

40. Verma, B.N., Sinha, R., Prasad, Prem K. and Das, H.N., 1990. Blue green algae from the paddy fields of North Bihar. *Phykos*, 29(1 and 2): 55–56.

41. Das, H.N. and Verma, B.N., 1990. Persistent effects of phosphamidon on survival, growth and heterocyst frequency of *Nostoc linckia. Environ. and Ecol.*, 8(4): 1142–1146.

42. Kumar, A. and Verma, B.N., 1990. Cellular and extracellular nitrogen content of two N_2 fixing cyanobacteria. *Phykos*, 29(1 and 2): 57–62.

43. Verma, B.N. and Sahay, A. P., 1991. Algal phenolics: A protocol. In: *Microbes and Environment*, (Eds.) A.B. Prasad and R.S. Bilgrami. Narendra Publishing House, Delhi, p. 51–74.

44. Das, H.N. and Verma, B.N., 1992. Effect of phosphamidon on non N_2 fixing blue-green algae, *Synechocystis aquatilis* Sauv. *Phykos*, 31(1 and 2): 151–157.

45. Sahay A.P., Das, P.K. and Verma, B.N., 1992. Studies on the algal flora of Nepal–I, Chlorophyceae. *Geophytology*, 20(2): 155–158.

46. Singh, M.P. and Verma, B.N., 1993. Interactions amongst the abiotic factors of a fresh water lake. *IBC*, 10(1).

47. Sahay, A.P., Das, P.K. and Verma, B.N., 1993. Studies on the algal flora of Nepal–II, Cyanophyceae and Euglenophyceae. *Geophytology*, 23(1): 181–183.

48. Das, H.N. and Verma, B.N., 1993. Limnology of a pond and a lake at Rajnagar (Madhubani), Bihar. *Biojournal*, 5(1 and 2): 41–46.

49. Das, H.N., Prasad, Prem K. and Verma, B.N., 1994. Methods for ascertaining relative spectrum frequency and rejuvenation potency of a soil algae. *Indian Bot. Soc.*, 73: 35–39.

50. Kund, K. and Verma, B.N., 1994. Biological assessment of algal bloom with special reference to phenolics. In: *Advances in Plant Sciences Research*, Vol. I, (Ed.) K.C. Sahni. International Book, Dehradun, India, 1: 165–178.

51. Das, H.N. and Verma, B.N., 1994. Post treatment impact of carbofuron on *N. linckia* and *S. aquatilis. National Bull. Env. Sci.*, 12: 26–28.

52. Kund, K. and Verma, B.N., 1994. Study on the phytoplankton diversity indices of a fish pond. *National Bull. Env. Sci.*, 12.

53. Sahay, A.P., Das, P.K. and Verma, B.N., 1995. Changes in the content of phenolic substances during growth of a cyanobacterium *Oscillatoria surviceps var. angusta*. *Phykos*, 34(1 and 2): 9–16.

54. Verma, B.N. and Das, S.N., 1997. Algal flora of Chitwan and Nawalparasi district of Nepal. *Phykos*, 35(1 and 2): 119–128.

55. Verma, B.N., 1998. Charophyta: Morphotaxonomy and cytotaxonomy. In: *Advances in Phycology*, (Eds.) B.N. Verma *et al.* APC Publication Pvt. Ltd., New Delhi, pp. 173–176.

56. Verma, B.N. and Vidyavati, 1999. Cladophorales: An overview over morphotaxonomy *vis-à-vis* cytotaxonomy. In: *Recent Trends in Algal Taxonomy*, (Ed.) Vidyavati. APC Publication Pvt. Ltd., New Delhi, pp. 101–126.

57. Verma, B.N. and Sinha, A.K., 2002. Studies on the interaction among soil algae *in vitro*. In: *Proceedings of International Symposium on Microbial Biotech. for Sustainable Development*, Jabalpur. Scientific Publisher, Jodhpur.

58. Das, S.N., Verma, B.N. and Vishwanathan, P.N., 2003. Environmental degradation and algae. In: *Phycology: Nature and Nurture*, (Eds.) B.N. Verma *et al.* Kalyani Publication, Delhi, pp. 287–299.

59. Prasad, Prem K. and Verma, B.N., 2003. Exploration of BGA as a possible mean of biological control against nematode disease incidence. In: *Phycology: Nature and Nurture*, (Eds.) B.N. Verma *et al.* Kalyani Publication, Delhi, pp. 197–206.

60. Verma, B.N., Singh, M.P. and Labh, L., 2003. Post-treatment effects of EMS on *Chara* Linn. *Phycology: Nature and Nurture*, (Eds.) B.N. Verma *et al.* Kalyani Publication, Delhi, pp. 273–285.

Foreword

I am happy to learn that a compendium is going to be published on superannuation of Prof. B. N. Verma to whom I have been associated for more than four decades interacting each other at various occasions of academic deliberations. We also worked together while reviewing the cytotaxonomical works done on *Cladophorales*.

The present compendium embodied 32 original papers and review articles divided in two volumes–*Phycocodiversity: Aspects and Prospects* and *Biosphere: Forms and Functions*. It is very thoughtful of the editor Dr. P.K. Prasad to dedicate his efforts to his teacher and guide Prof. B.N. Verma.

Prof. Verma is a committed researcher and dedicated teacher. Besides, he is popular among contemporary researchers. Involvement of more than 75 eminent research scientists contributing articles on different aspects of Phycology, Mycology, Microbiology, Biotechnology of higher plants, Ethnobotany, Radiation Botany, Limnology, Environmental Science etc. in the compendium is the testimony of wide contacts of Prof. Verma. It is his *tour-de-force* which initiated and established a team of young phycologists in present Bihar.

Phycocodiversity: Aspects and Prospects have 16 chapters written by eminent phycologists. Each chapter is well written and quite informative. I am confident that the students, researchers and teachers will be benefited from it.

I congratulate Dr. P.K. Prasad for successful publication of this compendium and thanks for the pain underlying tiring job of editor.

I wish healthy and prosperous life to Prof. B.N. Verma.

Prof. (Mrs.) Vidyavati

Ex Vice-Chancellor
Kakatiya University
Warangal, A.P.

Message

Professor M.N. Noor

UGC Visiting Professor, Department of Botany,
Ranchi University, Ranchi – 834 008
Email: m_n_noor@yahoo.com
Tel.: 0651-2491689, 9431597668 (M)

Some Glimpses from the Memory Lane

When I go down to the memory lane, I become nostalgic as certain wavelengths flash out automatically on my mental spectrum pertaining to Professor Bhola Nath Verma with whom I have been emotionally attached for over 4 decades from now. I shall try to catch up just a few flashes of the past golden moments in this brief write-up expecting the readers to kindly bear with me for a little bit historical backdrop of the laboratory where Prof. Verma initiated his scientific career in 1965 under the generous supervision of legendary Phycologist, Prof. Jwala Prasad Sinha.

After completing doctoral work on Cytology of Cladophorales and Oedogoniales in 1958 under the guidance of Professor M.B.E. Godward (London University), Prof. Sinha returned back to Patna University and finally shifted to *Ranchi University* in the year 1961. At that time, research activity in the Department of Botany was a far cry and the scientific infrastructure was virtually dismal. Bestowed with an ardent visionary aptitude, Prof. Sinha established Phycology Laboratory with full facilities in the Department which eventually emerged out as one of the Centres of Phycological Research in the country. In view of his substantial contributions in the field of algal studies., he may rightly be called as *'Father of Algology'* in erstwhile Bihar State of India.

With the facilities available at hand, Prof. Verma joined the laboratory and started working on 'Cytology of Cladophorales'. I have personally seen him virtually playing with chromosomes and I do perfectly recollect his musical vibrations whenever he prepared a well-quashed metaphasic plate in the lab. The moment was really exciting. Basically, he is an Algal Cytologist and his articles, published in reputed Journals, have been well received by the scientific community. I must congratulate him for his work on algal cytology which, my opinion, is quite a difficult proposition.

In the end, I do hope that this Commemoration Volume with around 32 articles published in 2 volumes in honour of Prof. Verma would be interesting and useful to Plant Scientists.

Message

A. Vaishampayan

Professor of Microbial Genetics and Biological Nitrogen Fixation,
B.H.U., Banaras, U.P.
Phone: 0542-6702542/2575358; +91-9415201138;
E-mail: vaishampavan geneticist@yahoo.co.in
prfssr.vaishampavan@gmail.com

Dear Dr. Prem Kumar,

I share my brief views about the academic accomplishments of Prof. B.N. Verma, as follows:

"I am pleased to place on record the glory of my profound academic associations with Professor B.N. Verma from as back as 1969,(when he was an Associate Lecturer at L.S. College Muzaffarpur and I was an I.Sc. student), till date, when he is not in an active service but I have travelled to my present position by virtue of the key of science enlightened in my mind as one of his favourite undergraduate and postgraduate disciples. In between, apart from being one of the most popular and painstaking teacher (Lecturer, and later as a Reader and Professor) of Phycology under the Bihar University services, Professor Verma was ever regarded as an excellent and innovative, researcher on the cytochemical, physico-chemical and biochemical studies of algae with the production of a large number of quality Ph.D. thesis and research publications in referred scientific journals with the receipt of due recognitions far and wide. I was always amazed to find him talking sweet tounged solely in and around the subject of research and nothing else, proving that adding to the knowledge jewel is the greatest earning of life. I am sure this chain of meaningful scientific addition will continue through his brilliant students and students' disciples for time immemorial to keep his name as a noted scholar of Botany, in general, and Algology, in particular. I wish for his long, healthy and active life."

Message

Dr. (Mrs.) Pushpa Srivastava

Emeritus Fellow–UGC,
Algal Biotechnology Laboratory,
Department of Botany,
University of Rajasthan, Jaipur

Prof. B.N. Verma of Department of Botany, University of Muzaffarpur (Bihar) is a close associate of mine. My association dates back to 1999, when he was delivering his invited lecture in the Conference organised in the honour of Prof. G.S. Venkataraman of Indian Agricultural Research Institute (PUSA) to mark his superannuation. Prof. Verma is a doyen of phycology and was engaged actively in his chosen field–The blue green algae, so dear to his heart. In addition to algae he has equal fascination for higher plants. His collection of plants including a wide variety of cacti stand a witness to his overall interest in plants as a whole. His publications have found a place in National and International Journals both. More than his contributions in the field of phycology, he is known for his courteous nature, open heartedness and ever smiling personality. In person, I have enjoyed his brotherly affection, timely advice and encouragement in the moments of pain and pleasure both. To be a scientist is different from being a noble person, Dr. Verma combines both these God given qualities. If it is true that behind every man there is a women, it is very much true for Mrs. Krishna Verma–simple, affectionate caring and sharing, ever smiling and back bone of her husband. I wish a very healthy prosperous and long life to the couple.

Message

Prof. N. Anand, D.Sc.,

Chief Editor, PHYKOS,
Vice Chancellor, Vels University
Pallavaram, Chennai – 600 117

Dear Dr. Prem,

I am very much delighted to know that a Commemoration Volume is being brought out in honour of my friend and senior Phycologist of our country Prof. B.N. Verma. Prof. Verma was well known to my teacher Prof. T.V. Desikachary , and the beginning of our association dates back to 1970's at the CAS in Botany, University of Madras, Chennai. Prof. Verma's contribution to phycological research especially cytological studies on algae are significant. His keen interest in the Biodiversity of green algae has revealed the presence of a number of new green algal taxa. He has also been one of the leading Limnologists of the country. I have known Prof. Verma earlier through his publications and later during phycological meetings. He has acquired the distinction of being a great teacher and a serious researcher. I congratulate the authors of the volume and Dr. Prem Prasad for their efforts in compiling the Commemorative publications.

Campus: Velan Nagar, P. V. Vaithiyalingam Road, Pallavaram, Chennai - 600 117, India Phone: (91-44) 2266 2500/01/02/03 Fax: (91-44) 22662513 Admn. Office: 521/2, Anna Salai. (Opp. G.R. Complex), Nandanam, Chennai - 600 035. Phone: 2431 5541 /42 Website: www.velsuniv.org E-mail: velscollege@gmail.com.

Preface

While celebrating superannuating of our revered teacher and guide Prof. B. N. Verma, sentiments and emotions broke down. An idea flashed down at the moment to present something special to our guru which could be memorial, a token of respect and befitting to his scientific temperament. Publication of a volume in his honor/ dedicated to him was instantly considered the only way to mark our respect to this mentor. The drive was on to collect contributions from contemporary phycologists with long working experience in the respective domain. It was a difficult time for the editor to select out of the contributions for the comprehensive volume. Participation of some close associates of Prof. Verma, however, could not be ignored. It was this reason that the conservative approach on Algae had to be dropped and a broader perspective was undertaken to accommodate these contributions.

The present compendium contains altogether nineteen research articles of which eleven are devoted to basic and fundamental aspects of Cyanobacteria and Algae. The biodiversity status of Chilka lake, algal flora in polluted environments and the botanical ramble among members of Cyanobacteria in natural and artificial symbioses speak of an in depth observations of expert in the field. The systematic study and phylogenetic treatment of the mixotrophic Euglenineae has also been discussed. Besides these topics of fundamental interest, biotechnological promises, soil reclamation and some experimental works concerning nitrogen metabolism, antibiotic and radiation effect of Cyanobacteria find a place in this volume. Remaining eight articles are concerned with higher plants and heterotrophic fungi. The ethno medicinal importance of some common vascular Cryptogams, the Pteridophytes, in the Aravalli range of Rajasthan have been discussed. Some horticultural crops and their biotechnological renaissance have been looked into. Spices cultivated in India and their known anti-microbial activities have been enumerated. Developing weevil resistance in an economic plant like Sweet Potato through tissue culture technique is an important achievement in biotechnology. This has been discussed at length. The seedling diseases due to storage fungi of the crop seeds, the associated metabolic disorder and the study of mycoflora in plants belonging to Umbelliferae have been well described. The structure and behavior of eucaryotic chromosomes in Radish and over exploitation of Victoria Park in Gujarat causing the degradation of the forest are some of the contributions in the present volume.

The present treatise is thus hetero-organic ranging from procaryotic algae to Vascular Cryptogams to higher plants on one hand and on the other from mixotrophic forms to heterotrophs.

Editor is indeed grateful to authors and experts in the field for generously contributing their papers and thoughtfully extending all out co-operation, which made this effort a success.

Prem Kumar Prasad

Contents

Chapter 1

Nitrogen Metabolism in Cyanobacteria

*Mayashree B. Syiem, Arvind Kumar Singh, Amar Nath Rai**

Department of Biochemistry, North-Eastern Hill University, Shillong – 793 022, Meghalaya

Introduction

Although, molecular nitrogen is abundant in the earth's atmosphere, it gets incorporated into the biosphere only through assimilation by microorganisms and plants. Wide range of nitrogenous compounds in inorganic ion forms such as nitrate, nitrite or ammonia as well as simple organic forms like urea, amino acids and some nitrogen containing bases etc. can be metabolized by various organisms. In addition, some prokaryotes are capable of fixing atmospheric nitrogen using nitrogenase enzyme.

Among them cyanobacteria are ancient and morphologically diverse group of gram negative eubacteria, many of which combine two seemingly mutually exclusive processes: O_2-evolving oxygenic photosynthesis and O_2-sensitive nitrogenase dependant nitrogen fixation.

The two processes are separated either temporally in non-heterocystous forms (where alternating cycles of nitrogen fixation and photosynthesis take care of the problem) or spatially, in heterocystous forms (in which case, the nitrogen fixation occurs in heterocysts and photosynthesis in the vegetative cells). Thus, cyanobacteria

* Corresponding Author E-mail: rai_amar@hotmail.com

have evolved an efficient way to protect O_2-sensitive nitrogenase enzyme from O_2 evolved during photosynthesis.

Even though, cyanobacteria are a self sufficient group of microbes that are capable of independent existence, many enter into symbiotic associations with a wide range of plants (from algae to angiosperms) and with some animals (Rai,1990; Schenk, 1992; Warner, 1992; Loffelhardt and Bonhert, 1994; Meeks, 1998; Rai *et al.*, 2000). Symbiosis, as a mode of existence, exerts considerable changes on the cyanobiont's metabolism. In cyanobacterial-plant symbioses, the cyanobiont is almost always diazotrophic, fixing and providing fixed nitrogen to its eukaryotic partner (Stewart *et al.*, 1983; Rai, 1990; Rai *et al.*, 1996). This necessitates adapting morphological, physiological and biochemical changes to bring about higher rate of nitrogen fixation.

Many nitrogen fixing cyanobacteria are also capable of metabolizing a wide range of combined nitrogen sources, the most common being nitrate, nitrite, ammonia urea and atmospheric nitrogen. Many are also capable of utilizing amino acids such as glutamine, arginine and asparagines as sole nitrogen source to various degrees(Herrero and Flores, 1990; Flores and Herrero, 1994 and Herrero *et al.*, 2001). In these organisms there exist systems of hierarchical preferences in the order in which the various inorganic nitrogen sources are assimilated when present in combination. For example, when ammonium is present in the growth medium the nitrate assimilating system gets expressed at very low levels.

Nitrogen Fixation in Heterocystous and Non-heterocystous Forms of Cyanobacteria

Morphologically cyanobacteria can be grouped into various categories (Table 1.1). In non-heterocystous forms of cyanobacteria, nitrogenase enzyme responsible for catalyzing conversion of molecular nitrogen to ammonium is expressed in all the cells while in heterocystous forms the enzyme is localized in the heterocysts. The differentiation of heterocysts from vegetative cells is a nitrogen-regulated process in cyanobacteria and is accompanied by morphological, biochemical and genetic changes (Wolk *et al.*, 1994). The presence of externally available fixed nitrogen sources such as ammonia, nitrate, nitrite, urea and some amino acids repress heterocyst differentiation (Wolk *et al.*, 1994), while their absence leads 5-15 per cent of the vegetative cells to differentiate into heterocysts. The enzyme nitrogenase which functions exclusively in heterocysts under aerobic conditions is the conventional *Mo*-dependent nitrogenase (*Nif1*) (Elhai and Wolk, 1990 and Thiel *et al.*, 1995) made up of two different proteins. The *Mo-Fe* protein (dinitrogenase) is a α_2 β_2 tetramer and its α and β subunits are encoded by *nifD* (Lammers *et al.*, 1983; Golden *et al.*, 1985) and *nifK* (Mazur and Chui, 1982) genes respectively. The other component of the nitrogenase complex is *Fe*-protein (dinitrogenase reductase), which is a dimer of two identical subunits coded by *nifH* gene. There are reports of significant DNA rearrangement during heterocyst differentiation from vegetative cells in *Anabaena* sp. strain PCC 7120. In heterocysts of this cyanobacterium *nifHDK* is contiguous but an interruption of 11kb DNA fragment is found in *nifD* gene in vegetative cells(Golden *et al.*, 1985). A second rearrangement is also known to occur in the process of heterocyst differentiation deleting a 55kb DNA fragment located in *fdxN* gene (bacterial type ferrodoxin gene whose function is not known in cyanobacteria) (Golden *et al.*, 1987). These rearrangements involve site-specific excisases encoded by *xisA* (Lammers *et al.*, 1986) and *xisF* (Carrasco *et al.*, 1994), respectively. Under anaerobic conditions *Anabaena variabilis* ATCC 29413 have also been shown to possess another *Mo*-dependent nitrogenase (*Nif*2) functioning in the vegetative cells (Thiel *et al.*, 1995 and Thiel and Pratte, 2001). Two other alternative nitrogenases–one vanadium-dependent nitrogenase encoded by *vnfDGK* genes and one *Fe*-only-nitrogenase have also been reported in *Anabaena variabilis* (Elhai and Wolk, 1990; Kentemich *et al.*, 1991). Nitrogenase enzyme shows extreme sensitivity to O_2 exposure which manifests both at the level of synthesis and activity (Weare and Benemann, 1974; Gallon and Chaplin, 1987; Gallon, 1992; Rai *et al.*, 1992). Studies on the strategies for O_2 protection during aerobic nitrogen fixation show that there is a temporal separation between N_2-fixation and photosynthesis (Wolk *et al.*, 1994). Under microaerobic to anaerobic conditions, non-heterocystous cyanobacterium *Plectonema boryanum* also shows a temporal separation of nitrogenase activity and net O_2 evolution due to photosynthesis (Weare and Benemann, 1974). In these cells nitrogenase protein is found to be irreversibly inactivated and degraded on exposure to air and nitrogenase activity regains only after fresh synthesis of the protein (Rai *et al.*, 1992). Even though the nitrogenase protein is ~15 fold higher in non-heterocystous cyanobacteria than in the heterocystous forms, the nitrogenase activities are found to be similar (Rai, 1998). This indicates that much of the nitrogenase protein in non-heterocystous cyanobacteria is not involved in N_2-fixation (Stal and Bergman, 1990; Rai *et al.*, 1992) and may be involved in reduction of O_2 as in *Azotobacter* (Thorneley and Ashby, 1989).

Table 1.1: Nitrogen-fixing Cyanobacterial Species

Sl.No.	*Type*	*Representative Species*
I	Unicellular species showing budding or binary fission:	
	(*i*) All strains are aerobic N_2-fixers	*Gloeothece, Synechocystis*
	(*ii*) Some strains are aerobic N_2-fixers	*Synechococcus, Cyanothece*
	(*iii*) Microaerobic or anaerobic N_2-fixer	*Gloeocapsa*
II	Unicellular species showing multiple fission: All strains show microaerobic/anaerobic N_2-fixation	*Dermocarpa, Pleurocarpa, Chroococcidiopsis, Xenococcus, Myxasarcina*
III	Nonheterocystous filamentous species:	
	(*i*) Anaerobic/microaerobic N_2-fixers	*Plectonema, Lyngbya, Phormiditim, Pseudoanabaena*
	(*ii*) Some strains capable of aerobic fixation	*Oscillatoria, Microcoleus, Trichodesmium*
IV	Heterocystous filamentous aerobic N_2-fixers dividing in one plane	*Anabaena, Nostoc, Scytonema, Cylindrospermum, Nodularia, Calothrix*
V	Heterocystous filamentous aerobic N_2-fixers dividing in more than one plane	*Chlorogleopsis, Fischerella*

The relationship between the heterocysts and the vegetative cells is of mutualistic metabolite interchange. The vegetative cells lack the capability to fix atmospheric nitrogen and hence rely on the heterocysts for supply of fixed nitrogen. On the other hand, the vegetative cells transfer carbohydrates to the heterocysts. The exact nature of the translocated carbohydrate is not yet established although glucose, sucrose, maltose etc. are likely candidates (Table 1.2). The carbohydrate metabolism in the heterocysts is through oxidative pentose phosphate pathway as well as via glycolytic pathway generating pyruvate (Gallon and Chaplin, 1987 and Haselkorn, 1978).

Table 1.2: Effect of Carbohydrates on Heterocyst Frequency and Nitrogenase Activity in *Nostoc* ANTH

Growth Medium	*Heterocyst Frequency (%)*	*Nitrogenase Activity (nmol C_2H_2 Reduced μg^{-1} Chl a h^{-1})*
N_2-medium	13.95	3.09
N_2-medium + 15 mM sucrose	17.50	8.65
N_2-medium + 50 mM fructose	19.02	9.05
N_2-medium + 50 mM glucose	21.64	12.85
N_2-medium + 50 mM maltose	21.02	4.19
N_2-medium + 50 mM lactose	18.77	4.79
N_2-medium+ 10 mM galactose	17.00	3.55

One of the characteristics of heterocystous cyanobacteria is that they enter into symbiotic associations with other plants (Table 1.3).

Table 1.3: N_2-Fixing Symbioses Involving Cyanobacteria

Host (Plants)	*Cyanobacteria*	*Comments*
Angiosperms *Gunnera*	*Nostoc* sp.	Cyanobacterium is located inside host cell in stem nodules. Symbiosis is intracellular.
Gymnospenns *Cycads (e.g. Cycas, Macrozamia)*	*Nostoc* sp.	Cyanobacterium is located in the cortical zone of the coralloid roots. Symbiosis is intercellular.
Pteridophytes *Azolla*	*Nostoc* sp. (usually called *Anabaena azollae)*	Cyanobacterium occupies the mucilage filled cavities on the ventral surface ofthe dorsal lobes of the leaves. Symbiosis is intercellular.
Bryophytes		
Hornworts *Anthoceros* *Notothylus* Liverworts:	*Nostoc*	Cyanobacterium occupies mucilage filled cavities on undersurface of hornwort/liverwort gametophytic thallus.
Blasia *Cavicularia*	*Nostoc*	Symbiosis is intercellular.
Mosses *Sphagnum*	*Nostoc*	Cyanobacterium occupies the hyaline cells of the moss.

Contd...

Table 1.3–Contd...

Host (Plants)	*Cyanobacteria*	*Comments*
Fungi: Lichens	*Nostoc,* *Scytonema,* *Fishcerella,* *Calothrix,* Unicellular cyanobacteria	N_2-fixing lichens consist of two-membered associations between fungi and cyanobacteria or three-membered associations containing a green alga as well. Symbioses are intercellular and are referred as cyanolichens. Of 18,000 documented species of lichens about 8 per cent are cyanolichens.
Algae Marine diatoms *(Rhizosolenis,* *Hemiaulus)*	*Richelia,* *Calothrix*	Cyanobacterium is located in the periplasmic space.
Freshwater diatom *(Rhopalodia)*	Unicellular cyanobacteria	Cyanobacterium appears as bluish green inclusions in the cytoplasm.
Marine sponges 38 genera of Calcarea and *Desmospongia* groups.	*Aphanocapsa* *Phormidium*	Cyanobacterium occurs inter or intracellularly through out the sponge tissue and/or superficial tissues. Symbioses are mainly restricted to the photic zone only.
Prokaryotes Various non-photosynthetic bacteria	Various non-heterocystous and heterocystous cyanobacteria	Cyanobacteria act as the host. The bacterium occupies the mucilage sheath. Bacteria are clustered around the heterocysts or heterocyst-vegetative cell junctions in heterocystous forms. The symbiont occurs intracellularly in *Pleurocapsa minor.*
Non-photosynthetic protists Amoeba *(Paulinella)* Glaucophyta *(Cyanophora)*	*Cyanelle*	Intracellular inclusions resembling unicellular thin-walled cyanobacteria referred as 'cyanelles'.

In symbiosis, the demand for fixed nitrogen is far higher than in the free-living state as the cyanobionts is almost entirely responsible for provision of the total fixed nitrogen requirement of the whole symbiosis. To meet such a demand, cyanobacteria in symbiosis differentiate into much higher heterocyst frequency (in some cases up to 80 per cent). In free-living conditions, nitrogenase activity of a cyanobacterium shows a linear increase with the increase in heterocyst frequency (Table 1.4).

Table 1.4: Linear Relation Between Heterocyst Frequency and Nitrogenase Activity of Few Cyanobacterial Species

Cyanobacterial sp.	*Heterocyst Frequency (%)*	*Nitrogenase Activity (nmol C_2H_2 Reduced μg^{-1} Chl a h^{-1})*
Nostoc linkia	4.52	1.21
Anabaena doliolum	4.75	1.52
Anabaena cycadae	7.69	2.46
Nostoc ANTH	13.95	3.09

Increase in heterocyst frequency of a cyanobionts depends on its role in the symbiosis and on the availability of carbon nutrition in the form of fixed carbon while in symbiosis. In those symbioses where the cyanobionts transfers fixed-N to its partner(s) and in return receive fixed-C, the heterocyst frequency is highest. Cyanobionts supplying fixed-N but not receiving any fixed-C in return come next followed by the cyanobionts that supply both fixed-N and fixed-C to its partner(s). In the last case, the heterocyst frequency of the cyanobionts is almost similar to its free-living counterpart. Microscopic studies have revealed that double and multiple heterocysts occur beyond a heterocyst frequency of 30 per cent. Studies have revealed that the nitrogenase activity of a cyanobiont shows a linear relationship with increase in single heterocysts (as in the case of free-living cyanobacteria) rather than to the total heterocyst frequency. Immunogold studies have shown that all heterocysts including the multiple ones possess nitrogenase protein (Rai *et al.*, 1989). As seen in the Table 1.5, in the case of *Anthoceros punctatus-Nostoc* symbiosis, the old *Nostoc* cyanobiont colonies isolated from the base of the *Anthoceros* gametophyte have maximum heterocyst frequency. Under the microscope they also reveal occurrence of maximum number of double and multiple heterocysts. Yet nitrogenase activity was much higher in the mature colonies where the ratio of single heterocysts to double or multiple heterocysts is much higher. The reason for lower nitrogenase activity in these colonies despite the existence of nitrogenase protein in all heterocysts could be due to restricted availability of carbohydrate from the vegetative cells. In the case of multiple heterocysts, the heterocysts in the middle of a row are not in contact with any vegetative cells at all and those at the end of the chain are connected only to one vegetative cell as against single heterocysts that are connected to two vegetative cells directly on both sides. Hence, the availability of proper amount of photosynthate is a predisposed criterion for efficient nitrogenase activity. The whole focus of the cyanobiont in symbiosis is towards higher rate of nitrogen fixation which creates a much higher demand for fixed-C as a source of energy and reductant for nitrogenase activity (Rai, 1998). The reductant in the heterocyst for N_2-fixation is ferrodoxin, although under iron deficient condition flavodoxin is used as a substitute for ferrodoxin (Gallon and Chaplin, 1987). Heterocysts lack photosystem II activity and hence there is no generation of reductant from water in the non-cyclic electron transport. Generation of reductant by cyclic electron transport involving light and photosystem I is possible, provided there is a suitable source of electron. H_2, via the action of hydrogenase is a possible source of reductant for N_2-fixation in heterocysts. Apart from that, by losing photosystem II, heterocysts protect nitrogenase from O_2 exposure that irreversibly inactivates nitrogenase.

Table 1.5: Correlation Between Heterocyst Frequency and Nitrogenase Activity with Age in the Cyanobiont *Nostoc* ANTH Isolated from Different Parts of the *Anthoceros* Gametophytic Thallus

Portion of the Gametophyte	*Heterocyst Frequency (%)*	*Nitrogenase Activity (nmol C_2H_2 Reduced mg^{-1} dry wt h^{-1})*
Tip (young colonies)	2.91	1.456
Middle (mature colonies)	19.20	2.018
Base (old colonies)	45.80	1.135

Nitrate Uptake and its Assimilation in Cyanobacteria

Nitrate is the most abundant form of fixed-N available to microorganisms and plants in nature and in cyanobacteria it is the most common form of nitrogen nutrition. Exogenous nitrate is efficiently transported into the cyanobacterial cells where it is then reduced first to nitrite and then to ammonia for assimilation by the combined effort of nitrate reductase (NR) and nitrite reductase (NIR) enzymes (Flores and Herrero, 1994). Heterocysts lack both nitrate uptake and nitrate reductase systems but both these processes are present in the vegetative cells (Kumar *et al.*, 1985 and Rai and Bergman, 1986). The transport of nitrate into the cyanobaeterial cells is energy dependent and needs participation of ATP to drive nitrate transport (Rai *et al.*, 1981, Flores *et al.*, 1983a and Bagchi *et al.*, 1989). Based on studies of positive effect of sodium ions on the intracellular levels of accumulation of nitrate, in nitrate reductase mutants of unicellular *Synechococcus* sp., it has been suggested that there exits a cotransport of nitrate and sodium with a downhill electrochemical gradient for sodium from outside the cells to inside which might be responsible for nitrate being actively taken up into the cells (Rodriguez *et al.*, 1992). Repression of NR in heterocysts is not due to the absence of *Mo*-cofactor but lack of apoprotein. This may be a strategy to eliminate competition for Molybdenum (Mo) and reductant between the two molybdo-enzymes–nitrogenase and nitrate reductase (Rai, 1998). Rai and Bergman in 1986 have also established that the loss of nitrate uptake and NR during heterocyst differentiation is irreversible and it is probably due to genomic rearrangement during heterocyst development resulting in *nif* gene co-transcription. Structural genes for nitrite reductase *(nirA)*, nitrate/nitrite uptake *(nrt ABCD)* and nitrate reductase *(nar B)*, have been found to be co-transcripted as a single operon in *Synechococcus* sp. strain PCC 7942 and *Anabaena* sp. strain 7120 (Omata *et al.*, 1993; Cai and Wolk, 1993 and Frias *et al.*, 1997). Assimilation of externally available nitrate depends on the status of externally available ammonium in the growth medium. Cyanobacterial cultures engaged in active nitrate uptake and assimilation show drastic inhibition of nitrate uptake on exposure to ammonium (Rai *et al.*, 1981; Ohmori *et al.*, 1977; Flores *et al.*, 1980 and Garbisu *et al.*, 1992). This inhibition is reversible and if the ammonium in the medium is exhausted then the nitrate uptake resumes within minutes. It is suggested that ammonium may inhibit nitrate uptake by lowering the ATP concentration which is required for nitrate uptake (Rai *et al.*, 1981 and Rai and Singh, 1982). However, it also seems possible that a product of ammonium assimilation via glutamine synthetase has a feedback role in inhibition of nitrate uptake on exposure to externally supplied ammonium (Flores *et al.*, 1983b). Nitrate uptake also show strict dependence on CO_2-fixation(Flores *et al.*, 1983c and Manzano *et al.*, 1976). Hence, there are two regulatory forces working on nitrate uptake–ammonium show negative regulation while CO_2-fixation shows a positive effect. The two processes have antagonic roles on nitrate uptake. It is possible that CO_2-fixation products combine with the negative effectors generated by ammonium assimilation, neutralizing the negative effectors (Flores *et al.*, 1983c). Furthermore, nitrate is reported to cause induction of nitrate reductase (Bagchi *et al.*, 1985b). Nitrite uptake is shown to take place both actively and passively. There could be a common transport system for both nitrate and nitrite as products of ammonium assimilation via GS also inhibit the nitrite uptake (Flores *et al.*, 1987 and Martin-Nieto *et al.*, 1989). Bhattacharya *et al* (2002) have recently shown that in *Nostoc* ANTH, there is a nitrite uptake system quite distinct from the common nitrate–nitrite uptake system.

Ammonia Uptake and its Utilization in Cyanobacteria

Ammonia occupies a unique biochemical position as it is the only form of inorganic nitrogen that can be directly incorporated into the organic molecules in the living beings. It occupies the status of the obligate intermediate in the utilization of other inorganic forms of nitrogen. Ammonia is the preferred source of nitrogen and it regulates most of the key enzymes of nitrogen metabolism such as nitrogenase, glutamine synthetase and nitrate reductase enzymes in living organisms. Ammonium transport system (ATS) in cyanobacteria has been extensively studied using ^{14}C methylammonium (an ammonium analogue) (Rai *et al.*, 1984 and Shehawy and Kleiner, 1999). Ammonium uptake in cyanobacteria has been shown to occur in two phases–an initial rapid phase of 2-3 minutes duration that is MSX insensitive and independent of methylammonium metabolism, followed by a slower second phase which is sensitive to MSX exposure and dependant on methylammonium metabolism (Singh *et al.*, 1985; Singh *et al.*, 1986 and Singh *et al.*, 1987). Further studies on GS defective mutant of *Anabaena cycadeae* suggests that the second phase of the ammonium transport may be a separate ATS and hence there could be two ATS present (Singh *et al.*, 1985). Recent studies show that in *Synechocystis* sp. PCC 6803 there are three putative *amt* (ammonium transport) genes that are under nitrogen control *i.e.* they are derepressed under nitrogen depleted conditions and repressed under nitrogen replete conditions, *amt 1* gene responsible tor a high affinity transporter with K_5 for methylammonium 2.7μM is expressed at higher rates than the other two *amt* genes, *amt l* transcription is also under control of transcription factor, NtcA (Montesinos *et al.*, 1998). Kleiner in 1985 has suggested that the importance of the ATS lies in the uptake of exogenous ammonium and retention of ammonium produced during nitrogen fixation. Ammonium taken up from external sources or generated during nitrite/nitrate assimilation and N_2-fixation is incorporated by glutamine synthetase-gtutamate synthase (GS-GOGAT) pathway into organic forms. The levels of GS (*glnA* gene product) enzyme vary in cyanobacteria on nitrogen nutrition (Merida *et al.*, 1991 and Flores and Herrero, 1994). Under N_2-fixing conditions GS levels are much higher than when the cyanobacterial cells are grown in the ammonium supplemented medium. Similarly, under nitrogen depleted conditions there is increase in *glnA* mRNA synthesis and GS activity. There are two distinct promoters for *glnA* gene: one *E. coli* type promoter and one *nif*- like promoter (Turner *et al.*, 1983). Under N_2-fixing conditions *glnA* transcription occurs through a *nif*- like promoter. Regulation of GS protein in cyanobacteria is not under control of adenylation/deadenylation (Merida *et al.*, 1991).

Amino Acids Nutrition and Transport in Cyanobacteria

Some nitrogen fixing cyanobacteria are able to utilize amino acids such as glutamine, arginine and asparagines as sole nitrogen source (Herrero and Flores, 1990; Flores and Herrero, 1994; Herrero *et al.*, 2001). On the other hand, amino acids like glutamate, histidine and lysine are reported to be inhibitory to growth (Chapman and Meeks, 1983; Flores and Muro-Pastor, 1990; Prakasham *et al.*, 1991). However, the relative efficiency of different cyanobacteria to utilize them as sole nitrogen source or the degree to which these amino acids can support cyanobacterial growth varies widely. *Synechococcus* PCC 6803 shows comparable growth on arginine to that of nitrate but *Anabaena* sp. PCC 7120 shows slower growth on arginine than on nitrate(Herrero and Flores, 1990; Prakasham *et al.*, 1991; Flores and Herrero, 1994; Herrero *et al.*, 2001). There are scanty and contradictory reports about the role of amino acids on cyanobacterial metabolism. Arginine has been reported to be a weak repressor of nitrogenase and nitrate reductase in *Anabaena* sp. PCC 7120 (Herrero and Flores, 1990), while there are reports of arginine inducing NR activity in *Oscillatoria chalybea* (Bednarz and Schmid, 1991). Similarly, glutamine has been found to repress nitrogenase activity but not heterocyst formation (Thiel and Lyons, 1986). At the same time, there are reports of the unicellular cyanobacterium *Synechococcus* sp. strain PCC 7002 being able to utilize a wide range of organic compounds such as urea, most amino acids and purines (especially hypoxanthine and xanthine as N-sources for growth and other functions (Kapp *et al.*, 1975).

In *Synechocystis* sp. strain PCC 6803 and in *Anabaena* sp. strain 7120 arginine was found to be taken up and concentrated within the cells by means of a transport system specific for basic amino acids (Labarre *et al.*, 1987; Flores and Muro-Pastor, 1990; Herrero and Flores, 1990; Flores and Herrero, 1994). Three amino acid transport systems have been reported in *Synechocystis* sp. strain PCC 6803, one specific for basic amino acids and glutamine, one for neutral amino acids excluding glutamine and one specifically for glutamine and glutamate (Labarre *et al.*, 1987). A high affinity and a low affinity transport system for both glutamine and glutamate (Chapman and Meeks, 1983) and a single transport system for leucine (Thiel, 1988) have been reported in *Anabaena variabilis* ATCC 29413. Strasser and Falkner in 1986 have reported a common transport system for aspartate and glutamate in *Nostoc* sp. Similarly, a common transport system for both glutamine and glutamate has also been reported in *Anabaena* PCC 7120 (Flores and Muro-Pastor, 1988). In addition, *Anabaena* 7120 possesses three ATP dependant high affinity amino acid transport system: two for neutral amino acids (Montesinos *et al.*, 1995) and one for basic amino acids (Herrero and Flores, 1990). Further, there are two low affinity transport systems: one specific for acidic amino acids (Montesinos *et al.*, 1995) and one for basic amino acids (Herrero and Flores, 1990). In *Anabaena* sp. exogenously supplied glutamine is converted to glutamate mainly by the action of GOGAT although glutaminase activity has also been detected (Rowell *et al.*, 1977; Haystead *et al.*, 1973; Montesinos *et al.*, 1995).

Studies involving amino acids uptake and utilization in *Nostoc* ANTH show growth in glutamine, asparagine and arginine supplemented media as sources of fixed-nitrogen (Table 1.6). Glutamine and asparagine support growth *of Nostoc* ANTH better than arginine. The heterocyst formation, nitrogenase and NR activities are completely repressed by glutamine and asparagine while the arginine exerts only partial inhibition. However, the glutamine synthetase (GS) activity remains unaffected in presence of these amino acids. These observations are similar to that *of Anabaena* PCC 7120 (Herrero and Flores, 1990). Unlike in *Anabaena variabilis* where glutamine inhibits nitrogenase only, in case *of Nostoc* ANTH glutamine inhibits both nitrogenase activity and heterocyst formation (Thiel and Lyons, 1986; Chen *et al.*, 1987; Bhattacharya *et al.*, 2002). The NR activity of *Nostoc* ANTH is nitrate inducible/ ammonium repressible and cultures grown in glutamine and asparagines show strong inhibition of NR activity than those grown in arginine (50 per cent). These results agree with partial repression of NR activity in *Anabaena* PCC 7120 by arginine (Herrero and Flores, 1990) but are in contrast to the arginine induction of NR activity in *Oscillatoria chalybea* (Bednarz and Schmid, 1991). The GS activity in *Nostoc* ANTH remains nearly similar to those grown in N_2 medium in presence of glutamine, arginine and asparagines. Similar results have been reported earlier in *Nostoc muscurum* (Singh *et al.*, 1991).

Table 1.6: Growth, Heterocyst Frequency (HF), Nitrogenase Activity (N_2ase), Nitrate Reductase Activity (NR) and Glutamine Synthetase/Transferase Activity (GS) of 4 Day Old *Nostoc* ANTH as a Function of Different Nitrogen Sources

Addition to $BG11_0$ medium (mM)	*Growth*	*Heterocyst Frequency (per cent)*	*N_2ase (nmol C_2H_4 Formed/µg Chl a/h)*	*NR (nm NO_2 Formed $min^{-1}mg^{-1}$ Protein)*	*GS (nmol γ glutamyl Hydroxamate Formed min^{-1} mg^{-1} Protein)*
Nil	1.2±0.1	5.3±0.2	12±0.6	1.8±0.11	610±7
NO_3(5)	1.3±0.1	0.0	0.0	4.2±0.15	598±3
NH_4^+ (2)	1.4±0.1	0.0	0.0	0.2±0.10	376±6
Glutamine(1)	2.1±0.2	0.2	0.1	0.3±0.06	595±8
Asparagine(1)	1.9±0.2	0.3±0.1	0.5±0.1	0.5±0.04	581±7
Arginine(1)	1.2±0.1	2.6±0.3	6.1±0.3	0.9±0.01	601±4

Studies on uptake of different amino acids in *Nostoc* ANTH as a function of different nitrogen sources in the media showed that glutamine uptake was biphasic in nature: an initial rapid phase of about 2 min followed by a slower second phase that lasted the rest of the experimental duration. N_2-grown *Nostoc* ANTH shows glutamine uptake rates of 55 and 11 nmol min^{-1} mg^{-1} Chl *a* during the first and the second phase, respectively. The glutamine uptake rates are much higher (93 and 44 nmol min^{-1} mg^{-1} Chl *a*) in glutamine-grown cultures. This increase in the rate of glutamine uptake is sensitive to chloramphenicol–a protein synthesis inhibitor. The activity and pattern of glutamine uptake in nitrate and ammonium grown cultures are similar to that of N_2-grown cultures.

The N_2-grown cultures show asparagine uptake rates of 30 and 11 nmol min^{-1} mg^{-1} Chl *a* during first and second phase and the rate of asparagine uptake is higher in asparagine-grown cultures (56 and 29.2 nmol $min^{-1}mg^{-1}$ Chl *a*).

The N_2-grown *Nostoc* ANTH cultures show arginine uptake rates of 34 and 21 nmol $min^{-1}mg^{-1}$ Chl *a* while the rates are 115 and 51 nmol min^{-1} mg^{-1} Chl *a* in arginine grown cultures during first and the second phase of uptake studies. Hence, the results suggest that the uptake of glutamine, arginine and asparagine in *Nostoc* ANTH are biphasic and substrate-inducible (Bhattacharya *et al.*, 2002). Also, such substrate induced amino acid uptake requires *de novo* protein synthesis as was evident from the fact that chloramphenicol inhibited uptake processes in all the amino acids studied. Unlike ammonium uptake, glutamine, asparagines and arginine uptake are not inhibited by presence of nitrate or ammonium and hence, uptake of these amino acids probably is not under direct control of nitrogen source present in the surrounding, but may be related to the overall nitrogen status of the cell. Such a lack of direct control by nitrate and ammonium has also been seen for leucine uptake in *Anabaena variabilis* (Thiel, 1988) and for arginine uptake in *Anabaena* PCC7120 (Herrero and Flores, 1990).

Studies involving effects of DCMU, the inhibitor of non-cyclic photosynthetic electron transport, CCCP (an uncoupler) and DCCP (an inhibitor of F_0 fraction of ATPase) on uptake rates of glutamine, asparagine and arginine found that presence of the protonophore CCCP and the ATPase inhibitor DCCP strongly inhibit the uptake of the amino acids studied. The DCMU (an inhibitor of photosynthetic O_2-evolution) partially inhibit the uptake of glutamine, asparagine and arginine. Hence, the uptake of the above mentioned amino acids are energy dependant as was already reported in the cases *of Anabaena variabilis Anabaena* PCC *7120* (Thiel, 1988; Herrero and Flores, 1990; Montesinos *et al.*, 1995).

Urea Uptake and Assimilation

Urea is continuously fed into the environment through biological activity and through addition as N-fertilizer for agricultural purposes. Many cyanobacteria belonging to different taxonomic groups can utilize urea as their N-source (Kratz and Myers,1955; Neilson and Larsson, 1980; Rawson, 1985). Urea uptake appears to be carrier mediated and it is taken up intact and actively, using energy derived mainly from cyclic photophosphorylation and then metabolized intracellularly (Healey, 1977; Jahns *et al.*, 1988). Urea transporters in *Anabaena doliolum* and *Anacystis nidulans* have been reported to possess K_m of 250 and 400 µM respectively at pH 7.0 (Rai and Singh, 1987a). Presence of ammonia in the medium does not seem to inhibit urea uptake although, simultaneous presence of urea in the medium depresses ammonia uptake (Healey, 1977; Rai and Singh, 1987a). Since intracellular accumulation of ammonia inhibits its own uptake, this inhibition by urea presumably needs hydrolysis of urea taken up by the cells to ammonia (Healey, 1977; Rai and Singh, 1987a). The enzyme urease degrades the intracellular urea in cyanobacteria releasing one molecule of CO_2 and two molecules of NH_4^+ (Berns *et al.*, 1966; Mackerras and Smith, 1986; Singh and Ahmad, 1989). Urease has been purified from *Spirulina maxima* (Carvajal *et al.*, 1982) and from *Anabaena cylindrical* (Argall *et al.*, 1992) and has been shown to require Ni_2^+ ions in the growth medium. Cyanobacterial urease has been speculated to be a scavenger of amino nitrogen from urea produced internally by degradation of purines and arginine (Ge *et al.*, 1990). Cyanobacterial cells possess cyanophycin granules which are used as nitrogen reserves and are mobilized during N-starvation. They consist of multi-L-arginyl-poly (L-aspartic acid) (Simon, 1971; Allen *et al.*, 1980; Allen and Weathers, 1980). Their degradation is carried out by cyanophycinase to aspartic acid-arginine dipeptide, which gets further degraded to free arginine and eventually to urea (Gupta and Carr, 1981a). Gupta and Carr in 1981 have reported the presence of cyanobacterial arginase activity, which may be an indication of urea producing pathways in cyanobacteria.

Concluding Remarks

Nitrogen fixing capability of cyanobacteria has catapulted this group of organisms into the midst of intense research. The behavioural changes of cyanobacteria in presence of various fixed nitrogen sources in terms of growth, physiology and different metabolic processes are of great interest as they modify nitrogen fixing ability of the cyanobacteria. The presence of ammonium in the cyanobacterial growth media represses the expressions of the proteins involved in the utilization of other nitrogen sources like nitrate or N_2. Apart from being the most preferred source of nitrogen, ammonium regulates various processes in N-metabolism and photosynthesis in cyanobacteria. GS mediated ammonium metabolism seem to be necessary for all the repressive effects on N-metabolism in presence of externally supplied ammonium. This suggests involvement of a common mechanism in these regulations. Use of amino acids as N-nutrition shows that relative efficiency with which amino acids can support growth in cyanobacteria varies. Amino acid uptake seems to be substrate inducible, energy dependent, requires *de novo* protein synthesis and is biphasic in nature. Nitrate and ammonium repress ammonium uptake but not amino acid uptake.

Studies involving N-metabolism lead to understanding the molecular mechanism involved in N_2-fixation and modifications in this process on exposure to various alternative N-sources. This knowledge may be important in developing high yielding N_2-fixers for N-fertilization in rice fields and viable, useful symbiotic associations of cyanobacteria with crop plants for enrichment of soil.

Acknowledgements

The authors are thankful to Department of Science and Technology, Government of India, for financial support under FIST programme.

References

Allen, M.M., Hutchison, F. and Weathers, P.J.,1980. Cyanophycin granule polypeptide formation and degradation in the *Cyanobacterium Aphanocapsa* 6308. *J. Bacteriol.*, 141: 687–693.

Allen, M.M. and Weathers, P.J., 1980. Structure and composition of cyanophycin granules in the *Cyanobacterium Aphanocapsa* 6308. *J. Bacterial.*, 141: 959–962.

Argall, M.E., Smith, G.D., Stamford, N.P.J. and Youbens, B.N., 1992. Purification and properties of urease from the *Cyanobacterium Anabaena cylindrica. Biochem. Intl.*, 6: 1027–1036.

Bagchi, S.N., Palod, A. and Chauhan, V.S., 1989. Photosynthetic control of nitrate metabolism in *Phormidium uncinatum:* A Cyanobacterium. *Curr. Microbiol.*, 19: 183–188.

Bagchi, S.N., Rai, U.N., Rai, A.N. and Singh, H.N., 1985b. Nitrate metabolism in Cyanobacterium *Anabaena cycadeae*: Regulation of nitrate uptake and reductase by ammonia. *Physiol. Plant.*, 68: 322–326.

Bednarz, J. and Schmid, G.H., 1991. Induction of nitrate reductase activity by arginine in the filamentous Cyanobacterium *Oscillatoria chalybea. Z Natueforsch*, 46c: 591–596.

Berns, D.S., Holohan, P. and Scott, E., 1966. Urease activity in blue-green algae. *Science*, 152: 1077–1078.

Bhattacharya, J., Singh, A.K. and Rai, A.N., 2002. Nitrogen nutrition in the Cyanobacterium *Nostoc* ANTH: A symbiotic isolate from *Anthoceros:* uptake and assimilation of inorganic-N and amino acids. *Ind. J. Biochem. Biophys.*, 39: 163–169.

Bhattacharya, J., Singh, A.K. and Rai, A.N., 2002. Isolation and characterization of a chlorate-resistant mutant (Clo-R) of the symbiotic cyanobacterium Nostoc ANTH: Heterocyst formation and N_2–fixation in the presence of nitrate, and evidence for separate nitrate and nitrite transport systems. *Curr. Microbiol.*, 45: 99–104.

Boussiba, S. and Gibson, J., 1987. Regulation of methylammonium-ammonium transport in the unicellular Cyanobacterium *Synechococcus* R-2, PCC 7942. *FEMS. Microbiol. Lett.*, 43: 289–294.

Boussiba, S., Resch, C.M. and Gibson, J., 1984. Ammonium uptake and retention in some cyanobacteria. *Arch. Microbiol.*,138: 287–293.

Cai, Y. and Wolk, C.P., 1993. Differential effects of a *hetR* mutation on the expression of genes of *Anabaena* PCC 7120 that are directly involved in heterocyst differentiation and those that are not. In: *Fourth Cyanobacterial Workshop in Molecular Genetics,* Abstracts, Asilomar (Pacific Grove, CA), p. 26.

Carrasco, C.D., Ramaswamy, K.S., Ramasubramaniam, T.S. and Golden, J.W., 1994. *Anabaena xisF* gene encodes a developmentally regulated sites-specific recombinase. *Gene Develop.*, 8: 74–83.

Carvajal, N., Fernandez, M., Rodriguez, J.P. and Donoso, M., 1982. Urease of *Spirullina maxima. Phytochem.*, 21: 2821–2823.

Chapman, J.S. and Meeks, J.C., 1983. Glutamine and glutamate transport by *Anabaena variabilis. J. Bacteriol.*, 156: 122–129.

Chen, C., Van Baalen, C. and Tabita, F.R., 1987. DL-7-azatryptophan and citrulline metabolism in the Cyanobacterium *Anabaena* sp. Strain 1 F. *J. Bacteriol.*, 169: 1114–1119.

Elhai, J. and Wolk, C.P., 1990. Developmental regulation and spatial pattern of expression of the structural genes for nitrogenase in the Cyanobacterium *Anabaena. EMBO J.*, 9: 3379–3388.

Flores, E. and Herrero, A., 1994. Assimilatory nitrogen metabolism and its regulation. In: *The Molecular Biology of Cyanobacteria*, (Ed.) D.A Bryant. Kluwer Academic Publishers, Dordrecht, The Netherlands, pp. 487–517.

Flores, E. and Muro-Pastor, M.I., 1990. Mutational and kinetic analysis of basic amino acid transport in the Cyanobacterium *Synechocystis* sp. PCC 6803. *Arch. Microbiol.*, 154: 521–527.

Flores, E. and Muro-Pastor, M.I., 1988. Uptake of glutamine and glutamate by the dinitrogen-fixing Cyanobacterium *Anabaena* sp. PCC 7120. *FEMS Microbiol. Lett.*, 56: 127–130.

Flores, E., Gurrero, M.G. and Losada, M., 1983a. Photosynthetic nature of nitrate uptake and reduction in the Cyanobacterium *Anacystis nidulans. Biochim. Biophys. Acta*, 722: 408–416.

Flores, E., Gurrero, M.G. and Losada, M., 1980. Short-term ammonium inhibition of nitrate utilization by *Anacystis nidulans* and other cyanobacteria. *Arch. Microbiol.*, 128: 137–144.

Flores, E., Herrero, A and Guerrero, M.G., 1987. Nitrate uptake and its regulation in the Cyanobacterium *Anacystis nidulam. Biochim. Biophys. Acta*, 896: 103–108.

Flores, E., Romero, J.L., Gurrero, M.G. and Losada, M., 1983c. Regulatory interaction of photosynthetic nitrate utilization and carbon dioxide fixation in the cyanobacterium *Anacystis nidulans. Biochim. Biophys. Acta*, 725: 529–532.

Flores, E., Romero, J.L., Herrero, A. and Gurrero, M.G., 1983b. Nitrate assimilation by cyanobacteria. In: *Photosynthetic Prokaryotes: Cell Differentiations and Functions*, (Eds.) G.C. Papageorgiou and L. Packer. Elsivier, New York, pp 363–387.

Frias, J.E., Flores, E. and Herrero, A., 1997. Nitrate assimilation gene cluster from the heterocyst forming Cyanobacterium *Anabaena* sp. Strain PCC 7120. *J. Bacterial.*, 179: 477–486.

Gallon, J.R., 1992. Reconciling the incompatible: N_2-fixation and O_2. *New Phytol.*, 122: 571–609.

Gallon, J.R. and Chaplin, A.E. (Ed), 1987. *An Introduction to Nitrogen Fixation.* Cassell, London.

Garbisu, C., Hall, D.O. and Serra, J.L., 1992. Nitrate and nitrite uptake by free-living and immobilized N-starved cells of *Phormidium laminosum. J. Appl. Phycol.*, 4: 139–148.

Ge, X., Cain, K. and Hirschberg, R., 1990. Urea metabolism and urease regulation in the Cyanobacterium *Anabaena variabilis. Can. J. Microbiol.*, 36: 218–222.

Golden, J.W., Mulligan, M.E. and Haselkom, R., 1987. Different recombination site specificity of two developmentally regulated genome rearrangements. *Nature*, 327: 526–529.

Golden, J.W., Robinson, S.J. and Haselkorn, R., 1985. Rearrangement of nitrogen fixation genes during heterocyst differentiation in *Cyanobacterium Anabaena. Nature*, 314: 419–423.

Gupta, M and Carr, N.G., 1981b. Enzymology of arginine metabolism in heterocyst-forming cyanobacteria. *FEMS Microbiol Lett.*, 12: 179–181.

Gupta, M. and Carr, N.G., 1981b. Detection of glutamate synthase in heterocysts of *Anabaena* sp. Strain 7120. *J. Bacteriol.*, 148: 980–982.

Haselkorn, R., 1978. Heterocysts. *Ann, Rev. Plant Physiol.*, 29: 319–344.

Haystead, A., Dharmawardene, M.W.N. and Stewart, W.D.P., 1973. Ammonia assimilation in a nitrogen-fixing blue-green alga. *Plant Sci. Lett.*, 1: 439–445.

Healey, F.P., 1977. Ammonium and urea uptake in some freshwater algae. *Can. J. Bot.*, 55: 61–69.

Herrero, A. and Flores, E., 1990. Transport of basic amino acids by the dinitrogen-fixing Cyanobacterium *Anabaena* PCC 7120. *J. Biol. Chem.*, 265: 3931–3935.

Herrero, A., Muro-Pastor, A.M. and Flores, E., 2001. Nitrogen control in cyanobacteria. *J. Bacteriol.*, 183: 411–425.

Jahns, T., Zobel, A. Kleiner, D. and Kaltwasser, H., 1988. Evidence for carrier-mediated, energy dependent uptake of urea in some bacteria. *Arch. Microbiol.*, 149: 377–383.

Kapp, R., Stevens, S.E. and Fox, J.L. 1975. A survey of available nitrogen sources for the growth of the blue-green alga, *Agmenellun quadruplicatum. Arch Microbiol.*, 104: 135–138.

Kentemich, T., Haverkamp, G and Bothe, H., 1991. The expression of a third nitrogenase in the Cyanobaeterium *Anabaena variabilis. Z. Natutforsh*, 46c: 217–222.

Kleiner, D., 1985. Bacterial ammonium transport. *FEMS. Microbiol. Lett.*, 32: 87–100.

Kratz, W.A. and Myers, J., 1955. Nutrition and growth of several blue-green algae. *Amer. J. Bot.*, 42: 282–287.

Kumar, A.P., Rai, A.N. and Singh, H.N. 1985. Nitrate reductase activity in isolated heterocysts of the Cyanobacterium *Nostoc muscorum. FEBS Lett.*, 179: 125–128.

Labarre, J., Thuriaux, P. and Chauvat, F., 1987. Genetic analysis of amino acid transport in the facultatively heterotrophic Cyanobacterium *Synechocystis* sp. Strain PCC 6803. *J. Bacteriol.*, 169: 4668–4673.

Lammers, P.J. and Haselkorn, R., 1983. Sequence of the *nifD* gene coding for a subunit of dinitrogenase from *Cyanobacterium Anabaena. Proc. Nat. Acad. Sci., USA*, 80: 4723–4727.

Lammers, P.J., Golden, J.W. and Haselkorn, R., 1986. Identification and sequence of a gene required for a developmentally regulated DNA excision in *Anabaena. Cell*, 44: 905–911.

Loffelhardt, W. and Bonhert, H.J., 1994. Molecular biology of cyanelles. In: *The Molecular Biology of Cyanobacteria*, (Ed.) D.A. Bryant. Kluwer Academic, Dordrecht, The Netherlands, pp 65–89.

Mackerras, A.H. and Smith, G.D., 1986. Urease activity of the Cyanobacterium *Anabaena cylindrical. J. Gen. Microbiol.*, 132: 2749–2752.

Manzano, C., Candau, P., Gomez-Monero, G., Relimpio, A.M. and Losada, M., 1976. Ferredoxin-dependant photosynthetic reduction of nitrate and nitrate by particles of *Anacystis nidulans. Mol. Cell Biochem.*, 10: 161–169.

Martin-Nieto, J., Herrero, A. and Flores, E., 1989. Regulation of nitrate and nitrate reductases in dinitrogen-fixing cyanobacteria and *Nif* mutants. *Arch. Microbiol.*, 151: 475–478.

Mazur, B.J. and Chui, C.F., 1982. Sequence of the gene coding for the β subunit of dinitrogenase from the blue-green alga *Anabaena. Proc. Nat. Acad. Sci., USA*, 79: 6782–6786.

Meeks, J.C., 1998. Symbiosis between nitrogen-fixing cyanobacteria and plants. *Biosciences*, 48: 266–276.

Merida, A., Candau, P. and Florencio, F.J., 1991. Regulation of glutamine synthetase activity in the unicellular Cyanobacterium *Synachocystis* sp. Strain PCC 6803 by the nitrogen source: Effect of ammonium. *J. Bacterial.*, 173: 4095–4100.

Montesinos, M.L., Herrero, A. and Flores, E., 1995. Amino acid transport required for diazotrophic growth in the Cyanobacterium *Anabaena* sp. Strain PCC 7120. *J. Bacteriol.*, 177: 3150–3157.

Montesinos, M.L., Muro-Pastor, A.M., Herrero, A. and Flores, E., 1998. Ammonium/ methylammonium permeases of a Cyanobacterium identification and analysis of three itrogen-regulated *amt* genes in *Synachocystis* sp.strain 6803. *J. Biol. Chem.*, 273: 31463–31470.

Neilson, A.H. and Larsson, T., 1980. The utilization of organic nitrogen for growth of algae: Physiological aspects. *Physiol. Plant.*, 48: 542–553.

Ohmori, M., Ohmori, K. and Strotmann, H., 1977. Inhibition of nitrate uptake by ammonia in a blue-green alga, *Anabaena cylindrical. Arch. Microbiol.*, 114: 225–229.

Omata T., Andriesse, X. and Hirano, A., 1993. Identification and characterization of a gene cluster involved in nitrate transport of the *Cyanobacterium Synechococcus* sp. PCC 7942. *Mol. Gen. Genet.*, 236: 193–202.

Prakasham, R., Singh, A.K., Singh, H.N. and Rai, A.N., 1991. Inorganic nitrogen regulation of glutamate uptake in the Cyanobacterium *Nostoc muscorum. Physiol. Plant.*, 82: 257–260.

Rai, A.N., Borthakur, M. and Paul, D., 1996. Symbiotic cyanobacteria: Biotechnological applications. *J. Sci Ind. Res.*, 55: 742–752.

Rai, A.N., 1998. Cyanobacterial nitrogen metabolism: relevance in agricultural Biotechnology. In: *Cyanobacterial Biotechnology*, (Eds.) G. Subramanian, B.D. Kaushik and G.S. Venkataraman. Oxford and IBH Publishing Co. Pvt. Ltd. New Delhi, India, pp 223–231.

Rai, A.K. and Singh, S., 1987a. Kinetics and regulation of urea uptake in *Anabaena doliolum* and *Anacystis nidulans. J. Gen. Appl. Microbiol.*, 33: 471–479.

Rai, A.N. and Bergman, B., 1986. Modification of NO_3 metabolism in heterocysts of the N_2–fixing Cyanobacterium *Anabaena* 7120 (ATCC27893). *FEMS Microbiol. Lett.*, 36: 133–137.

Rai, A.N. and Singh, S., 1982. Regulation of nitrate uptake in *Nostoc mmcorum* by glutamine synthetase. *FEMS. Microbiol. Lett.*, 14: 303–306.

Rai, A.N. (Ed.), 1990. *Handbook of Symbiotic Cyanobacteria.* CRC Press, Boca Raton, Florida, USA.

Rai, A.N., Borthakur, M. and Bergman, B., 1992. Nitrogenase derepression, its regulation and metabolic changes associated with diazotrophy in the nonheterocystous Cyanobacterium *Plectonema boryanum* PCC 73110. *J. Gen. Microbiol.*, 138: 481–491.

Rai, A.N., Borthakur, M., Singh, S. and Bergman, B., 1989. *Anthoceros-Nostoc* symbiosis: Immunoelectronmicroscopic localization of nitrogenase, glutamine synthetase, phycoerythrin and ribulose-1, 5-bisphosphate carboxylase/oxygenase in the cyanobionts and the cultured (free-living) isolate 7801. *J. Gen. Microbiol.*, 135: 385–395.

Rai, A.N., Kashyap, A.K. and Gupta, S.L., 1981. ATP-dependent uptake of nitrate in *Nostoc muscorum* and inhibition by ammonium ions. *Biochem. Biophys. Ada.*, 674: 78–86.

Rai, A.N., Rowell, P. and Stewart, W.D.P., 1984. Evidence for an ammonium transport system in free-living and symbiotic Cyanobacteria. *Arch. Microbiol.*, 137: 241–246.

Rai, A.N., Soderback, E. and Bergman, B., 2000. Transley Review: Cyanobacterium-plant symbioses. *New Phytol.*, 147: 449–481.

Rawson, D.M., 1985. The effects of exogenous amino acids on growth and nitrogenase activity in the Cyanobacterium *Anabaena cylindrica* PCC 7122. *J. Gen. Microbiol.*, 131: 2549–2554.

Rodriguez, R., Lara, C. and Gurrero, M.G., 1992. Nitrate transport in the Cyanobacterium *Anacystis nidulans* R2. Kinetic and energetic aspects. *Biochem. J.*, 282: 639–643.

Rowell, P., Enticott, S. and Stewart, W.D.P., 1977. Glutamine synthetase and nitrogenase activity in the blue-green alga *Anabaena cylindrica. New Phytol.*, 79: 41–54.

Schenk, H.E.A., 1992. Cyanobacterial symbioses. In: *The Prokaryotes*, (Eds.) A. Balows, H.G. Truper, M. Dworkin W. Harder and K.H. Schleiter. Springer-Verlag, Germany, 4: 3819–3854.

Shehawy, R.M. and Kleiner, D., 1999. Ammonium (methylammonium) transport by heterocysts and vegetative cells of *Anabaena variabilis. FEMS Microbiol. Lett.*, 181: 303–306.

Simon, R.D., 1971. Cyanophycin granules from the blue-green alga *Anabaena cylindrical:* A reserve material consisting of copolymers of aspartic acid and arginine. *Proc. Nat. Acad. Set., USA*, 68: 265–267.

Singh, A.K., Singh, H.N. and Rai, A.N., 1991. Evidence for a role of glutamine synthetase in assimilation of amino acids as nitrogen source in the Cyanobacterium *Nostoc muscorum*. *Biochem. Intl.*, 25: 887–894.

Singh, D.T., Ghose, R. and Singh, H.N., 1987. Physiological characterization of the ammonium transport system in the free-living diazotrophic Cyanobacterium *Anabaena cycadeae*. *J. Plant Physiol.*, 127: 231–239.

Singh, D.T., Modi, D.R. and Singh, H.N., 1986. Evidence of glutamine synthetase and methylammonium (ammonium) transport as two distinct targets of methionine sulfoximine inhibitory action in the Cyanobacterium *Anabaena doliolum*. *FEMS. Microbiol. Lett.*, 37: 95–98.

Singh, D.T., Rai, A.N. and Singh, H.N., 1985. Methylammonium (ammonium) uptake in a glutamine auxotroph of the Cyanobacterium *Anabaena cycadae*. *FEBS Lett.*, 186: 51–53.

Singh, S. and Ahmad S., 1989. Regulation of urea uptake by ammonia in the Cyanobacterium *Anabaena doliolum*. *FEMS Microbiol. Lett.*, 61: 199–202.

Stal, L. and Bergman, B., 1990. Immunological characterization of nitrogenase in the filamentous nonheterocystous Cyanobaeterium *Oacillatoria limosa*. *Planta*, 182: 287–291.

Stewart, W.D.P., Rowell, P. and Rai, A.N., 1983. Cyanobacteria-eukaryotic plant symbioses. *Ann. Microbiol. (inst. Parteur)*, 134B: 205–228.

Strasser, P. and Falkner, G., 1986. Characterization of the giutamate/aspartate transport system in a symbiotic *Nostoc* sp. *Planta*, 168: 381–385.

Thiel, T. and Lyons, E., 1986. Effect of glutamine on growth and heterocyst differentiation in the Cyanobacterium *Anabaena variabilis*. *J. Bacteriol.*, 168: 769–774.

Thiel, T. and Pratte, B., 2001. Effect on heterocyst differentiation of nitrogen fixation in vegetative cells of the Cyanobacterium *Anabaena variabilis* ATCC 29413. *J. Bacteriol.*, 183: 280–286.

Thiel, T., 1993. Characterization of genes for an alternative nitrogenase in the Cyanobacterium *Anabaena variabilis*. *J. Bacteriol.*, 175: 6276–6286.

Thiel, T., Lyons, E.M., Erker, J.C. and Ernst, A., 1995. A second nitrogenase in vegetative cells of a heterocyst-forming Cyanobacterium. *Proc. Nat. Acad. Sci., USA*, 92: 9358–9362.

Thiel, T., 1988. Phosphate transport and arsenate resistance in the Cyanobacterium *Anabaena variabilis*. *J. Bacteriol.*, 170: 1143–1147.

Thorneley, R.N.F. and Ashby, G.A., 1989. Oxidation of nitrogenase iron protein by dioxygen without inactivation could contribute to high respiration rates of *Azotobacter* species and facilitate nitrogen in aerobic environment. *Biochem. J.*, 261: 181–187.

Turner, N.E., Robinson, S.J. and Haselkorn, R., 1983. Different promoters for the *Anabaena* glutamine synthetase gene during growth using molecular or fixed nitrogen. *Nature,* 306: 337–342.

Warner, D., 1992. *Symbiosis of Plants and Microbes.* Chapman and Hall, London, UK.

Weare, N.M. and Benemann, J.R. Nitrogenase activity and photosynthesis in *Plectonema boryanum. J. Bacteriol.,* 119: 258–265.

Wolk, C.P., Ernst, A. and Elhai, J., 1994. Heterocyst metabolism and development. In: *The Molecular Biology of Cyanobacteria,* (Ed.) D.A. Bryant. Kluwer Academic, Dordrecht, The Netherlands, pp 769–823.

Chapter 2

A Botanical Ramble Among Blue-Green Algae in Diverse Natural and Artificial Symbioses

A. Vaishampayan and A.K. Awasthi*

*Photobiological Nitrogen Fixation Research Laboratory,
Department of Genetics and Plants Breeding.
Institute of Agricultural Sciences,
Banaras Hindu University, Varanasi – 221 005*

Introduction

Blue-green algae [BGA], also popularly known as cyanobacteria, are prokaryotic organisms with a typical "higher plant-type" oxygen-evolving photosynthesis (Fogg *et al.*, 1973). Chlorophyll-a, phycobiliproteins (c-phycocyanin and c-phycoerythrin) and allophycocyanin are the basic pigments which provide the characteristic blue-green colour to these microorganisms (Stewart, 1977). Being ubiquitous,the BGA are distributed the world over under widely varying conditions and display unicellular to branched filamentous morphology, represented by the different genera and species (Fay and Van Baalen, 1987). The trichomes of the filamentous forms differentiate into vegetative cells and akinetes, as also into heterocysts in the majority of nitrogen-fixing species (Stewart, 1980). This is because heterocysts create anaerobic condition to avoid the detrimental effects of high oxygen concentration on the nitrogen-fixing enzyme, nitrogenase (Fleming and Haselkorn,1973). Therefore, the two vital life processes, *i.e.*, photosynthesis and diazotrophy (dinitrogen fixation) in the heterocystous BGA are spatially separated (photosynthesis in vegetative cells and nitrogen fixation in heterocysts) while the same are temporally controlled in the non-heterocystous forms (with alternate cycles of photosynthesis and nitrogen fixation) for protecting nitrogenase against the damaging effect of photochemically released oxygen inside the vegetative cells (Apte, 1992). Although the BGA generally reproduced by fission, budding, trichome breakage, hormogonia formation or akinete germination (Carr and Whitton, 1982; Fay, 1983; Vaishampayan and Mishra, 1990), yet transformation of genetic material has also been well established in this group of organisms (Stewart and Singh, 1975; Singh *et al.*, 1987; Vaishampayan, 1984a, 1988a,b, 1994a, 1995a).

The focus in this article is on the unique structural, physiological, biochemical and genetic adaptations which some of the BGA genera and species display in associative environment in symbiosis with different categories of specific plants. Exploratory researches during the recent years in this exciting area of study have rather encouraged successful efforts in constructing an artificial symbiosis using BGA as the dinitrogen fixing microsymbiotic components with the diverse groups of plants, in order to enrich the latter with natural N supply of BGA origin (Gusev and Korzhenevskaya, 1990, 1995; Vaishamapayan, 1995b; Vaishampayan and Awasthi, 1995a). The present emphasis, therefore, is on some of the salient recent research accomplishments with the BGA occurring in natural or artificial symbiosis.

Blue-Green Algae in Natural Symbioses

"Symbiosis" in terms of biological science has been variously defined from time to time by a set of workers (Scott, 1969; Ahmadjian and Paracer, 1986; Smith and Douglas, 1987; Rai, 1990a; Vaishampayan, 1995b) to mean the infinite life association of two dissimilar organisms, involving exchange of metabolites between the two partners, generally designated as micro–and macro–symbionts. In these associations, microsymbiont is the one which enlodges itself in the intercellular or intracellular space of the other partner, *i.e.*, the macrosymbiont. Such mutualistic associations, involving BGA as the microsymbionts, have been well known to occur with plants, animals (echiruoid worms and marine sponges), hollow shafts of hairs of polar bears, non-photosynthetic protists (belonging to the group of Glaucophyta) and bacteria (Smith and Douglas, 1987). The plants which display BGA symbioses in nature are algae (diatoms), fungi (lichens), bryophtes (liverworts, hornworts and mosses), pteridophyte (*Azolla*), gymnosperms (cycads) and angiosperm (*Gunnera*).

Among algae, *Rhizosolenia* is the one diatom which forms symbiosis with the BGA microsymbiont *Richelia intracellularis*, while fungal associations (lichens) have a wide range of BGA microsymbionts starting from the unicellular forms, like *Gloeocapsa*, *Gloeothece*, *Synechocystis* and *Hyella*, to the unbranched filamentous/heterocystous forms (*Calothrix*, *Nostoc* and *Scytonema*) and branched filamentous/heterocystous forms, *i.e.*, *Fischerella* (Rippka *et al.*,1979). *Nostoc* is generally the only BGA genus forming symbioses with bryophytes, gymnosperms (cycads) and angiosperms (*Gunnera*). The symbiotic association of the BGA microsymbiont *Anabaena azollae* with the water fern *Azolla* is most thoroughly known, as nitrogen fixation by this complex is extremely efficient in the water-logged paddy fields of tropical countries, particularly South and South-East Asia (Peters and Meeks, 1989).

As compared to the other categories of heterotrophic nitrogen fixers, *i.e., Rhizobia* and *Frankia*, the range of phototrophic BGA symbioses in nature is much wider. This is the reason why BGA have been considered as potential autotrophic nitrogen-fixing partner in establishing artificial symbioses with plants. Therefore, the role of BGA in supplying dinitrogen-derived fixed N (ammonium) to the host plant in an artificial symbiosis is well discernible. However, the form in which fixed N moves from microsymbiotic BGA to the macrosymbiont in the cases of both natural and artificial symbioses may vary (Gusev and Korzhenevskaya, 1990).

In plant-BGA symbiosis, heterocyst frequency of the filamentous microsymbiont is significantly higher over that possessed by the free-living BGA, and this forms a good basis for a remarkably higher rate of nitrogen fixation by the symbiotic BGA (Fay, 1983). The BGA in natural symbioses have relatively low levels of the ammonia-assmilating enzyme, glutamine synthetase (GS), as compared to the free-living BGA, possibly due to gene regulation at the transcription or activity level. Thus, a situation is created whereby the newly fixed ammonia is mostly liberated by the microsymbiotic BGA to the host system. The other remarkable feature of plant-BGA symbiosis is that, in contrast to the free-living BGA where each heterocyst contains GS which is double in quantity as compared to each vegetative cell, the distribution of GS is quite uniform in all types of cells in the microsymbiotic BGA (Rai, 1990a). So far as transport of photosynthetically fixed carbon is concerned, the plant-BGA symbioses fall into three categories, *i.e.*, (a) where the BGA transfers both fixed C and fixed N to the plant (in the case of lichens having BGA as the only microsymbiont), (b) where only fixed N is transferred by the BGA to the plant while the former fixes its own C to meet its requirement (in the case of lichens enlodging BGA and a green alga as dual microsymbiotic partners, as also in the case of *Azolla*), and (c) where the BGA supplies fixed N to the host plant and in return derives fixed C from the latter due to non-functional photosyntesis of the BGA microsymbiont (in the case of liverworts, cycads and *Gunnera*). An updated account of the different categories of naturally occurring plant-BGA symbioses is presented herewith.

Blue-Green Algae in Fungal Symbioses

Symbiotic associations of certain fungi with algae are generally referred to as lichens which conspicuously differ from the component partners with respect to thallus morpholgy. Nearly 8 per cent of lichen species are known to enlodge BGA as the extracellularly developed microsymbionts (Rai, 1990b). However, *Geosiphon pyriforme* is an exceptional lichen species in which the cell wall of the parenchymatous fungus lyse to enclose the microsymbiotic (*Nostoc*) cells by invagination of the plasmalemma (Knapp, 1933; Schnepf, 1964). If BGA is the only microsymbiotic component (bipartite lichen), it is found either spread throughout the lichen thallus, *e.g.*, in *Collema* species (homoiomerous type lichen thallus), or is restricted to a distinct symbiont layer in the upper cortex, *e.g.*, in *Peltigera canina* (heteromerous type lichen thallus). On the other hand, if a green alga as well as a BGA jointly form the microsymbiotic component of a lichen species (tripartite lichens), the BGA cells localize themselves in special structures known as cephalodia which may develop either outside the lichen thallus (*e.g.*, in *Peltigera aphthosa*) or within the thallus (*e.g.*, in *Lobaria pulmonaria*). Moreover, in the tripartite lichens, the heterocyst frequency of the BGA microsymbiont is significantly higher as compared to the free-living BGA or those present in the bipartite lichens (Rai, 1990b). This may be either due to movement of photosynthate from green alga to BGA, or under influence of an effector from the macrosymbiont (fungus) in the case of tripartite lichens (Hitch and Millbank, 1975; Rai, 1990b). However, experimental support lacks for either of these two propositions.

So far only four of the heterocystous BGA genera are known to be the definite microsymbiotic components of lichen species, *i.e.*, (i) *Calothrix* in association with *Lichina* species (Galloe, 1939; Ahmadjian, 1962), *Placynthium nigrum* (Geitler, 1934), and species of *Calotrichopsis* and *Proscyphus* (Henssen, 1967); (ii) *Nostoc* in association with *Collema* species (Degelius, 1954), *Peltigera* species (Linkola, 1923; Watanabe and Kiyohara,1973), *Pannaria* species (Shiman, 1957), *Hydrothyria venosa* (Kele, 1964), *Leptogium issatshenku* (Danilova, 1927) and *Stereocaulon* species (Lamb, 1951); (iii) *Fischerella* in association with species of *Ephebe, Spilonema* and *Stereocaulon* (Bornet, 1873; Hitch and Millbank, 1975; Huss-Danell, 1979); and (iv) *Scytonema* in association with species of *Pannaria, Coccocarpia, Erioderma, Dictyonema, Zahlbrucknerella, Lichinodium, Polychydium, Thurmutis, Lichenotahrix, Heppia, Lasioloma, Lopadium,* and *Stereocaulon* (Santesson, 1952; Henssen, 1964; Marton and Glaun, 1976; Rai, 1990b). Of these, the last one, *i.e., Scytonema* has been shown to forego its filamentous and heterocystous chatacter to appear as unicells in symbiotic association with the lichen *Heppia echinulata,* while the same cultured in laboratory resumes its filamentous and heterocystous nature (Marton and Galun, 1976), as reported earlier for the BGA species of *Calothrix/Dichothrix* in association with the lichen species *Placynthium nigrum* (Geitler, 1934).

The unicellular BGA, *Hyella* species has been demonstrated to be symbiotically associated with the lichen *Arthopyrenia halodytes* collected from the British Isles (Swinscow, 1965). The possibilities of other unicellular BGA genera, *e.g., Gloeocapsa, Gloeothece* and *Synechocystis* as the microsymbiotic components with the species of lichens, like *Pyrenopsidium, Phylliscum, Pterygiopsis, Synnalisa, Psorotichia, Pyrenopsis, Peccania, Stereocaulon, Peltula, Gloeoheppia, Gonohymenia, Pyrenopsidium, Lichinella* and *Thyrea,* have also been shown from time to time (Geitler, 1937; Jaag, 1945; Bubrick and Galun, 1984; Rai, 1990b).

Nostoc-containing lichens have shown the selective role of lectins in recognition events between micro- and macro-symbionts. In *Peltigera canina* and *P. polydactyla* the lectins of macrosymbiotic (fungal) origin were found to tightly bind the cultured BGA component (Lokhart *et al.*, 1978), suggesting that such lectins may be involved in establishing the initial stages of nitrogen-fixing symbiosis. Further, lectins produced by *Peltigera canina* and *P. horizontalis* macrosymbionts have been shown to bind to a cultured *Nostoc* isolated from another species of *Peltigera* (Petit, 1982; Petit *et al.*, 1983), suggesting their involvement in recognition and initial interaction between compatible lichen symbionts. Lectin binding sites in *Peltigera praetextata* has been demonstrated using *P. canina* lectin conjugated with fluoroceine isothiocyanate, which could label only 20 per cent of BGA microsymbiont, while 80 per cent of the endophytic BGA trichomes had their binding sites occupied with endogenous lectins (Lallement and Savoy, 1985), thereby suggesting the major role played by lectins in recognition process between BGA and fungi.

Using acetylene reduction technique or radioisotopic N incorporation (Stewart, 1980) the heterocystous BGA in symbiotic association with all the lichen species, except *Dendriscocaulon umhausence, Collema crustatum, Leptogium tremelloides, Lobaria amplissima, Nephroma perlie, Pannaria microphylla, Peltigera venosa, Psoroma hypnorum, Solarina saccata, S. spongiosa, Stereocaulon vesuvianum, Sticta dufourii* and *S. canariensis*, have been demonstrated to perform significant dinitrogen-fixing activity (Kershaw, 1985) with rates varying from 8 to 21 n mol acetylene reduced per microgram chlorophyll-a per hour in accordance with pH, temperature, light intensity, thallus miosture content, season of lichen collection, environmental condition prevailing before time of collection and the period of pre-treatment and/or storage (Millbank, 1984; Kershaw, 1985). Role of exogenous molybdenum on sharply enhancing the nitrogenase (Horstmann *et al.*, 1982) and combined N incompletely repressing the same (Rai *et al.*, 1983) in BGA fungal association has been established. Moreover, while nitrogen fixation in BGA occurs at the expense of ATP and reductant made available through photosynthetic assimilation of inorganic carbon. Certain BGA in association with lichens, like *Peltigera aphthosa*, also show dark nitrogenase activity which is supported by the catabolism of glycogen reserves accumulated in light (Rai *et al.*, 1981). The significance of continuous/uninterrupted nitrogen fixation by BGA microsymbionts in both light and dark may be in order to meet the total N requirement of both fungal and green algal partners in the case of tripartite lichens. Since the level of ammonia-assimilating enzymes in microsymbiotic BGA is only 5 per cent as compared to the free-living BGA, the fixed N in the form of ammonia is mostly liberated for use(assimilation) by the other partners of the bipartite/tripartite lichen(s).

Blue-Green Algae in Bryophytic Symbioses

In natural habitats, BGA as well as eukaryotic algae grow in association with bryophtes, comprising of mosses, liverworts and hornworts (Clymo and Hayward, 1982). The first proof of the nitrogen-fixing symbiosis of a BGA *Nostoc* species with the liverwort *Blasia pusilla* was given by Bond and Scott (1955) in which transfer of fixed nitrogen from BGA to the liverwort was established using the tracer technique. Generally, the BGA form epiphytic associations with mosses and are endophytic with liverworts and hornworts (Meeks, 1990). However, the mosses like *Sphagnum riparium* and *S. linderbergii* are exceptions which may enlodge BGA even endophytically (Granhall and Selander, 1973; Basilier *et al.*, 1978). Similarly, among the liverworts, *Marchantia berteroama* and *Porela navicularis* are exceptions to associate BGA epiphytically (Dalton and Chatfield, 1985; Brassel *et al.*, 1986). Species of *Nostoc* are primarily involved as a BGA microsymbiont in bryophytic associations, confined to the gametophytes of the latter. When associated epiphytically, these look like the free-living *Nostoc*, but as an endosymbiont these display distorted vegetative cell shape of higher frquencies of heterocysts (Meeks, 1990). However, another BGA genus, *Hapalosiphon* species has also been reported to be associated endosymbiotically with the moss *Sphagnum* species (Stewart, 1966). The per annum rates of nitrogen fixation in sub-arctic myers, shown by free-living *Nostoc*, and microsymbiotic *Nostoc* associated with *Sphagnum* endophytically and epiphytically were estimated to be 0.2, 0.2 and 9.4 gm per square meter. Similar rates of nitrogen fixation have been displayed by *Nostoc* in association with the mosses *Drapanocladus trichophyllus* (Granhall and Selander, 1973) and *D. exannulates* (Basilier *et al.*, 1978). Metabolic interaction between micro–and marcro- symbionts of a BGA-moss symbiotic complex was established with the help of radioisotopic N, and it was resolved that symbiotic *Nostoc* in this case was 14-22 times more active in dinitrogen fixation commensurate with the frequencies of heterocysts as compared to the free-living *Nostoc* in general (Basilier,1980; Meeks, 1990). However, the details of metabolic relation between such symbioses are still to be worked out. In grasslands and slash-burn sites too, nitrogen-fixing symbiotic association of *Nostoc* and / or *Anabaena* have been reported to occur with the mosses *Ceratodon purpureus* and *Funaria hygrometrica* (Brasell *et al.*, 1986), established through autofluorescence and epifluorescence microscopy.

The *Nostoc* communities are macroscopically visible as dark round spots (0.4-0.7 mm in diameter) in the gametophytic tissue of liverworts and hornworts with a distinctly organized common structure in both the cases (Enderlin and Meeks, 1983). Physiological and taxonomic details of the epiphytic associations of BGA with liverworts are lacking. However, in accordance with the available pieces of information, the epiphytic associations with bryophytes, in general, apparently occur more often then thought earlier (Meeks, 1990). *Marchantia berteroava* exhibited approximately 33 per cent nitrogen-fixing activity as compared to *Ceratodon purpureus* (Brasell *et al.*,1986) or *Porella navicularis* (Dalton and Chatfield, 1985).

Studies on endophytic associations of *Nostoc* species with liverworts and hornworts have been more intensive in view of the facts that these easily grow in laboratory culture and the associations are capable to be reconstituted from the separately cultured partners.Moreover, *Nostoc* is distintly localized within 'slime cavities' in the gametophyte thallus which is quite favourable to select out tissue for structural and ultrastructural studies, since there has been great biological interest in screening the pronounced morphological changes in this microsymbiont in *ex-planta* culture. Researches done during the last decade on the symbiotically associated *Nostoc* species with the hornwort *Anthoceros punctatus* have shown that interactions between the partners can be categorized into two major stages, each with two distinct sub-stages (Meeks, 1990). The first stage involves initial establishment of the association which includes formation of *Nostoc* infective units and infection, while the second stage comprises of the activities related with regulation of growth and metabolism and differentiation of associated *Nostoc*.

Nostoc strains capable of fanning symbiosis convert more than half of their vegetative cells into hormogonia within 24 h of their co-culture with *Anthoceros punctatus*. The most important characteristics of the symbiotically competent *Nostoc* strains are the gliding motility and chemotactic response of their hormogonia which help in locating the 'slime pores' or cavity openings in the gametophyte thallus for colonization (Enderlin and Meeks, 1983; Campbell and Meeks, 1989). These authors have found that *Nostoc* strains isolated from even any other plant-BGA symbiosis could associate with *Anthoceros punctatus*. The broad genetic diversity among *Nostoc* strains ATCC 29133, 27895 and 29150 in this context has been observed based on DNA-DNA reassociation (Lachance, 1981).

Pigment characteristics and spontaneously arising mutants resistant to metabolic analogues were used to assess the competitive relationships between three *Nostoc* strains infecting *Anthoceros punctatus*. Resistance to analogues (*e.g.*, L-methionine-DL-sulfoximine) or periplasmic enzymes (5-fluorocytosine) cause an increase in Infectiveness of the strain relative to its parental culture (Enderlin and Meeks, 1983). Rodgers and Stewart (1977) identified the symbiotically competent *Nostoc* isolates from *Anthoceros punctatus* and *Blasia pusilla*, and described these as *Nostoc sphaericum* on the basis of culture morphology. The microsymbionts isolated from *Anthoceros husnotii* and *A. laevis* too were similarly described to be *Nostoc sphaericum* (Duckett *et al.*, 1977). On the other hand, these authors identified *Nostoc calcicola* to be associated with *Blasia pusilla*. However, Watanabe and Kiyohara (1973) earlier described the nitrogen-fixing microsymbiotic BGA from *B. pusilla* and *Cavicularia densa* as *Nostoc sphaericum*. In fact the morphology of the organisms kept on changing in laboratory culture (Meeks, 1990). In view of this, as well as the systematic problems enumerated by Rippka *et al.* (1979), along with the genotypic and phenotypic diversity of symbiotically competent *Nostoc* strains and the non-exclusive/non-colinear competitive relationships between competent strains, it is difficult to assign the species of *Nostoc* on the basis of morphology in the different liverwort and hornwort associations.

The symbiotic *Nostoc* in bryophytic association converts 12.5 to 45 per cent of vegetative cells into heterocysts (Meeks, 1990). Rodgers and Stewart (1977) reported specific activities of symbiotic nitrogen fixation (acetylene reduction; of 70 n mol of acetylene reduced per mg of *Nostoc* protein per hour, a value which is 3.2-fold higher than their free-living counterparts. The proof of a host photosynthesis-dependent nitrogenase activity was obtained from a well-planned acetylene reduction experiment in which *Nostoc* colonies and gametophyte tissue of *Anthoceros punctatus* were darkened by a layer of fine charcoal and the sporophytes exposed to light. Almost nil acetylene was reduced in these experiments if the sporophytes were also darkened or excized from the gametophytes (Stewart and Rodgers, 1977). However, it has been implied that the micro-symbiotic *Nostoc* species (strain 7801) directly maintains a pool of photosynthate in vegetative cells that supports nitrogenase activity in heterocysts. The pool of photosynthate is most probably supplemented by extracellular products of *Anthoceros* metabolism (Meeks, 1990).

The Blue-Green Alga *Anabaena azollae* in Symbiosis with *Azolla*

Of the various categories of BGA symbiosis with plants, the mutualistic association of the nitrogen-fixing endosymbiont *Anabaena azollae* with the water-fern *Azolla* is extremely attractive and more thoroughly worked out. Nitrogen fixed by this complex is very efficient, contributing nearly 2-7 Kg N/ha/d which is equivalent to that fixed by forage leguminous crops, and significantly higher than the agronomically recommended rates of nitrogen fixation (Vaishampayan, 1993a,b, I994a,b; Vaishampayan and Banerjee, 1995). But despite studies of the science of nitrogen fixation and related physiological, biochemical, genetic and somaclonal mutagenesis aspects of this complex (Peters and Meeks, 1989; Vaishampayan *et al.*, 1992b; Vaishampayan, 1995c-e), coupled with field investigations on its agronomics and economics in relation to rice culture (Singh, 1979; Singh, 1994; Kannaiyan, 1994; Vaishampayan and Awasthi, 1995b; Vaishampayan and Singh, 1995), a universally acceptable/sustainable *ex-planta* culture of the nitrogen-fixing microsymbiont *Anabaena azollae* is not yet available for experimental work from any of the *Azolla* species known so far.

Azolla is heterosporous, originally established by Lamarck in 1873 (See Svenson, 1944), and belongs to the family Azollaceae (Schulthorpe, 1967; Moore, 1969; Konar and Kapoor, 1974; Chatterjee and Dutta, 1980; Lumpkin and Plucknett, 1980; Chatterjee and Maiti, 1981). The genus is considered to include six extant species in two sections, based primarily upon features of their reproductive structures (Peters and Calvert, 1982a). The sub-genus *Euazolla* (characterized by three megaspore floats) includes *A. filiculoides* Lamarck, *A. caroliniana* Willdenow, *A. mexicana* Presl. and *A. microphylla* Kaulfuss, while the sub-genus *Rhizosperma* (characterized by nine megaspore floats) includes *A. pinnata* R. Brown and *A. nilotica* De Caisne (Peters, 1984). It is important to note here that a variety of *A. filiculoides*, earlier known as *A. filiculoides var. rubra*, is currently recognized as a distinct species. *i.e.*, *A. rubra* (also known as *A. japonicum*), while *A. pinnata* has been described at two different times by two different schools of thought as *A. pinnata* var. *imbricata* and *A. pinnata var. pinnata*, based on little or sometimes no noticeable difference in their leaf branching morphology. The nitrogen-fixing micro-symbiotic BGA *Anabaena azollae* Strasburger is associated with all these *Azolla* species, confined to the specialized cavities inside the dorsal leaf lobes of the fern. Actually, the sporophytes of *Azolla* plant consists of prostrate multi-branched floating stem which bears deeply bilobed leaves and adventitious roots. While the ventral leaf lobes are only one-cell thick, the dorsal lobes are thicker and aerial. At maturity, the dorsal lobes develop a well-defined cavity which is characterized by the presence of epidermal trichomes (Calvert *et al.*, 1985). It is internally occupied by the BGA *Anabaena azollae* (Gunning *et al.*, 1978). The synchronous development of *Anabaena* with that of *Azolla* leaves and the development profile occurring along a main stem axis were initially reported by Hill (1975). The stem apices and the undifferentiated *Anabaena* cells associated with these are protected by an overarching of the differentiated dorsal and ventral leaf lobes. So these are completely isolated from the environment. Autoradiographic and electron microscopic studies have given a clear concept of the association (Sun *et al.*, 1984a,b). Establishment of *Anabaena* filaments in the cavity of each leaf begins in the developing leaves contiguous with the apical *Anabaena* colony. The partitioning of *Anabaena* filaments normally involves a specialized epidermal trichome, termed as primary branched hair (Calvert and Peters, 1981; Peters and Calvert, 1982b; Peters and Meeks, 1989). This hair originates from the axil of the leaf primordium on the apical meristem and is rapidly differentiated into a single stalk cell and several elongated terminal cells which grow towards the *Anabaena* colony (Uheda, 1986). In due course, the cavities are closed up, engulfing the primary branched hairs and *Anabaena* filaments. During the closure of the development of leaf cavity, *Anabaena* filaments differentiate heterocysts and additional epidermal hairs are formed on the cavity wall. These hairs are located at the back of the cavity depression and are known as secondary branched hair (Braun-Howland and Nierzwicki-Bauer, 1990). Apart from primary and secondary branched hairs there are some other hairs consisting of only two cells and are known as simple hairs. The epidermal hairs of the leaf cavities exhibit transfer cell structure (Pate and Gunning, 1972; Duckett *et al.*, 1975; Peters *et al.*, 1978). Filaments of apical *Anabaena* colonies lack heterocysts (Hill, 1977). During leaf maturation, cell division in the endophyte decreases, the individual cells enlarge and differentiation of heterocysts continues. Sometimes in the mature cavities heterocyst frequency is about 30 per cent as recorded in *Azolla caroliniana* (Peters, 1984: Patra and Singh, 1984). Heterocyst frequency in *Anabaena azollae* increases with the leaf age of *Azolla*. The synchronous development of the micro- and macro-symbionts was first demonstrated in *Azolla filiculoides* (Hill, 1975, 1977), followed by *A. caroliniana* (Becking, 1979; Kaplan *et al.*, 1986), *A. pinnata* (Kaplan *et al.*, 1986) and *A. mexicana* (Patra and Singh, 1984).

The microsymbiont *Anabaena azollae* contains chlorophyll-a, phycobiliproteins and carotenoids, while the *Azolla*, macrosymbiont possesses chlorophyll-a and -b as well as carotenoids (Peters and Mayne, 1974a; Tyagi *et al.*, 1980). The association and the individual partners exhibit, the typicals C-3 cycle (Peters, 1984). In *Azolla caroliniana*, the endophyte accounts for about 15 per cent of the total chlorophyll and proteins (Ray *et al.*, 1978), The microsymbiont contributes as little as 2 per cent of the association's total photosynthetic capability (Plazinski, 1990). Sucrose that is synthesized by *Azolla* is transferred to *Anabaena* (Peters, 1975).

The close relationship between photosynthesis and nitrogen fixation in *Azolla-Anabaena* symbiosis has been thoroughly worked out. (Peters and Mayne, 1974b; Peters, 1975; Peters *et al.*, 1979, 1980a; Tyagi *et al.*, 1981; Kaplan *et al.*, 1986; Kaplan and Peters, 1988). Photosynthesis is, of course, the ultimate source of all the ATP and reductant used in nitrogen fixation. Dark aerobic nitrogenase catalyzed reduction of substrate is dependent upon endogenous reserves of photosynthates and dark rates are always lesser than one-half of those obtained in light. Endogenous reserves of photosynthates and photophosphorylation are the primary driving forces of nitrogenase activity in light. However, dark respiration driven activities may be ATP-limited (Peters, 1975). Although phycobiliproteins are generally associated with photosystem-II, the relative rate of acetylene reduction is greater in the phycobiliprotein region. Moreover, heterocysts of the endophyte are found to contain phycobiliproteins. Kaplan *et al.* (1982, 1986) have presented information on the total phycobiliproteins as a function of leaf cavity and age of *Anabaena azollae*.

A number of studies have been conducted on nitrogenase activity of *Azolla-Anabaena* complex, using acetylene reduction, with rates expressed on the basis of fresh weight, dry weight, total nitrogen, chlorophyll contents or protein content (Ashton and Walmsley, 1976; Becking, 1976, 1979; Peters, 1976; Peters *et al.*, 1976; Watanabe *et al.*, 1977; Talley and Rains, 1980a,b; Watanabe, 1982; Nierzwicki-Bauer, 1990; Vaishampayan *et al.*, 1992b). Optimum culture conditions for this nitrogen-fixing BGA-fern symbiosis were worked out by Peters *et al.* (1980b). In *Azolla caroliniana*, the nitrogenase catalyzed reduction of acetylene, radioisotopic N and protons, as well as their interrelationships are best defined for *Anabaena azollae* and the entire symbiotic complex (Newton and Calvins, 1976; Peters, 1976, 1977; Peters *et al.*, 1976, 1977, 1979, 1980a, 1981a,b).

Increase in microsymbiotic heterocyst frequency has been found to be directly related with nitrogenase activity of *Anabaena azollae* filaments (Kaplan and Peters, 1988; Peters and Meeks, 1989). Few reports are available on the genetics of nitrogen fixation in *Anabaena azollae* (Franche and Cohen-Bazire, 1985; Nierzwicki-Bauer and Haselkorn, 1986; Franche *et al.*, 1987). Cloned DNA from *Anabaena* species (designated as PCC 7120) has been used to determine the distribution of restriction sites around the *nif HDK* genes of *Anabaena azollae* extracted from four *Azolla* species (*A. caroliniana, A. filiculoides, A. microphylla* and *A. mexicana*). Although many restriction sites of the symbiotic *Anabaena nif HDK* genes differ from those of the free-living *Anabaena* species PCC 7120, three apparently identical restriction sites have been found in and around the *nif D* region, the arrangement of *Anabaena azollae* and *Anabaena* species PCC 7120 *nif HDK* genes is apparently similar (with *nif H* and *nif D* linked together and *nif K* being some distance away from *nif D*). The *Anabaena azollae nif HDK* genes appear to be strongly conserved among all the *Azolla* species examined so far, regardless of the geographic origin of the fern (Franche *et al.*, 1987).

Physiological and biochemical studies of the whole association, endophyte-free *Azolla* macrosymbiont and populations of microsymbiotic *Anabaena azollae* isolated from all stages of development provide a more refined approach to an understanding of the structure, functions and host-symbiont interactions in *Azolla* (Peters and Meeks, 1989). Nitrogen fixed by *Anabaena* in mature leaf cavities Is transported to the apical region meeting the nitrogen requirements of both *Azolla* and *Anabaena* (Kaplan and Peters, 1988). Shifting of nitrogen is related with the fact that *Anabaena* located in mature leaf cavities has diminished capability for utilizing and metabolizing the ammonia (Ray *et al.*, 1978). In fact glutamine synthetase (GS), glutamate synthase (GOGAT) and glutamine dehydrogenase (GDH) are the ammonia-assimilating enzymes, and the macrosymbiont alone accounts for association's 90 per cent activities of these enzymes while actively participating to the regulation of the metabolic processes in *Anabaena azollae* (Ray *et al.*, 1978). Other symbiotic BGA and bacteria have also shown low levels of GS compared to the free-living organisms (Stewart, 1977; Haselkorn, 1978). Comparative biochemical studies with some free-living BGA and *Anabaena azollae* suggest that GS activity of the latter decreases as heterocyst frequency and population of epidermal hairs in the leaf cavities increase (Haselkorn *et al.*, 1980; Orr and Haselkorn, 1982; Haselkorn, 1992). These authors have also indicated about the lower levels of GS antigen in *Anabaena azollae*, as compared to the free-living BGA. Peters *et al.* (1980b) suggested that regulation of GS in the association is at the level of activity rather than synthesis. It has been observed that GS m-RNA transcript levels in *Anabaena azollae* is one-tenth of that in free-living BGA (Nierzwicki-Bauer and Haselkorn, 1986). GDH has a regulatory role in reassimilation of the released ammonia by GS at high intra-cavity ammonia concentrations (Braun-Howland and Nierzwicki-Bauer, 1990).

Anabaena azollae exhibits photoheterotrophic (Peters and Mayne, 1974a) or mixotrophic (Peters, 1975) metabolism. Sucrose, glucose and fructose from *Azolla* might serve as reduced carbon sources for the endophyte in mature leaf cavities, providing reductant for nitrogen fixation (Ray *et al.*, 1978). The specialized leaf cavities of *Azolla* provides physical protection, requisite minerals and other nutrients as well as adequate moisture to the microsymbiont (Peters, 1984).

Although lectins (phytohaemagglutinins) have been known to play an important role in recognition process between two symbiotic partners, *e.g.*, legume-*Rhizobium* symbiosis (Sindhu and Dadarwal, 1994), yet it has been a subject of critical discussion in *Azolla-Anabaena* symbiosis because the symbiotic state in this case is maintained all through the life-cycle (Braun-Howland and Nierzwicki-Bauer, 1990). However, conventional haemagglutinin techniques have been used to demonstrate the presence of lectins in *Anabaena azollae* freshly isolated from three *Azolla* species (Kobiler *et al.*, 1982). Yet, consensus is that the lectins of macro-symbiotic (*Azolla*) origin are specifically involved with the maintenance of fern-BGA symbiosis (Ladha and Watanabe, 1984; McCowen *et al.*, 1987). It is important to mention here that in contrast to the results of immuno-fluorescence studies, data on lectin studies indicate that haemagglutinating activity could be inhibited by pre-adsorption of the *Azolla pinnata* lectin to microsymbiotic BGA freshly isolated from *A. pinnata*, but not to the microsymbiont isolated from *Azolla filiculoides*'and vice-versa (McCowen *et al.*, 1987). Subsequently, it was established through serological tests that lectins of macrosymbiotic origin are different in the diferent *Azolla* species, while *Anabaena azollae* maintains a uniform type of lectin of microsymbiotic origin (freshly isolated from any *Azolla* species). Of course, the phytohaemagglutinins isolated from *Anabaena azollae* were sharply distinguishable in molecular weights and cross-reacting properties, as compared to the free-living BGA (Vaishampayan and Awasthi, 1995b). In view of the recent serological studies the *ex-planta Anabaena azollae* culture of Newton and Herman (1979), which shows lectins more like those of a free-living *Anabaena* species, has yet to be unequivocally confirmed to be either the actual endosymbiont of *Azolla* or an epiphytic contaminant to be mistaken as *Anabaena azollae* (Vaishampayan, 1994a). Although the possibilities of the occurrence of *Anabaena azollae* In free-living condition does exist, but it has not been possible to reassociate it with the fern (Peters and Meeks, 1989). Even efforts on the isolation of *Anabaena azollae* from the various *Azolla* species have been well documented (Huneke, 1933; Venkataraman, 1962; Ashton and Walmsley, 1976; Becking, 1976; Newton and Herman, 1979; Tel-Or *et al.*, 1983; Berliner and Fisher, 1987; Meeks *et al.*, 1988), but none of these has been reconstituted with the endophyte-free *Azolla*. In this context, the only report of Lin *et al.* (1988) is worth mentioning on an artificial reconstruction of *Azolla-Anabaena* symbiosis from heterologous megasporocarps. Actually, when the apical cap and indusium were excized from a fertile megasporocarp apparatus, the plants emerging from such megaspores were BGA-free. However, if an apical cap from a different fertile megasporocarp was placed on the top of the excized megasporocarp, reinfection with the BGA associated with the transplanted apical cap could be demonstrated and verified by SEM, monoclonal antibodies and nitrogen fixation assay (Lin *et al.*, 1988). However, using this technique when free-living form of *Anabaena azollae* or any other filamenmtous, heterocystous and nitrogen-fixing BGA was allowed to infect on endophyte free *Azolla* megaspore apparatus, the difference in growth rate eventually resulted in one partner outgrowing the other

(Peters and Meeks, 1989; Vaishampayan, 1993a,b; 1995b). Therefore, efforts to reconstitute the functional symbiosis by replacing the BGA microsymbiotic chamber with a cultured BGA isolate have yet to be strengthened for a conclusive output in this direction.

Blue-Green Algae in Cycad Symbiosis

BGA symbiosis with gymnosperms is exemplified by Cycad plants only. There are nearly 150 living cycad species on record right now (Giddy, 1984) belonging to the families Cycadaceae (*Cycas*), Stangeriaceae (*Stangeria*) and Zamiaceae (*Bowenia, Ceratozamia, Dioon, Encephalartos, Lepidozamia, Macrozamia, Microcycas* and *Zamia*). With the exception of *Zamia pseudoparasitica* (which grows epiphytically on trees in the rain forest of Panama), all cycads are terrestrial plants. Their coralloid roots developing out of the tap roots (other than lateral roots and apogeotropical roots) need special mention as these carry distinct cylindrical zones of nitrogen-fixing BGA. This is quite apparent as a pronounced dark blue-green band in a cross- or longitudinal sectioning of a mature coralloid root. There are elongated cycad cells traversing the intercellular BGA zone which help in the exchange of metabolites between the micro- and macro-symbionts (Lindblad and Bergman, 1990). However, there are also reports on the occurrence of intracellular micro-symbiotic BGA in the coralloid roots of *Cycas revoluta* (Obukowicz *et al.*, 1981) and *Macrozamia communis* (Nathanielsz and Staff, 1975).

In thirteen different cycad species, involving the genera *Bowenia* (Bowyer and Skerman, 1968), *Cycas* (Grill Caiola, 1980), *Dioon* (Grilli Caiola, 1980), *Encephalartos* (Grobbellar *et al.*, 1987). *Macrozamia* (Grilli Caiola, 1980) and *Zamia* (Lindblad *et al.*, 1985), the BGA *Nostoc* was identified as the (microsymbiont. In *Cycas revoluta* the microsymbiont *Anabaena* species has been shown to occur (Zhu, 1982), while *Microzamia communis* has been demonstrated to associate both *Nostoc* and *Anabaena* as microsymbiotic partners (Nanthanielsz and Staff, 1975). In contrast to these *Nostoc* and/or *Anabaena* associations, *Calothrix* species has been found to be associated with a South African species of *Encephalartos* (Grobbellar *et al.*, 1987). Restriction fragment length polymorphism analyses of the chromosomal DNA preparations from microsymbiotic BGA (freshly isolated from the coralloid roots of *Ceratozamia mexicana, Dioon edule, Zamia furfuraceae* and *Z. skinneri*) showed differences in sizes of kDNA fragments hybridizing to specific BGA probes, thereby suggesting the existence of different BGA species/strains with the cycads (Lindblad *et al.*, 1989). This study explored further that although a mixture of *Nostoc* strains may associate with a single cycad species, yet a single BGA type predominates in the roots of a single cycad plant. Involvement of specific recognition signals) has been suggested since only heterocystous and nitrogen-fixing BGA have been associated with the cycads (Lindblad and Bergman, 1990). The microsymbionts *Nostoc* 7422, *Nostoc* 73102 and *Nostoc* species, isolated from *Cycas, Macrozamia* and *Zamia* have been shown to have numerous fimbriae which are proteinaceous structures speculated to be involved in *Nostoc-Peltigera* symbiosis (Dick and Stewart, 1980; Lindblad and Bergman, 1990). However, whether this has a definite/specific and sole role to play in establishing recognition between BGA-cycad association remains to be ascertained, since many free-living BGA, *e.g.*, *Anabaena cylindrica, A. variabilis* and *Nostoc muscorum* do also possess fimbriae.

During infection, studied so far in a solitary species, *Macrozamia communis*, the BGA filaments enter through a break in the dermal layer and reach the specified cylindrical zone across the continuous intercellular or even intracellular cortical channel (Nathanielsz and Staff, 1975). Separation and reunion of cycad-BGA symbiotic partners have been scarcely studied, but of course, both the partners have shown to grow independently in laboratory culture (Lindblad and Bergman, 1990).

The BGA microsymbionts of *Cycas revoluta* (Zhu, 1982), *Dioon edule* (Grilli Caiola, 1975), *Macrozamia communis* (Grilli Caiola, 1974) and *Zamia skinneri* (Lindblad *et al.*, 1985) have shown the occurrence of akinetes. In other cases akinetes have been rarely found. Heterocyst frequency of the endosymbiotic BGA has been found to vary from 5-6 per cent in growing tips to nearly 46 per cent in the basal older part (Lindblad *et al.*, 1985; Lindblad and Bergman, 1986), in sharp contrast to the free-living BGA which uniformly have 5-7 per cent heterocysts. These authors have also observed double, triple or quadruple heterocysts with higher frequencies in microsymbiotic BGA of *Cycas revoluta* and *Zamia skinneri*, a feature which is rarely found in the free-living BGA.

Nearly 50 to 150 cycad species with the BGA endosymbionts, like *Nostoc*, *Anabaena* or *Calothrix*, have been demonstrated through acetylene reduction and tracer techiques to perform nitrogen fixation. These include all the 33 species of *Encephalartos*, tested so far (Grobbelaar *et al.*, 1986). The rate of nitrogen fixation by BGA-cycad symbiosis has been found to be nearly 7-times higher than that of the free-living BGA (Kumar *et al.*, 1986). A remarkable feature noted in the corraloid roots of *Cycas* and *Zamia* is the fact that while nitrogenase activity is highest at the root tip, and is considerably reduced at the basal end, the heterocyst frequency was noted to be in the reverse order, *i.e.*, lowest at the root tip and enhanced to as great as 45 per cent at the base (Lindblad and Bergman, 1986). This clearly indicates that the majority of multiple heterocysts may not be functional due to the hampered transport of important compounds from distant vegetative cells, because otherwise in experiments using antibodies directed towards the *Fe*-protein of the nitrogenase complex, the nitrogen-fixing enzyme was found in all heterocysts (Bergman *et al.*, 1986). Even the possibility of nitrogenase being synthesized in the vegetative cells of the microsymbiotic BGA has been suggested, because a lowered oxygen tension may prevail in the BGA zone. Since the zone is completely surrounded by respiratory cycad root tissue, and since photolysis of water and intracellular production of oxygen will be inhibited in the absence of light, there shall be hardly any oxygen left to damage nitrogenase in the vegetative cells (Lindblad and Bergman, 1990).

In contrast to the situation in BGA-fungal symbioses or *Azolla-Anabaena* symbiosis, the microsymbiont of *Cycas revoluta, Ceratozamia mexicana* and *Zamia skinneri* exhibit a very high *in vitro* activity of the primary ammonia-assimilating enzyme GS in relation to the free-living BGA (Lindblad and Bergman, 1986). These authors have also shown that the cellular levels and sub-cellular distribution of the GS protein in the microsymbiont of *Cycas revoluta* and *Zamia skinneri* were quite comparable to those of the free-living BGA. The other ammonia-assimilating enzyme GOGAT has also been demonstrated to occur in *Cycas revoluta* (Lindblad *et al.*, 1987).

In the free-living BGA, a biphasic uptake of ammonium has been reported to occur (Rai *et al.*, 1984). While the first phase represents transport into the cells, the second phase is dependent upon sustained assimilation of ammonia via GS. However, in *Cycas revoluta* microsymbiont, the second phase of ammonia assimilation was not detectable, although the BGA component displayed substantial *in vitro* GS activity (Rai *et al.*, 1986; Lindblad and Bergman, 1990).

The Blue-Green Alga *Nostoc* Species in Symbiosis with *Gunnera*

Gunnera is the only angiospermic genus to invariably display a nitrogen fixing symbiosis with the BGA *Nostoc* species, distributed all through the New Zealand, South-East Asia, Central America and South America (Bonnet, 1990). Its all ten species from New Zealand (Silvester and Mc Namara, 1976), as well as one each from Indonesia, *i.e.*, *Gunnera macrophylla* (Becking, 1976), Hawaii, *i.e.*, *G. kaalensis* (Towata, 1985) and Germany, *i.e.*, *G. chilensis* (Von Neumann *et al.*, 1970) have been tested to harbour *Nostoc*. The specific attributes of *Gunnera* which help its association with *Nostoc* species as the sole natural BGA-angiosperm symbiosis, have been discussed earlier in detail by the previous reviewers (Becking, 1977; Silvester, 1977; Bonnet, 1990), but, of course, it can be pointed out here that it is the inherent property of *Gunnera* plant to make glands with unique features and an unusual structure of the cell wall. The cells are quite tolerant to BGA introduction and replication. *Gunnera* maintains a favourable microenvironment for growth of *Nostoc* as an epiphyte on the apex, and the intracellular levels of fixed nitrogen are low enough to favour retention of *Nostoc* by inducing heterocyst differentiation and nitrogen fixation. *Nostoc* cells colonize only the internal portion of a *Gunnera* plant where glands are produced. However, in the absence of glands, this microsymbiotic association of *Nostoc* is found on extracellular locations of *Gunnera*, particularly in the mucus covering the shoot apex.

Hoffman (1985) has demonstrated that the glands of *Gunnera chilensis* possesses a typical central papilla with 5-7 circumscribing papillae, thereby producing 4-6 channel junctures. Through the independent entry of *Nostoc* at different channel junctures, several separate clusters of BGA cells are formed that are frequently colonized inside the mature glands. *Nostoc* cells in healthy and recently colonized glands are surrounded by a membrane of *Gunnera* origin which is similar to the plasma membrane with respect to staining and thickness. Although there is less agreement that the cell wall material of *Gunnera* origin surrounds the *Nostoc* cells, yet Von Neumann *et al.*, (1970) is of the opinion that wall material comes to surround *Nostoc* cells as these penetrate the cells of *Gunnera*. In this context, Silvester and Mc Namara (1976) viewed the presence of wall material as an indicator of senescence, and proposed that wall material was present only around degenerating cells and hence produced by the *Gunnera* cell wall after colonization was completed; this was later confirmed by Towata (1985).

Nostoc has been generally treated to be an endosymbiont with *Gunnera*. But, if it is true, what should be the role of the shoot apex mucus which colonizes *Nostoc* cells as an epiphyte, appears to be a question mark. It has been hypothesized that the mucus helps in (a) protection of the exposed glandular tissue wherefrom the epidermis has been degraded, (b) prevention of desiccation of the population of *Nostoc* growing epiphytically on the shoot apex, (c) nutrition of *Nostoc* cells living epiphytically on the shoot apex, (d) inhibition of growth of microbes which might otherwise cause disease and decay of the exposed *Gunnera* tissue, (e) provision of a viscous medium through which the *Nostoc* cells migrate on their way through the channel junctures towards interior regions of the gland, and (f) prevention of excessive differentiation of *Nostoc* cells into heterocysts which might fragment the filaments or otherwise impair their ability to penetrate *Gunnera* cell wall. The cells of *Nostoc* (both newly colonized and the older ones) exist as members of motile filaments both within the channel junctions towards interior of glands as well as on the surface of the shoot apex of *Gunnera* (Sprent, 1979). The frequency of heterocysts in the total intracellular population increases up to 65 per cent with the age of glands (Silvester, 1976).

It is important to point out here that both *Nostoc* and *Gunnera* are capable to thrive independently of each other, yet *Nostoc* forms intracellular symbiosis with the latter. This, indeed, reflects a unique potential of both the partners in this respect, since *Gunnera* cell sustains with a living foreign organism multiplying inside, while *Nostoc* cells continue multiplying well at the expense of molecular nitrogen even being apart from the outer world (confined inside the *Gunnera* membrane). In order to locate the factor(s) responsible for imparting such unique potential to the microsymbiotic BGA, a large number of free-living *Nostoc* and *Anabaena* species were screened for their potential to establish symbiosis with *Gunnera maniculata* and *G. chilensis*. While *G. maniculata* could associate with six different *Nostoc* isolates (Bonnet and Silvester, 1981), *G. chilensis* could colonize none (Von Neumann *et al.*, 1970). *Anabaena* was not at all acceptable as a microsymbiont by the *Gunnera* system (Bonnet, 1990). However, investigations on identical reconstitution of *Nostoc* using various isolates of *Gunnera* are completely lacking. Thus, one thing is clear that the specificity of symbiosis is restricted to only certain *Nostoc* species, though the recognition factor is yet to be worked out.

The *Gunnera-Nostoc* symbiosis performs a substantial nitrogen fixation to meet its own needs. The rate of nitrogen fixation has been estimated to be 72 kg N per ha per year by *Gunnera dentata*. extrapolated from laboratory derived estimates (Silvester and Smith, 1969). However, direct field assessment is in support of 12-21 kg N fixed per ha per annum by *Gunnera macrophylla-Nostoc* symbiosis (Becking, 1976). Remarkably, in the gland tissues of *Gunnera* the rate of nitrogen fixation by BGA microsymbiont Is extremely high (Bonnet, 1990).

Nostoc cells isolated from *Gunnera macrophylla* have been shown to grow heterotrophically (Harder, 1917). Even within the gland these fix very low quantity of inorganic carbon and lack the detectable phycocyanin and photosystem-II activity (Silvester, 1976). However, with the increase in light intensity on *Gunnera* glands colonized by *Nostoc* cells, the acetylene reducing activity of the gland slices have been shown to increase under possible influence of the photosystem-I mediated phosphorylation operative inside microsymbiotic heterocysts (Bonnet, 1990). Although biochemical factors and regulatory mechanisms which promote heterocyst differentiation in endosymbiotic *Nostoc* and facilitate transfer of fixed nitrogen from *Nostoc* heterocysts to *Gunnera* are not yet fully known. Radioisotopic N fixed by *Nostoc* has been shown, however, to be detectable in alcohol-insoluble extracts of *Gunnera* tissues, indicating thereby that nitrogen fixed by BGA microsymbiont is transferred and assimilated into *Gunnera* protein (Silvester and Smith, 1969).

Blue-Green Algae in Artificial Symbioses

The BGA are diazotrophic (Fay, 1983) and their significant role in improving the soil N fertility of wet fields at the expense of photosynthetic energy produced by their own is well documented in India and other tropical countries (Singh, 1961; Roger and Kulasooriya, 1980; Venkataraman, 1981). However, in view of their delicate form and structural organization, the viability and nitrogen-fixing potentials of BGA are adversely affected on exposure to the various abiotic stress factors of the modern agronomic practices (Singh and Vaishampayan, 1978; Vaishampayan *et al.*, 1978; Singh *et al.*, 1979; Vaishampayan and Hemantaranjan, 1983; Vaishampayan and Mishra, 1989). This is in view of the fact that the modern agricultural fields are generally treated with the heavy doses of synthetic agrochemicals, *i.e.*, herbicides and nitrogenous fertilizers to maintain the high yielding properties of the new crop varieties. Among these, the herbicides are generally designed to be either photosynthetic inhibitors or photosynthetic energy deviators (Dodge, 1975), thereby killing the target weeds as well as the BGA as non-target biological species owing to their identical oxygenic photosynthesis (Vaishampayan, 1983a, 1984b-f, 1985a-d; Vaishampayan *et al.*, 1992a). On other face of the coin, the combined N/synthetic N fertilizers repress nirogen fixation at a threshold level in view of the inherent properties of these organisms to switch off their dinitrogen-fixing machinery no sooner the concentration of fixed nitrogen (ammonium) in the external, medium/soil environment exceeds its biological consumption (Singh *et al.*, 1978a,b; Vaishampayan, 1981, 1982a-f, 1983b-f, 1984g,h; Vaishampayan and Singh, 1981a,b; Vaishampayan and Hemantaranjan, 1984; Vaishampayan and Sahay, 1984a-c; Prasad and Vaishampayan, 1987), thereby, paradoxically converting a nitrogen fixer into an ammonia consumer for maintaining rather its own survival. The net result is the fixation of no greater than 27-30 kg N per ha per year as an average by the different free-living BGA species (Singh and Singh, 1989).

The above situation has been seriously considered by the concerned scientific community to improve these photobiological nitrogen fixers by exploiting their characteristics of fast growth rate, simple genome organization, somewhat response to DNA transformation and an efficient homologous recombination, apart from their "higher plant-type" oxygen-evolving photosynthesis, chlorophyll biosynthesis and amino acid synthesis (Vaishampayan, 1983g,h, 1984i; Haselkorn, 1992). Accordingly, considerable phycotechnological developments have been made during the last couple of decades to establish the acquisition of herbicide-resistance (Vaishampayan and Prasad, 1980, 1981a-c, 1982, 1984a,b; Vaishampayan, 1983i, 1984j; Prasad and Vaishampayan, 1984, 1994; Prasad *et al.*, 1986; Mishra *et al.*, 1991a; Vaishampayan, 1995a) and derepression of nitrogen fixation for the constitutive synthesis and activity of the enzyme nitrogenase (Singh and Singh, 1978; Mishra *et al.*, 1991b; Prasad *et al.*,

1991) in model nitrogen-fixing systems under Indian agro-climatic conditions. However, on reinoculation to the fields such genetically improved BGA species undergo a competition problem with the pre-dominant native species. This natural contest comes in way of their adaptation to field condition. And here comes the rationale and ethics for attempting establishment of artificial plant-BGA symbiosis for offering protection to the natural as well as genetically improved nitrogen-fixing BGA against direct exposure to the detrimental abiotic stress factors described above, and facilitating a congenial situation whereby nitrogen is fixed in close proximity to the macrosymbiont (the selected host plant) for absolute and easy-to-reach consumption by the latter. This is a highly fascinating area of modern biological science for enabling our fruit, vegetable and grain crops to have in-built N supply of BGA origin.

Prior to attempting creation of an artificial plant-BGA symbiosis at the intracellular level it is a pre-requisite to ascertain the possibility of introducing a BGA cell inside the cell of target host plant and assess the fate of either of such artificially associated partners. It is also important at the same time to assess whether the nitrogen-fixing BGA cell introduced inside a plant protoplast will supply ammonium to the plant cell, and if so, whether it is possible to regenerate a stable nitrogen-fixing plant from a cell (protoplast with endosymbiotic BGA). Rather, the latter is to be considered even in the case of intercellular plant-BGA interactions, where in BGA cells are mixed with plant callus to occupy intercellular space.

The plant seldom contains a microsymbiont inside the cells (with the exception of a few cells) in the case of intercellular interaction, and thus nitrogen fixation in such an association takes place only in individual zones or organs of the plant (Vose, 1983; Gusev and Korzhenevskaya, 1990), like the natural symbiotic systems where the association constitutes only 1-2 per cent of microsymbiotic cells (Stewart *et al.*, 1983). Contrary to this, the positive aspects of introducing the intact cells of nitrogen-fixing BGA into plant cells or tissues are plentiful in the case of intracellular symbiosis. The most important benefit of such an association is that the oxyphobic nitrogen-fixing enzyme, *i.e.*, nitrogenase, will remain protected from oxygen which is regularly generated by the BGA as a byproduct of BGA photosynthesis but would be absorbed by the plant during respiration. Moreover, the total photosynthetic energy of BGA origin would be channelized to nitrogenase, since the BGA under symbiosis would be able to otherwise receive the bulk photosynthate of plant origin. The same energy would also be useful for the photoassimilation of fixed nitrogen by its incorporation into organic compounds. BGA, in fact, possess a number of specific properties by virtue of which these could be identified as the most suitable partners of plant cells in artificial associations, *viz.*, (a) BGA are believed to have played a role in the formation of eukaryotic cells in the evolutionary process through symbiosis with non-photosynthetic organisms (Margulis, 1981); (b) BGA are the most prominent phototrophs, with the exception of green algae, to enter into symbiotic relations with other organisms in nature, forming associations of both exo- and endo-symbiotic types, thus presenting valid grounds for their introduction into both isolated protoplasts (Reisser, 1984) and populations of cultured cells (Fay, 1983); (c) apart from acting as a nitrogen-fixing component, the BGA perform a number of metabolic functions in symbiosis with plants which involve excretion of carbohydrates, polysaccharides, peptides, amino acids, organic acids, glycerol, "gibberelin-like" growth promoting substances and vitamins (Shah and Vaidya, 1977; Gibson and Smith, 1982; Yagodina, 1983) to help plant cell culture; (d) during formation of symbiosis much importance is attached to the coincidence in growth rates of partners. In this context, the growth rates of BGA are comparable to those of higher plant cells in batch cultures (Street, 1977; Suleimanova and Mineyeva, 1981).

But, in spite of the attributes referred to above in favour of establishing an artificial BGA-plant symbiosis, there are certain well-identified limitations which many times come in way of developing/establishing desired associations, *viz.* (a) only certain BGA species belonging to the genera *Nostoc* and *Anabaena* generally occur in natural associations with only a limited number of higher plants, displaying a high degree of host-symbiont specificity (Stewart *et al.*, 1983; Gusev and Korzhenevskaya, 1990); (b) toxins produced by many of the BGA hamper plant cell culture (Gibson and Smith, 1982; Yagodina, 1983), and thus selection of host plant becomes important which may be reluctant/immune to the effects of such toxins; (c) the optimum pH values are acidic (5.0-5.5) in plant, and neutral to alkaline (7-10) in BGA; (d) the optimum temperature for the growth of plant cells is 30-40°C which is higher than that for BGA (24-27°C); (e) the concentration of mineral salts in the medium for growing plant cells is higher than that used for BGA culture; (f) BGA do not require any organic source for growth, while plant cells generally require sucrose and other organic compounds for callus culture and regeneration (Stanier *et al.*, 1971; Vasil *et al.*, 1980).

With quite a careful consideration of the non-congenial factors, noted above, the isolated plant protoplasts have been used to introduce various species of BGA, *viz.* *Anacystis nidulans*, *Gloeocapsa* species, *Anabaena variabilis* and *Nostoc muscorum* (Hughes *et al.*, 1978; Gusev and Korzhenevskaya, 1990, 1995: Vaishampayan, 1995b), of which the latter three are nitrogen-fixing. It is important to mention here that *Anabaena variabilis* is a facultative phototroph and is thus also capable of chemo-heterotrophic growth in the dark. Further, an auxotrophic mutant of *Anabaena variabilis* has also been introduced into plant protoplasts to establish a possible dependence of the microsymbiotic BGA on the growth and metabolism of macrosymbiont (the selected host plant) (Wolk, 1973). The filaments of *Anabaena variabilis* are fragmented and cells are treated with lysozyme to obtain spheroplasts (Gusev *et al.*, 1979; Semenova *et al.*, 1982), while *Gloeocapsa* cells are sometimes subjected to sonication to remove the outer sheath (Gusev and Korzhenevskaya, 1990) for introduction to the plant protoplasts to establish an artificial symbiosis.

Nearly 25 artificial plant-BGA associations are on record so far, including the symbioses of (a) *Medicago sativa* with *Anabaena variabilis*, *Chlorogloeopsis fritschii*, *Calothrix elenkinii* and *Nostoc muscorum*; (b) *Daucus carota* with *Anabaene flos-aquae*, *Plectonema boryanum* and *Gloeothece* species; (c) *Solanum laciniatum* with *Anabaena variabilis* and *Chlorogloeopsis fritschii*; (d) *Nicotiana tobacum* with *Anacystis nidulans*, *Anabaena variabilis* and *Nostoc* species; (e) *Papaver somniferum* with *Anabena variabilis*; (f) *Panax ginseng* with *Chlorogloeopsis fritschii*, *Anabaena variabilis*, *Calothrix elenkinii* and *Nostoc muscorum*; and (g) *Dioscorea deltoidea* with *Anacystis nidulans*.

Localization of BGA and relation of partners in artificial associations have well-marked similarities with the natural symbioses, *e.g.*, the stimulating effect of plant cells and tissues on the formation of heterocysts (the nitrogen-fixing apparatus and nitrogenase activity of BGA); supply of fixed N by the BGA to support the growth of plant tissues; transfer of carbon compounds from plant tissues to BGA as an added source of photosynthetic energy to perform the physiological and biochemical functions of dinitrogen fixation more efficiently for a higher output of ammonium. Nevertheless, there do exist considerable differences between natural and artificial symbioses between plant and BGA, *i.e.*, (a) in the case of natural associations, microsymbiotic growth is inhibited or reduced in a plant host, whereas in the artificial association it is stimulated by metabolites from the eukaryotlc partner; (b) heterocyst frequency of the microsymbiont is not as high as that in the endophyte of natural symbioses; and (c) in the absence of a specialized space for the BGA inside the host plant body (which is otherwise well-defined in the case of natural symbioses) the microsymbiont of artificial symbioses make their own habitats in the population of plant cell/tissue cultures. Probably in view of these facts, many other attempts to introduce BGA into isolated plant protoplasts failed to produce viable systems. Inadequacy of introduction techniques has also been a major factor behind the failure. But, of course, wherever success has been achieved, attempts have been made to establish interactions for nitrogenous metabolites of BGA with the complete exclusion of combined N from the medium in order to facilitate gain of fixed N by plant cells in the presence of the artificially introduced BGA (Gorelova *et al.*, 1985).

Suspension and callus cultures as well as inoculation of plant cuttings and regeneration of plants from mixed cultures are in practice to establish nitrogen-fixing associations with BGA (Gusev and Korzhenevskaya, 1995; Vaishampayan, 1995b) and the time appears to be not far off when we shall be able to combine our desired nitrogen-fixing BGA species and rare mutant isolates (which otherwise fail to compete with the native species on exposure to the field environment) in artificial, but functional, symbiosis with a wide array of fruit, vegetable and cereal plants for an ascertained viability and applicability of the associated BGA strain to significantly exert its beneficial effect on the concerned host either through supplying fixed N or supporting with other valuable metabolic excretory biochemicals.

Concluding Remarks

Despite enormous investigations on record on the biology of blue-green algal symbiosis with diverse groups of plants, there still exist a few important pitfalls and gap areas to be systematically worked out for harvesting the actual applied potential of symbiotic BGA, particularly in view of the fact that these fix nitrogen at a much higher rate than the free-living counterparts.

For most of the plant-BGA symbiotic complexes the mechanisms of increase in heterocyst and regulation of carbon supply for nitrogenase activity, as well as control mechanism of N-assimilating enzyme systems are not yet well known. Such worth investigating aspects will not only be important for improving any BGA-plant symbiotic complex, but would also provide insight for modifying our free-living BGA community with additional acquisitions, as the latter fixes at the most a very low quantum of 27-30 kg N per ha per annum.

One of the major limitations in identifying putatively unique regulatory mechanisms/systems of nitrogen-fixing BGA has been the inability to biologically complement mutations by experimentally controlled transfer of DNA between recipient and donor cells. Many of such complementation barriers have been removed by the construction of basic shuttle vectors capable of replication in *Escherichia coli* and certain strains of free-living and symbiotic *Nostoc* and *Anabaena* (Wolk *et al.*, 1984; Flores and Wolk, 1985; Meeks, 1990), yet far remains to be done to achieve some definite information on the regulatory aspects of the various physiological and biochemical processes of BGA in symbioses with all categories of plants except those in association with the water-fern *Azolla*. Although lot of progress has been made during the last few years on the detailed biology of *Azolla-Anabaena* symbiotic nitrogen-fixing complex, but aspects of its molecular biology are mostly reported to be of U.S. origin (Peters and Meeks, 1989; Braun-Howland and Nierzwicki-Bauer, 1990). Possibly this is in view of the high sophistications related with this area of study and owing to the truth that work on the agronomics of this complex attracts the western scientific teams less than those who are its actual users in the tropical belts of South-East and Far-East Asia. However, despite the progress made so far, the recognition process between *Azolla* and *Anabaena* still remain to be thoroughly worked out. Probably the inability to establish *ex-planta* culture of *Anabaena azollae* beyond all shadows of antigenic and serological tests have been a major limitation which needs to be resolved for a clear understanding of the symbiotic process and a routine genetic manipulation of micro- and macro-symbionts, as well as a sustainable interchange with an improved microsymbiotic BGA partner.

Similarly, the *Nostoc-Gunnera* symbiosis merits a thorough investigation because the host in this case is an angiosperm; *Nostoc* occurs as an endosymbiont in *Gunnera*: *Nostoc* heterocysts which differentiate within *Gunnera* cells, are able to synthesize or assemble all factors necessary for nitrogen fixation and to regulate their own internal oxygen tension; and *Nostoc* enables *Gunnera* to grow as a luxuriant herb in soils deficient in fixed nitrogen. Naturally, therefore, future concerted research in this area may provide useful clues to convert our grain crop species (which are all angiospermic) into nitrogen-fixing plants either by making possible novel endosymbiotic associations of nitrogen-fixing BGA with crop species or by introducing *nif* (nitrogen-fixing) genes of BGA origin into the plant chloroplasts through microprojectile bombardment. Obviously, this area of work does not now appear to be hypothetical only, since we have already described in this article at least six angiospermic genera which display an artificial symbiosis with the four nitrogen-fixing blue-green algae. Though it is difficult right now to predict as to whether an artificial symbiosis of a nitrogen-fixing BGA at the intercellular or intracellular level (in context with the target host plant cells) will be superior for the formation and survival of such laboratory-made novel associations, and whether the initial intracellular localization of the symbiont will be preserved as such during regeneration. But, of course, soon after we get resolved these issues it shall not only be possible to construct a large variety of artificial symbioses of nitrogen-fixing BGA with the economically useful fruit, vegetable and cereal plants, but it would also be feasible to use the rare herbicide- and heavy metal-resistant and derepressed BGA mutants as viable and constitutive nitrogen-fixing microsymbionts in such artificial associations for getting relieved of the problem of strain competition in fields.

Acknowledgements

The help rendered by the team members of the Photobiological Nitrogen Fixation Research Laboratory of this Department in literature survey and discussion, and by the Head of the Department for extending requisite facilities for accomplishing a major portion of authors' own work incorporated in this article is gratefully acknowledged. The authors also feel indebted to the Indian Council of Agricultural Research for supporting this Department through Recurrent Plan Projects as one of the main recognized centres of All-India Co-ordinated Biological Nitrogen Fixation Research, as well as the University Grants Commission for sanctioning the Career Award Project to one of us (A.V.) with the financial assistance from whom a part of fresh work, included in this article, has been conducted during the recent past.

References

Ahmadjian, V., 1962. Lichens. In: *Physiology and Biochemistry of Algae,* (Ed.) R.A. Lewin, Academic Press, New York, pp. 817–832.

Ahmadjian, V. and Paracer, S., 1986. *Symbiosis.* University Press of New England, Hanover, NH, pp. 1–342.

Apte, S.K., 1992. Molecular biology of cyanobacterial nitrogen fixation: Recent advances. *Indian Journal of Microbiology,* 32: 103–126.

Ashton, P.J. and Walmsley, R.D., 1976. The aquatic fern *Azolla* and its *Anabaena* symbiont. *Endeavour,* 35: 39–43.

Basilier, K., 1980. Fixation and uptake of nitrogen in *Sphagnum* blue-green algal associations. *Oikos,* 34: 239–247.

Basilier, K., Granhall, U. and Stenstrom, T.A., 1978. Nitrogen fixation in wet minerotrophic moss communities of a subarctic mire. *Oikos,* 31: 236–251..

Becking, J.H., 1976. Contributions of plant algal associations. In: *Proc. First Internat. Symp. Nitrogen Fixation,* Vol. II (Eds.) W.E. Newton and C.J. Nyman. Washington State University Press, Pullman, Washington, USA, pp. 581–591.

Becking, J.H., 1977. Dinitrogen-fixing associations in higher plants other than legumes. In: *A Treatise on Dinitrogen Fixation, Section III: Biology,* (Eds.) R.W.F. Hardy and W.S. Silver. John Wiley and Sons, New York, USA, pp.185–211.

Becking, J.H., 1979. Environmental requirements of *Azolla* for use in tropical rice production. In: *Nitrogen and Rice.* International Rice Research Institute, Los Banos, Laguna, Philippines, pp. 345–373.

Bergman, B., Lindblad, P. and Rai, A.N., 1986. The Cycas revolution Nitrogenase in free-living and symbiotic cyanobacteria: Immuncelectron microscopic localization. *FEMS Microbiol. Lett.,* 35: 75–87.

Berliner, M.D. and Fisher, R.W., 1987. Surface lectin binding to *Anabaena variabilis* end to cultured and freshly isolated *Anabaena azollae. Curr. Microbiol.*, 16: 150–152.

Bond, G. and Scott, G.D., 1955. An examination of some symbiotic systems for fixation of nitrogen. *Ann. Bot.*, 19: 67–75.

Bonnet, H.T., 1990. The *Nostoc–Gunnera* association. In: *Handbook of Symbiotic Cyanobacteria*, (Ed.) A.N. Rai, CRC Press, Inc., Boca Raton, Florida, USA, pp. 161–171.

Bonnet, H.T. and Silvester, W.B., 1981. Specificity in the *Gunnera–Nostoc* endosymbiosis. *Hew Phytol.*, 89: 121–139.

Bornet, E., 1873. Recherches sur les gonidies des lichens. *Ann. Sci. Net. Bot.*, (Ser.5) 17: 145–147.

Bowyer, J.W. and Skerman, V.B.D., 1968. Production at axenic cultures of soil-borne and endophytic blue-green algae. *J. Gen. Microbiol.*, 54: 299–315.

Brasell, H.M., Davies, S.K. and Mattay, J.P., 1986. Nitrogen fixation associated with bryophytes colonizing burnt sites in Southern Tasmania, Australia. *J. Bryol.*, 14: 139–150.

Braun-Howland, E.B. and Nierzwicki-Bauer, S.A., 1990. *Azolla–Anabaena* symbiosis: Biochemistry, physiology, ultrastructure and molecular biology. In: *Handbook of Symbiotic Cyanobacteria*, (Ed.) A.N. Rai. CRC Press Inc., Boca Raton, Florida, USA, pp. 65–117.

Bibrick, P. and Galun, M., 1984. Cyanobiont diversity in the Lichinaceae and Reppiaceae. *Lichenologist*, 16: 279–287.

Calvert, H.E., Pence, M.K. and Peters, G.A., 1985. Ultrastructural ontogeny of leaf cavity trichome in Azolla implies and functional role in metabolite exchange. *Protoplasms*, 129: 10–27.

Calvert, H.E. and Peters, G.A., 1981. The *Azolla–Anabaena* azollae relationship. IX. Morphological analysis of leaf cavity hair populations. *New Phytol.*, 89: 327–335.

Campbell, E.L. and Meeks, J.C., 1989. Characteristics of hormogonia formation by symbiotic *Nostoc* spp. in response to the presence of *Anthoceros punctatus* or its extracellular products. *Appl. Environ. Microbiol.*, 55: 125–137.

Carr, N.G. and Whitton, B.A., 1982. (eds.) *The Biology of Blue-Green Algae*. Blackwell Scientific Publications, London.

Chatterjee, P.B. and Dutta, S., 1980. Snails: A new pest of *Azolla. Internat. Rice Res. News Let.*, 5: 24–25.

Chatterjee, B.N. and Maiti, S., 1981. *Azolla* plants as fertilizers. In: *The Principles and Practice of Rice Growing*, (Eds.) B.N. Chatterjee and S. Maiti. Oxford and IBH Publ. Co., Calcutta, India, pp. 165–167.

Clymo, R.S. and Hayward, P.M., 1982. *The Ecology of Sphagnum*, (Ed.) A.J.E. Smith. Chapman and Hall Ltd., London, pp. 229–254.

Dalton, D.A. and Chatfield, J.M., 1985. A new nitrogen-fixing cyano-phytehepatic association: *Nostoc and Porella. Am. J. Bot.*, 72: 781–798.

Danilova, A.V., 1927. Le *Nostoc* en etat de symbiose. *Russ. Arkh. Protist.*, 6: 83–92.

Degelius, G., 1954. The lichen genus *Collema* in Europe: Morphology, taxonomy, ecology. *Symp. Bot. Ups.*, 13: 499–522.

Dick, H. and Stewart, W.D.P., 1980. The occurrence of fimbriae on a nitrogen-fixing cyanobacterium which occurs in lichen symbiosis. *Arch. Microbiol.*, 124: 107–128.

Dodge, A.D., 1975. Some mechanisms of herbicide action. *Science Prog.*, 7: 132–154.

Duckett, J.G., Prasad, A.K.S.K., Davies, D.A. and Walker, S., 1977. A cytological analysis of the *Nostoc*-bryophyte relationship. *New Phytol.*, 79: 349–377.

Duckett, J.G., Toth, R. and Soni, S.L, 1975. An ultrastructural study of the *Azolla-Anabaena azollae* relationship. *New Phytol.*, 75: 111–118.

Enderlin, C.S. and Meeks, J.C., 1983. Pure culture and reconstitution of the *Anthoceros–Nostoc* symbiotic association. *Planta*, 158: 157–175.

Fay, P., 1983. *The Blue-Greens*. Edward Arnold Publ., London.

Fay, P. and Van Baalen, C., 1987. (Eds.) *The Cyanobacteria*. Elsevier Scientific Publ., Amsterdam.

Fleming, H. and Haselkorn, R., 1973. Differentiation in *Nostoc muscorum*: Nitrogenase is synthesized in heterocysts. *Proc. Nat. Acad. Sci., USA*, 70: 2727–2731.

Flores, E. and Wolk, C.P., 1985. Identification of facultative heterotrophic, nitrogen-fixing cyanobacteria able to receive plasmid vectors from *Escherichia coli* by conjugation. *J. Bacteriol.*, 162: 1339–1344.

Fogg, G.E., Stewart, W.D.P., Fay, P.and Walsby, A.E., 1973. *The Blue-Green Algae*. Academic Press, London.

Franche, C. and Cohen-Bazire, G., 1985. The structural *nif* genes of a symbiotic *Anabaena-azollae* show a highly conserved physical arrangement. *Plant Sci.*, 39: 125–132.

Franche, C., Gunning, B.E.G., Rolfe, B.G. and Plaziski, J., 1987. Use of heterologous hybridization in phylogenetic studies of symbiotic *Anabaena* strains. In: *Current Plant Science and Biotechnology in Agriculture*, (Eds.) D.P.S. Varma and N. Brisson. Internat. Microb. Assocciation, Montreal, Quebec, Canada, pp. 305–306.

Galloe, O., 1939. *Natural History of the Danish Lichens (Part 6)*, Muksgaard, Copenhagen.

Geitler, L., 1934. Beitrage zur Kenntnis der Flechtensymbiose IV–V. *Arch. Protistenkd*, 82: 51–59.

Geitler, L., 1937. Beitrage zur Kenntnis der Flechtensymbiose VI. *Arch. Protistenkd*, 88: 161–167.

Gibson, C.E. and Smith, R.V., 1982. Freshwater plankton. In: *Botanical Monographs*, Vol. 19 (Eds.) N.G. Carr and B.A. Whitton. Blackwell Scientific Publ., Oxford, U.K., pp. 484–487.

Giddy, C., 1984. *Cycads of South Africa*, 2nd edn. (Ed.) C. Struik. Cape Town, pp.1–18.

Gorelova, O.A., Rerabek, J., Korzhenevskaya, T.G., Butenko, R.G. and Guesv, M.V., 1985. Influence of the cyanobacterium *Chlorogloeopsis fritschii* on growth and biosynthetic activity of *Solanum laciniatum* cells in mixed culture,. *Fiziol. Rast.*, 32: 1158–1169.

Granhall, U. and Selander, H., 1973. Nitrogen fixation in subarctic mire. *Oikos*, 24: 8–24.

Grilli Caiola, M., 1974. A light and electron microscopic study of the blue-green algae living either in the coralloid roots of *Macrozamia communis* or in isolated culture. *G. Bot. Ital.*, 108: 161–175.

Grilli Caiola, M., 1975. A light and electron microscopic study of blue-green algae growing in the coralloid roots of *Encephalartos altensteinii* and in culture. *Phyclogia*, 14: 25–33.

Grilli Caiola, M., 1980. On the phycobiont of cycad coralloid roots. *New Phytol.*, 85: 737–552.

Grobbelaar, N., Hatting, W. and Marshall, J., 1986. The occurrence of coralloid roots on the South African species of the Cycadales and their ability to fix nitrogen symbiotically. *S. Afr. J. Bot.*, 52: 467–475.

Grobbelaar, N., Scott, W.E., Hattingh, W. and Marshall, J., 1987. The identification of the coralloid root endophytes of the Southern African cycads and the ability of the cycads to fix dinitrogen. *S. Afr. J. Bot.*, 53: 111–129.

Gunning, B.E.S., Huges, J.E. and Hardham, J.R., 1978. Formative and proliferative cell divisions, cell differentiation and developmental changes in the meristem of *Azolla* roots. *Planta*, 143: 121–144.

Gusev, M.V., Wineyeva, L.A. and Semenova, L.R., 1979. The effect of lysozyme, EDTA, magnesium and mannitol on spheroplast formation by Anacystis nidulans. *Mikrobiologiya*, 48: 693–714.

Gusev, M.V. and Korzhenevskaya, T.G., 1990. Artificial associations. In: *Handbook of Symbiotic Cyanobacteria*, (Ed.) A.N. Rai. CRC Press Inc., Boca Raton, Florida, USA, pp. 173–230.

Gusev, M.V. and Korzhenevskaya, T.G., 1995. Artificial associations of nitrogen-fixing cyanobacteria with plants. In: *Proc. 10th Internat. Cong. Nitropgen Fixation*, (Ed.) Igor A. Tikhonovich. All-Russia Res. Inst., Russia (In press).

Harder, R., 1917. Ernahrungsphysiologische Untersuchungen an Cyanophyoeen, hauptsachlich dem endophytischen, *Nostoc punctiforme*. *Z. Bot.*, 9: 145–157.

Haselkorn, R., 1978. Heterocysts. *Ann. Rev. Pl. Physiol.*, 29: 319–344.

Haselkorn, R., 1992. Developmentally regulated gene rearrangements in prokaryctes. *Ann. Rev. Genet.*, 26: 113–130.

Haselkorn, R., Mazur, B., Orr, J., Rice, D., Wood, N and Rippka, R., 1980. Heterocyst differentiation and nitrogen fixation in cyanobacteria (blue-green algae). In: *Nitrogen Fixation*, (Eds.) W.E. Newton and W.H. Orme-Jounson. University Park Press, Baltimore, USA, pp. 259–278.

Henssen, A., 1964. Was ist *Pyrenothrix nigra*? *Ber. Dtsch. Bot. Ges.*, 77: 317–322.

Henssen, A., 1967. Eine Revision der Flechtenfamilien Lichinaceae und Ephebaceae. *Symb. Bot. Ups.*, 18: 123–141.

Hill, D.J., 1975. The pattern of development of *Anabaena* in the *Azolla-Anabaena* symbiosis. *Planta*, 133: 237–242.

Hill, D.J., 1977. The role of *Anabaena* in the *Azolla-Anabaena* symbiosis. *New Phytol.*, 78: 611–616.

Hitch, C.J.B. and Millbank, J.W., 1975. Nitrogen metabolism in lichens. VII. Nitrogenase activity and heterocyst frequency in lichens with blue-green phycobionts. *New Phytol.*, 75: 239–256.

Hoffman, E., 1985. The *Gunnera-Nostoc* symbiosis: A developmental study of the uninfected gland, *Master's Thesis*, University of Oregon, Eugene.

Horstmann, J.L., Denison, W.C. and Silvester, W.B., 1932. 15-dinitrogen fixation and molybdenum enhancement of acetylene reduction by *Lobaria* spp. *New Phytol.*, 92: 235–251.

Hughes, B.G., White, F.G., Vernon, L.P. and Smith, M.A., 1978. Uptake of cyanelles and blue-green algae into barley and tobacco protoplasts. *Plant Physiol.*, 61: 54–66.

Huneke, A., 1933. Beitrage zur Kenntnis des Symbiose zwischen Azolla and *Anabaena*. *Beitr. Biol. Pflanzen* 20: 315–343.

Huss–Danell, K., 1979. The cephalodia and their nitrogenase activity in the lichen *Stereocaulon paschale*. *Z. Pflanzenphysiol.*, 95: 431–442.

Jaag, O., 1945. Untersuchungen uber die Vegetation und Biologie der Algen des nackten Gesteins in den Alpen im Jura und im schweize-rrischen Mitteland. *Beitr. Kryptogamenflora Schweiz*, 9: 3–20.

Kannaiyan, S., 1994. Sporulation and spore germination in symbiotic nitrogen-fixing water-fern *Azolla*. In: *Biology and Application of Nitrogen-Fixing Organisms: Problems and Prospects*, (Eds.) A.B. Prasad and A. Vaishampayan. Scientific Publishers, Jodhpur, India, pp. 71–94.

Kaplan, D., Calvert, H.E. and Peters, G.A., 1982. Phycobiliproteins in the *Azolla* endophyte as a function of leaf age and cell type. *Plant Physiol.*, 69: 132–156.

Kaplan, D., Calvert, H.E. and Peters, G.A., 1986. The *Azolla* and *Anabaena-azollae* relationship. XII. Nitrogenase activity and phycobili-proteins of the endophyte as a function of leaf age and cell type. *Plant Physiol.*, 80: 884–890.

Kaplan, D. and Peters, G.A., 1988. Interaction of carbon metabolism in the *Azolla-Anabaena* symbiosis. *Symbiosis,* 6: 53–68.

Kele, R.A., 1964. Isolation of the mycobiont and phycobiont on an underwater lichen, *Hydrothyrea venosa. Honours Thesis,* Biology Department, Clark University, Worcester, MA.

Kershaw, K.A., 1985. *Physiological Ecology of Lichens.* Cambridge University Press, Cambridge, U.K.

Knapp, E., 1933. Uber *Geosiphon pyriforme* Fr. Wettst., eine intrazellulare Pilz-Algen Symbiose. *Ber. Dtsch. Bot. Ges.,* 51: 210–216.

Kobiler, D., Cohen-Sharon, A. and Abdulkadir, S., 1982. Lectins are involved in the recognition between *Anabaene* and *Azolla. Israel J. Bot.,* 31: 324–336.

Konar, R.N. and Kapoor, R.K., 1974. Embryology of *Azolla pinnata. Phytomorphology,* 24: 228–261.

Kumar, A.P., Perraju, B.T.V.V. and Singh, H.N., 1986. Carbon nutrition and regulation of uptake hydrogenase activity in free-living and symbiotic *Anabaena cycadeae. New Phytol.,* 104: 115–123.

Lachance, M.A., 1981. Genetic relatedness of heterocystous cyanobacterie by deoxyribonucleic acid-deoxyribonucleic acid reasssociation. *Int. J. Syst. Bacterial.,* 31: 139–151.

Ladha, J.K. and Watanabe, I., 1984. Antigenic analysis of *Anabaena-azollae* and the role of lectin in the *Azolla-Anabaena* symbiosis. *New Phytol.,* 98: 295–298.

Lallemant, R. and Savoye, D., 1985. Lectins and morphogenesis: Facts and outlooks. In: *Lichen Physiology and Cell Biology,* (Ed.) D.H. Brown. Plenum Press, New York, pp. 335–367.

Lamb, I.M., 1951. On the morphology, physiology and taxonomy of the genus *Stereocaulun. Can. J. Bot.,* 29: 522–544.

Lin, C., Watanabe, I., Tan, L.F. and Liu, C.C., 1988. Re-establishment of symbiosis to *Anabaena*-free *Azolla. Zhunguo Kexue (Scientia Sinica),* 30 B.

Lindblad, P. and Bergman, B., 1986. Glutamine synthetase: Activity and localization in cyanobacteria of the cycads *Cycas revoluta* and *Zamia skinneri. Planta,* 169: 1–12.

Lindblad, P. and Bergman, B., 1990. The cycad-cyanobacterial symbiosis. In: *Handbook of Symbiotic: Cyanobacteria,* (Ed.) A.N. Rai, CRC Press Inc., Boca Raton, Florida, USA, pp.137–159.

Lindblad, P., Hallbom, L. and Bergman, B., 1985. The cyanobacterium-*Zamia* symbiosis: Acetylene reduction and heterocyst frequency. *Symbiosis,* 1: 19–32.

Lindblad, P., Rai, A.N., and Bergman, B., 1987. The *Cycas revoluta-Nostoc* symbiosis: Enzyme activities in nitrogen and carbon metabolism in the cyanobiont. *J. Gen. Microbiol.,* 133: 1695–1702.

Lindblad, P., Haselkorn, R., Bergman. B. and Nierzwicki-Bauer, S.A., 1989. Comparison of DNA restriction fragment lengths of *Nostoc* in and from cycads. *Arch. Microbiol.,* 152: 20–27.

Linkola, K., 1923. Kulturen *mit Nostoc*–Gonidien der *Peltigera* Arten. *Ann. Zool. Soc. Bot. Fenn, Vanamo,* 1: 1–3.

Lokhart, C.K., Rowell, P. and Stewart, W.D.P., 1978. Phytohaemagglutinins from the nitrogen-fixing lichens *Peltigera canina* and *Peltigera polydactyla. FEMS Mcrobiol. Lett.,* 3: 127–136.

Lumpkin, T.A. and Plucknett, D.L., 1980. *Azolla*: Botany, physiology and use as a green manure. *Econ. Bot.,* 34: 111–153.

Margulis, L., 1981. *Symbiosis in Cell Evolution.* W.H. Freeman Publ., San Francisco, USA.

Marton, K. and Galun, M., 1976. *In vitro* dissociation and reassociation of the symbionts in the lichen *Heppia echinulata. Protoplasma,* 87: 135–151.

McCowen, S.W., MacArthur, L. and Gates, J.E., 1987. *Azolla* fern lectins that specifically recognize endosymbiotic cyanobacteria. *Curr. Microbiol.,* 14: 329–333.

Meeks, J.C., 1990. Cyanobacterial-bryophyte associations. In: *Handbook of Symbiotic Cyanobacteria,* (Ed.) A.N. Rai, CRC Press Inc., Boca Raton, Florida, USA, pp. 43–63.

Meeks, J.C., Joseph, C.M. and Haselkorn, R., 1988. Organization of *nif* genes in cyanobacteria in symbiotic association with *Azolla* and *Anthoceros. Arch. Microbiol.,* 150: 61–71.

Millbank, J.W., 1984. Nitrogen fixation by lichens. In: *Current Developments in Biological Nitrogen Fixation,* (Ed.) N.S. Subba Rao. Oxford and IBH Publ. Co., New Delhi, India, pp. 197–218.

Mishra, A.K., Vaishampayan, A. and Singh, V.P., 1991a. Isolation and physiological characterization of a herbicide-resistant mutant of a nitrogen-fixing cyanobacterium. In *Physiological Strategies for Crop Improvement,* (Eds.) D.N. Tyagi, B. Bose, A. Hemantaranjan and T.M. Devi. B.H.U., India, pp. 63–68.

Mishra, A.K., Vaishampayan, A. and Singh, V.P., 1991b. A herbicide-resistant cyanobacterial mutant derepressed for nitrogen fixation under ammonium–mediated condition. In: *Proc.Golden J. Symp. Ind. Soc. Genet. Pl. Breeding.* Indian Agricultural, Research Institute, New Delhi. India, pp. 98–100.

Moore, A.W., 1969. *Azolla*: Biology and agronomic significance. *Bot. Rev.,* 35: 17–34.

Nathanielsz, C.P. and Staff, I.A., 1975. On the occurrence of intracellular blue-green algae in cortical cells of the apogeotropic roots of *Macrozamia communis* L. Johnson. *Ann. Bot.,* 39: 363–371.

Newton, J.W. and Calvins, J.F., 1976. Altered nitrogenase pools induced by the *Azolla-Anabaena* symbiosis. *Plant Physiol.,* 58: 798–799.

Newton, J.W. and Herman, I., 1979. Isolation of cyanobacteria from the aquatic fern, *Azolla. Arch. Microbiol.,* 120: 161–165.

Nierzwicki-Bauer, S.A., 1990. *Azolla-Anabaena* symbiosis: Use in agriculture. In: *Handbook of Symbiotic Cyanobacteria,* (Ed.) A.N. Rai. CRC Press Inc., Boca Raton, Florida, USA, pp. 119–136.

Nierzwicki-Bauer, S.A. and Haselkorn, R., 1986. Difference in *mRNA* levels in *Anabaena* living freely or in symbiotic association with *Azolla. EMBO J.,* 5: 29–35.

Obukowicz, M., Schaller, M. and Kennedy, G.S., 1981. Ultrastructure and phenolic histochemistry of the *Cycas revoluta-Anabaena* symbiosis. *Sew Phytol.,* 87: 751–757.

Orr, J. and Haselkorn, R., 1982. Regulation of glutamine synthetase activity and synthesis of free-living and symbiotic *Anabaena* spp. *J. Bacteriol.,* 152: 626–635.

Pate, J.S. and Gunning, B.E.S., 1972. Transfer cells. *Ann. Rev. Pl. Physiol.,* 23: 173–196.

Patra, R.N. and Singh, P.K., 1984. Symbiotic algal heterocyst frequency in three species of *Azolla. Proc. Ind. Nat. Sci. Acad.* (Part B: Biol. Sci.), 50: 128–131.

Peters, G.A., 1975. The *Azolla-Anabaena azollae* relationships. III. Studies on metabolic capabilities and a further characterization of the symbiont. *Arch. Microbiol.,* 103: 113–122.

Peters, G.A., 1976. Studies on the *Azolla-Anabaena azollae* symbiosis. In: *Proc. First Internet. Symp. Nitrogen Fixation,* (Eds.) W.E. Newton and C.J. Nyman. Washington State University Press, Pullman, Washington, USA, pp. 592–600.

Peters, G.A., 1977. The *Azolla-Anabaena* symbiosis. In: *Genetic Engineering for Nitrogen Fixation,* (Ed.) A. Hollaender. Plenum Press, New York, USA, pp. 231–258.

Peters, G.A., 1984. *Azolla-Anabaena* symbiosis: Basic biology, use and prospects for the future. In: *Practical Applications of Azolla for Rice Production,* (Eds.) W.S. Silver and E.C. Schroder. Martinus Nijhoff/Dr. W. Junk Publishers, The Netherlands, pp. 1–14.

Peters, G.A. and Calvert, H.E., 1982a. The *Azolla-Anabaena* symbiosis. In: *Advances in Agricultural Microbiology,* (Ed.) N.S. Subba Rao. Oxford and IBH Publ. Co., New Delhi, India, pp. 191–218.

Peters, G.A. and Calvert, H.E., 1982b. The *Azolla-Anabaena* azollae symbiosis. In: *Algal Symbiosis,* (Ed.) L.F. Goff. Cambridge University Press, New York, USA, pp. 109–145.

Peters, G.A. and Mayne, B.C., 1974a. The *Azolla-Anabaena azollae* relationship. I. Initial characterization of the association. *Plant Physiol.,* 58: 813–819.

Peters, G.A. and Mayne, B.C., 1974b. The *Azolla-Anabaena azollae* relationship. II. Localization of nitrogenase activity as assayed by acetylene reduction. *Plant Physiol.,* 58: 820–824.

Peters, G.A. and Meeks, J.C., 1989. The *Azolla-Anabaena* symbiosis: Basic biology. *Arm. Rev. Pl. Physiol. Pl. Mol. Biol.,* 40: 193–210.

Peters, G.A., Evans, W.R. and Toia, R.E. Jr., 1976. *Azolla–Anabaena azollae* relationship. IV. Photosynthetically-driven nitrogenase-catalyzed hydrogen production. *Plant Physiol.*, 58: 119–126.

Peters, G.A., Toia, R.E. Jr. and Lough, S.M., 1977. *Azolla-Anabaena azollae* relationship. V. 15-dinitrogen fixation, acetylene reduction and hydrogen production. *Plant Physiol.*, 59: 1021–1025.

Peters, G.A., Toia, R.E. Jr., Raveed, D. and Levine, N.J., 1978. The *Azolla-Anabaena* azollae relationship. VI. Morphological aspects of the association. *New Phytol.*, 80: 583–593.

Peters, G.A., Mayne, B.C., Ray, T.B. and Toia, R.E. Jr., 1979. Physiology and biochemistry of the *Azolla-Anabaena* symbiosis. In: *Nitrogen and Rice.* International Rice Reseasrch Institute, Los Banos, Laguna, Philippines,. pp. 325–344.

Peters, G.A., Ray, T.B., Mayne, B.C. and Toia, R.E. Jr., 1980a. *Azolla-Anabaena* association: Morphological and physiological studies. In: *Nitrogen Fixation* (Eds.) W.E. Newton and W.H. Orme-Johnson. University Park Press, Baltimore, USA, Vol. II: 293–309.

Peters, G.A., Toia, R.E., Jr., Evans, W.R., Crist, D.K., Mayne, B.C. and Poole, R.E., 1980b. Characterization and comparisons of five nitrogen-fixing *Azolla-Anabaena* associations. I. Optimization of growth conditions for biomass increase and N content in a controlled environment. *Plant, Cell and Environ.* 3: 261–269.

Peters, G.A., Ito, O., Tyagi, V.V.S. and Kaplan, D., 1981a. Physiological studies on nitrogen-fixing *Azolla*. In: *Genetic Engineering of Symbiotic Nitrogen Fixation and Conservation of Fixed Nitrogen,* (Eds.) J.M. Lyons *et al.* Plenum Press, New York, USA, pp. 343–362.

Peters, G.A., Ito, O., Tyagi, V.V.S., Mayne, B.C., Kaplan, D. and Calvert, H.E., 1981b. Photosynthesis and nitrogen fixation in the *Azolla-Anabaena* symbiosis. In: *Current Perspectives in Nitrogen Fixation,* (Eds.) A.H. Gibson and W.E. Newton. Australian Academy of Sciences, Canberra, Australia, pp. 121–124.

Petit, P., 1982. Phytolectins from the nitrogen-fixing lichen *Peltigera horizontalis*: The binding pattern of primary protein extract. *New Phytol.*, 91: 705–721.

Petit, P., Lallemant, R. and Savoy, D., 1933. Purified phytolectin from the lichen *Peltigera canina* var. *canina* which binds to the phycobiont cell walls and its use as cytochemical marker in situ. *Hew Phytol.*, 94: 103–111.

Planzinski, J., 1990. The *Azolla-Anabaena* symbiosis. In: *Molecular Biology of Symbiotic Nitrogen Fixation,* (Ed.) P.M. Gresshoff. CRC Press Inc., Boca Raton, Florida, USA, pp. 51–76.

Prasad, A.B. and Vaishampayan, A., 1984. Genetic recombination in *Nostoc muscorum* strains. In: *Perspectives in Cytology and Genetics,* (Eds.) G.K. Manna and U. Sinha. Typographers Rashtravani Press, New Delhi, India, pp. 179–184.

Prasad, A.B. and Vaishampayan, A., 1987. Genetic dissections in cyanobacteria. In: *Mutagenesis: Basic and Applied,* (Ed.) A.B. Prasad. Print House, Lucknow, India, pp. 253–284.

Prasad, A.B. and Vaishampayan, A., 1994. Prospects of using pesticide-resistant cyanobacterial mutants as a viable bio-N fertilizer and a source material for genetic engineering studies. In: *Biology and Application of Nitrogen-Fixing Organisms: Problems and Prospects,* (Eds.) A.B. Prasad and A. Vaishampayan. Scientific Publishers, Jodhpur, India, pp. 47–70.

Prasad, A.B., Samanta, R., Vishwakarma, M.L. and Vaishampayan, A., 1986. Biological effects of a mercury fungicide on a nitrogen-fixing blue-green alga *Nostoc muscorum*: Isolation and preliminary characterization of a Hg-resistant mutant. *New Phytol.,* 102: 45–50.

Prasad, A.B., Vaishampayan, A., Vishwakarma, M.L. and Shrivastava, D.K., 1991. Development of fungicide-resistance coupled with nitrogen–fixing ability during ammonium-mediated growth, as a mutational event, in a cyanobacterium *Nostoc muscorum. Biomed. Lett. (Camb.),* 46: 245–251.

Rai, A.K., 1990a. Cyanobacteria in symbiosis. In: *Handbook of Symbiotic Cyanobacteria,* (Ed.) A.N. Rai. CRC Press Inc., Boca Raton, Florida, USA, pp. 1–7.

Rai, A.K., 1990b. Cyanobacterial-fungal symbioses: The cyanolichens. In: *Handbook of Symbiotic Cyanobacteria,* (Ed.) A.N. Rai. CRC Press Inc., Boca Raton, Florida, USA, pp. 9–41.

Rai, A.N., Rowell, P. and Stewart, W.D.P., 1983. Interactions between cyanobacterium and fungus during 15-dinitrogen incorporation and metabolism in the lichen *Peltigera canina. Arch. Microbiol.,* 134: 136–148.

Rai, A.N., Rowell, P. and Stewart, W.D.P., 1984. Evidence for an ammonium transport system in free-living and symbiotic cyanobacteria. *Arch. Microbiol.,* 137: 241–262.

Rai, A.N., Lindblad, P. and Bergman, B., 1984. Absence of the glutamine synthetase-linked methylammonium (ammonium) transport system in the cyanobiont of the *Cycas*-cyanobacterial symbiosis. *Planta,* 169: 379–393.

Ray, T.B., Peters, G.A., Toia, R.E. Jr. and Wayne, B.C., 1978. *Azolla-Anabaena* relationship. VII. Distribution of ammonia assimilating enzymes, protein and chlorophyll between host and symbiont. *Plant Physiol.,* 62: 463–467.

Reisser, W., 1984. Endosymbiotic cyanobacteria and cyanellae. InL *Encyclopedia of Plant Physiolology,* (Eds.) H. Linskens and J. Heslop-Harrison. Hew Series, Vol. 17. Springer-Verlag, Berlin, pp. 91–122.

Rippka, R., Deruelles, J., Waterbury, J.B., Herdman, M. and Stanier, R.Y., 1979. Generic assignments, strain histories and properties of pure cultures of cyanobacteria. *J. Gen. Microbiol.,* 111: 1–12.

Rodgers, G.A. and Stewart, W.D.P., 1977. The cyanophyte-hepatic symbiosis. I. Morphology and physiology. *New Phytol.,* 78: 441–451.

Roger, P.A. and Kulasooriya, S.A., 1980. *Blue-green Algae and Rice.* International Rice Research Institute, Los Banos, Laguna, Philippines.

Satensson, R., 1952. Follicolous lichens. *Symb. Bot. Ups.,* 12: 590–595.

Semenova, L.R., Mineyeva, L.A. and Gusev, M.V., 1982. The effect of osmotic stabilizing agents on the formation of spheroplasts in cyanobacteria and their photosynthetic activity. *Mikrobiologiya*, 51 259–271.

Schnepf, E., 1964. Zur Feinstruktur von *Geosiphon pyriforme*, ein Versuch zur Deutung cytoplasmatischer Membranen und Kompartimente. *Arch. Microbiol.*, 49: 112–118.

Schulthorpe, C.D., 1967. *The Biology of Aquatic Vascular Plants*. Edward Arnold Publ. Ltd., London, U.K.

Scott, G.D., 1969. *Plant Symbioses*. Edward Arnold, London, U.K.

Shah, A.K. and Vaidya, B.S., 1977. Detection of vitamin B-12 and pantothenic acid in cell exudates of blue-green algae. *Biol. Plant*, 19: 426–439.

Shiman, H., 1957. Beitrage zur Lebelsgeschichte homoeomerer Cyanophyceen-Flechten. *Osterr. Bot. Z.*, 104: 409–417.

Silvester, W.B., 1976. Endophyte adaptation in *Gunnera-Nostoc* symbiosis. In: *Symbiotic Nitrogen Fixation in Plants*, (Ed.) P.S. Nutman. Cambridge University Press, Cambridge, U.K., pp. 521–217.

Silvester, W.B., 1977. Dinitrogen fixation by plant, associations excluding legumes. In: *A Treatise on Dinitrogen Fixation, Section IV: Agronomy and Ecology*, (Eds.) R.W.F. Hardy and A.H. Gibson. John Wiley and Sons, New York, USA, pp. 141–165.

Silvester, W.B. and McNamara, G.S., 1976. The infection process and ultrastructure of the *Gunnera-Nostoc* symbiosis. *New Phytol.*, 77: 135–139.

Silvester, W.B. and Smith, D.R., 1969. Nitrogen fixation by *Gunnera-Nostoc* symbiosis. *Nature (London)*, 224: 1231–1232.

Sindhu, S.S. and Dadarwal, K.R., 1994. Biosynthesis of exopolysaccharides and its role during nodule development in Rhizobium-legume symbiosis. In: *Biology and Application of Nitrogen-Fixing Organisms: Problems and Prospects*, (Eds.) A.B. Prasad and A. Vaishampayan. Scientific Publ., Jodhpur, India, pp. 189–221.

Singh, A.L., 1994. Use of *Azolla*-cyanobacterial symbiosis in rice culture. In: *Biology and Application of Nitrogen-Fixing Organisms: Problems and Prospects*, (Eds.) A.B. Prasad and A. Vaishampayan. Scientific Publ., Jodhpur, India, pp. 95–118.

Singh, A.L. and Singh, P.L., 1989. *Nitrogen Fixation in Indian Rice Fields (Azolla and Blue-green Algae)*. Agro-Botanical Publ., Bikaner, India.

Singh, D.T., Nirmala, A.K., Modi, D.R., Katiyar, S. and Singh, H.N., 1987. Genetic transfer of herbicide-resistance gene(s) from *Gloeo capsa* spp. to *Nostoc muscorum*. *Molec. Gen. Genet.*, 208: 436–438.

Singh, H.N. and Singh, H.N. Jr., 1978. An azide-resistant mutant of the blue-green alga *Nostoc muscorum* producing heterocysts and nitrogenase in the presence of fixed inorganic nitrogen source. *Arch. Microbiol.*, 119: 197–201.

Singh, H.N., Singh, H.R. and Vaishampayan, A., 1979. Biological effects of rice-field herbicide *lasso* (alachlor) on various strains of the nitrogen-fixing blue-green alga *Nostoc muscorum* and characterization of herbicide-induced mutants resistant to methylamine and L-methiontae-DL-sulfoxinimine. *Environ. Exp. Bot.*, 19: 5–12.

Singh, H.N. and Vaishampayan, A., 1978. Biological effects of rice-field herbicide *machete* on various strains of the nitrogen-fixing blue-green alga *Nostoc muscorum*, 18: 87–94.

Singh, H.N., Vaishampayan, A. and Singh, R.K., 1978a. Evidence for the involvement of a genetic determinant controlling functional specificity of group VI B elements in the metabolism of molecular nitrogen and nitrate in a blue-green alga *Nostoc muscorum. Biochem. Biophys. Res. Commun.*, 81; 67–74.

Singh, H.N., Vaishampayan, A. and Sonie, K.C., 1978b. Mutation from mlybdenum-dependent growth to tungsten-dependent growth and further evidence for a genetic determinant common to nitrogenase and nitrate reductase in a blue-green alga *Nostoc muscorum. Hut. Res.*, 50: 427–432.

Singh, P.K., 1979. Use of *Azolla* in rice production in India. In: *Nitrogen and Rice.* International Rice Research Institute, Los Banos, Laguna, Philippines, pp. 125–137.

Singh, R.N., 1961. *Role of Blue-Green Algae in Nitrogen Economy of Indian Agriculture.* Indian Council of Agricultural Research, New Delhi, India.

Smith, D.C. and Douglas, A.E., 1987. *The Biology of Symbiosis.* Edward Arnold, London, U.K.

Sprent, J.I., 1979. *The Biology of Nitrogen-Fixing Organisms.* McGraw Hill, London, U.K.

Stanier, R.Y., Kunisawa, R., Mandel, M. and Cohen-Bazire, G., 1971. Purification and properties of unicellular blue-green algae (Order Chroococcales). *Bacterial. Rev.*, 35: 171–183.

Stewart, W.D.P., 1966. *Nitrogen Fixation in Plants.* Athlone Press, London, U.K.

Stewart, W.D.P., 1977. A botanical ramble among the blue-green algae. *Britist Phycol. J.*, 12: 89–115.

Stewart, W.D.P., 1980. Systems involving blue-green algae (cyanobacteria). In: *Methods for Evaluating Biological Nitrogen Fixation*, (Eds.) W.D.P. Stewart and J.R. Gallon. Academic Press, New York, USA, pp. 239–277.

Stewart, W.D.P. and Singh, H.N., 1975. Transfer of nitrogen-fixing (*Nif*) genes in the blue-green alga *Nostoc muscorum. Biochem. Biophys. Res. Commun.*, 62: 62–69.

Stewart, W.D.P., Rowell, P. and Rai, A.N., 1983. Cyanobacteria-eukaryotic plant symbioses. *Ann. Microbiol. (Inst. Pasteur)*, 134B: 205–214.

Street, H.E., 1977. Cell (suspension) culture-techniques. In: *Botanical Monographs* 2nd edn, (Ed.) H.E. Street. Blackwell Scientific Publ., Oxford, U.K., 11: 61–89

Suleimanova, Sh S. and Mineyeva, L.A., 1981. Effects of highlight intensity on growth and pigment content of cyanobacteria under conditions of photoauto- and photohetero-trophic cultivation. *Vestn. Mosk. Univ. Ser. Biol.*, 1: 42–53.

Sun, J.S., Chen, W.L., Zhu, Z.Q. and Zhu, Y.M., 1984a. An autoradiograpic study of the *Azolla* and *Anabaena azollae* relationship. *Acta Bot. Sci.*, 26: 115–119.

Sun, J.S., Zhu, Z.Q., Chen, W.L. and Li, S.Q., 1984b. Electron microscopic observation of the *Azolla-Anabaena azollae* relationship. *Acta Bot. Sci.*, 26: 343–348.

Svenson, H.K., 1944. The new world species of *Azolla. Amer. Fern J.*, 34: 69–85.

Swinscow, T.D.V., 1965. The marine species of *Arthopyrenia* in the British Isles. *Lichenologist*, 3: 55–59.

Talley, S.N. and Rains, D.W., 1980a. *Azolla* as a nitrogen source for temperate rice. In: *Nitrogen Fixation*, (Eds.) W.E. Newton and W.H. Orme-Johnson. University Park Press, Baltimore, USA, 2: 311–320 .

Talley, S.N. and Rains, D.W., 1980b. *Azolla filiculoides* Lam. as a fallow season green manure for rice in temperate climate. *Agron. J.*, 72: 11–18.

Tel-Or, E., Sandovsky, T., Kobiler, D., Arad, C. and Weinberg, R., 1983. The unique symbiotic properties of *Anabaena* in the water fern *Azolla*. In: *Photosynthetic Prokaryotes: Cell Differentiation and Function*, (Eds.) G.C. Papageorgiou and L.P. Packer. Elsevier Sci. Publ., New York, USA, pp. 303–314.

Towata, E.M., 1985. Morphometric and cytochemical ultrastructural analyses of the *Gunnera kaalensis/Nostoc symbiosis. Bot. Gaz.*, 146: 293–298.

Tyagi, V.V.S., Ray, T.B., Wayne, B.C. and Peters, G.A., 1981. The *Azolla-Anabaena azollae* relationship. XI. Phycobiliproteins in the action spectrum for nitrogenase-catalyzed acetylene reduction. *Plant Physiol.*, 68: 1479–1484.

Uheda, 1986. Isolation of empty packets from *Anabaena*-free *Azolla. Plant Cell Physiol.*, 27: 1187–1190.

Vaishampayan, A., 1981. Nutritional screening of amino acids in an auxotroph mutant of the cyanobacterium *Nostoc muscorum. Microbios Lett.*, 18: 111–116.

Vaishampayan, A., 1982a. A methylamine-resistant mutant of the blue-green alga, *Nostoc muscorum*: possible involvement of glutamine synthetase in methylamine metabolism. *New Phytol.*, 91: 607–613.

Vaishampayan, A., 1982b. A carbon auxotroph of the cyanobacterium *Nostoc muscorum*. In: *Proc. XIII Internat. Cong. Microbiol.*, (Ed.) American Soc. Microbiol. Boston, Massachusetts, USA, pp. 177–178.

Vaishampayan, A., 1982c. Cu–Fe interactions in the nitrogen-fixing cyanobacterium *Nostoc muscorum. Microbios. Lett.*, 21: 17–23.

Vaishampayan, A., 1982d. *Molybdenum-vanadium* biology in the cyanobacterium *Nostoc muscorum. Proc. First Internat. Phycol. Cong.* St. John's, Newfoundland, Canada, p.52–53.

Vaishampayan, A., 1982e. Amino acid nutrition in the blue-green alga *Nostoc muscorum. New Phytol.*, 90: 545–549.

Vaishampayan, A., 1982f. Amino acids' metabolism in the cyanobacterium *Nostoc muscorum*: A kinetic study of varying amino acid concentrations on a carbon and nitrogen auxotroph. In: *Proc. XII Internat. Cong. Biochemistry.* Perth, Western Australia, p. 252–253.

Vaishampayan, A., 1983a. Heterocyst differentiation and speciation in the nitrogen-fixing cyanobacterium *Nostoc muscorum* as influenced by a bipyridylium ion. In: *Proc. XXIX Internat. Cong. Physiol.,* Sydney (Australia) 6: 47–236.

Vaishampayan, A., 1983b. Genetics of methylamine-resistance leading to carbon auxotrophy in the cyanobacterium *Nostoc muscorum.* In: *XV Internat. Cong. Genet. (Abst.),* p. 812 (1462), New Delhi, India.

Vaishampayan, A., 1983c. UV-induced wide-range mutations in a cyanobacteriurm *Nostoc muscorum.* In: *Proc. VII Internat. Cong. Radiation Res.,* Rijswijk (The Netherlands), B: 4–41.

Vaishampayan, A., 1983d. Mo-V interactions during molecular nitrogen and nitrate metabolism in a nitrogen-fixing blue-green alga *Nostoc muscorum. Experientia,* 39: 358–360.

Vaishampayan, A., 1983e. Vanadium as a trace element in the blue-green alga, *Nostoc muscorum*: Influence of nitrogenase and nitrate reductase. *New Phytol.,* 95: 55–60.

Vaishampayan, A., 1983f. An ammonia Liberating strain of the nitrogen-fixing cyanobacterium *Nostoc muscorum. Transactions of Meeting on Biological Processes and Soil Fertility.* University of Reading, London, U.K., p. 130–131.

Vaishampayan, A., 1983g. Metabolic utilization of C-1 compound in a nitrogen-fixing cyanobacterium *Nostoc* sp. In: *Proc. VI Internat. Symp. Environ. Biogeochem.,* New Mexico, USA, 2: 3–15.

Vaishampayan, A., 1983h. Isolation of a carbon-requiring strain of a nitrogen-fixing cyanobacterium *Nostoc muscorum.* In: *Proc. III Internat. Symp. Microb. Ecol.,* Michigan State Univ., USA, V: 5–81.

Vaishampayan, A., 1983i. Pesticide-resistance in a nitrogen-fixing cyanobacterium. In: *Proc. Internat. Cong. Environ. Hazards Agrochem.,* Alexandria (Egypt), 3: 7–102.

Vaishampayan, A., 1984a. Genetic manipulation and recombination leading to pesticide-resistance in a nitrogen-fixing cyanobacterium. In: *Proc. III Europ. Cong. Biotechnol.,* Munich (Germany), 3: 353–358.

Vaishampayan, A., 1984b. Strong mutagenicity of a bipyridylium herbicide in a nitrogen-fixing blue-green alga. *Experientia,* 40: 1016–1019.

Vaishampayan, A., 1984c. Powerful mutagenicity of a bipyridylium herbicide in a nitrogen-fixing blue-green alga *Nostoc muscorum. Mut. Res.,* 138: 39–46.

Vaishampayan, A., 1984d. Mutagenicity of a bipyridylium herbicide in a nitrogen-fixing cyanobacterium *Nostoc muscorum. Environ. Internet.,* 10: 13–27.

Vaishampayan, A., 1984e. Biological effects of a herbicide in a nitrogen fixing cyanobacterium (blue-green alga): an attempt for introducing herbicide-resistance. *New Phytol.,* 96: 7–11.

Vaishampayan, A., 1984f. Studies on diuron uptake in a blue-green alga *Nostoc muscorum*. *J. Exp. Bot.*, 35: 897–904.

Vaishampayan, A., 1984g. Copper-iron interactions in a nitrogen-fixing cyanobacterium *Nostoc muscorum*. *J. Pl. Nut.*, 7: 567–573.

Vaishampayan, A., 1984h. Screening of amino acids for carbon and nitrogen enrichments in different strains of a cyanobacterium *Nostoc muscorum*. *Biochem. Physiol. Pfalzen*, 179: 411–417.

Vaishampayan, A., 1984i. Collection of a carbon-requiring strain of a nitrogen-fixing cyanobacterium. In: *Proc. V Internat. Cong. Culture Collec.*, (Ed.) Thailand Council of Scientific Indust. Res.), Bangkok, Thailand, pp. 63–68.

Vaishampayan, A., 1984j. Biological effects of the rice-field herbicide monuron on a nitrogen-fixing cyanobacterium *Nostoc muscorum*. *Microbios Lett.*, 28: 105–111.

Vaishampayan, A., 1985a. Mutagenicity of bipyridylium salts in a nitrogen-fixing cyanobacterium. *Microbios*, 43: 53–65.

Vaishampayan, A., 1985b. Potent mutagenicity of a bipyridylium herbicide in a nitrogen-fixing blue-green alga, *Nostoc muscorum*. *Biochem. Physiol. Pflanzen*, 180: 327–331.

Vaishampayan, A., 1985c. Mutagenic activity of alachlor, butachlor and carbaryl to a nitrogen-fixing cyanobacterium *Nostoc muscorum*. *J. Agric. Sci.*, 104: 571–576.

Vaishampayan, A., 1985d. Growth of *Nostoc muscorum* mutants with diuron (DCMU) and L-methionine-DL-sulfoximine (MSO). *Experientia*, 41: 137–139.

Vaishampayan, A., 1988a. Evidence for an inter-strain conjugational transfer of *nif* (nitrogen-fixing) gene(s) in a cyanobacterium *Nostoc muscorum*. In: *XVI Internat. Cong. Genet. (Abst.)*. National Research Council, Toranto, Ontario, Canada, p. 352.

Vaishampayan, A., 1988b. Collection and reconstruction of a non-heterocystous filamentous cyanobacterial culture from a sugar factory effluent fixing nitrogen with a remarkably higher rate under a simple micro-aerophilic condition. In: *Proc. VI Internat. Cong. Culture Collec. Microb. Gene Pool.* University of Maryland, USA, pp. 79–122.

Vaishampayan, A., 1993a. Mutation breeding for phosphate independence in the nitrogen-fixing water fern *Azolla* sp. In: *Proc. XV Internat. Bot. Cong.* Yokohama, Japan, pp. 522–534.

Vaishampayan, A., 1993b. Two-step mutation for reduced phosphorus requirement in *Azolla-Anabaena* symbiotic nitrogen-fixing complex. In: *XVII Internat. Cong. Genet. (Abst.)*, Birmingham, U.K., 18: 819.

Vaishampayan, A., 1994a. Microsymbiotic cyanobiont of the low phosphate-requiring *Azolla-Anabaena* symbiotic nitrogen-fixing complex. In: *Proc. Cyanobacterial Workshop: The Use of Cyanobacteria to Explore Basic Biol. Processes* (Eds.) A.R. Grossman and M.R. Schaefer. Asilomer, California, pp. 11–12.

Vaishampayan, A., 1994b. Recent advances in the molecular biology of *Azolla-Anabaena* symbiotic nitrogen-fixing complex and its use in agriculture. In: *Microbes and Environment,* (Eds.) A.B. Prasad and R.S. Bilgrami. Narendra Publishing House, New Delhi, India, pp. 121–144.

Vaishampayan, A., 1994c. Molecular genetics of nitrogen fixation in cyanobactoria. In: *Biology and Application of Nitrogen-Fixing Organisms: Problems and Prospects,* (Eds.) A.B. Prasad and A. Vaishampayan. Scientific Publishers, Jodhpur, India, pp. 1–18.

Vaishampayan, A., 1995a. Molecular biology of herbicide-resistant cyanobacterial mutants in relation to photosynthesis, dinitrogen fixation and N aasimilation. In: *Proc. Cyanobacterial Molec. Biol. Workshop,* (Ed.) D.A. Bryant. Asilomer, California, USA (In Press).

Vaishampayan, A., 1995b. Monoclonal mutagenesis in *Azolla-Anabaena* symbiotic nitrogen-fixing complex for agronomically important traits. In: *Proc. X Internat. Cong. Nitrogen Fixation,* (Ed.) Igor A. Tikhonovich. All–Russia Research Institute on Agricultural Microbiol., Podbelsky-Shesse, Saint Petereburgh, Pushkin, Russia (In Press).

Vaishampayan, A., 1995c. Siderophore-mediated iron uptake in nitrogen-fixing cyanobacteria. In: *Advancements in Iron Nutrition Research,* (Ed.) A. Hemantaranjan. Scientific Publ., Jodhpur, India, pp. 233–254.

Vaishampayan, A., 1995d. Mineral requirements of the free-living and symbiotic cyanobacteria. In: *Advancements in Micronutrient Research,* (Ed.) A. Hemantaranjan. Scientific Publ. Jodhpur, India, pp. 103–126.

Vaishampayan, A. and Awasthi, A.K., 1995a. Advances in molecular biology, agronomics and somaclonal mutagenesis of *Azolla-Anabaena* symbiotic nitrogen-fixing complex. In: *Microbiology in India: History, Progress and Applications,* (Ed.) Ajit Varma. Oxford and IBH Publ., New Delhi, India (In Press).

Vaishampayan, A. and Awasthi, A.K., 1995b. Biotechnological approaches to develop plant-cyanobacterial symbiotic somaclones as a potential bio-N fertilizer for wet agriculture. In: *Recent Excitments in Agricultural Biotechnology,* (Eds.) R.S. Bilgrami and D.N. Shukla. Narendra Publishing House, New Delhi, India (In Press).

Vaishampayan, A. and Banerjee, A.K., 1995. Genetic approaches to accomplish reduced phosphate-dependence of *Azolla-Anabaena* symbiotic nitrogen-fixing complex in wet agricultural fields. In: *Algal Ecology: An Overview,* (Eds.) A.N. Kargupta and B.N. Siddiqui. International Publications, Dehradun (In Press).

Vaishampayan, A. and Hemantaranjan, A., 1983. Inhibition of photosynthetic assimilation of inorganic carbon by monuron in a nitrogen-fixing cyanobacterium. In: *Proc. Internat. Cong. Environ. Hazards Agrochem. Developing Count.,* 2: 5–20.

Vaishampayan, A. and Hemantaranjan, A., 1984. Physiological significance of vanadium uptake during molecular nitrogen and nitrate metabolism in various strains of a nitrogen-fixing cyanobacterium *Nostoc muscorum. Pl. Cell Physiol.,* 25: 845–850.

Vaishampayan, A. and Mishra, A.K., 1989. Mutation for herbicide-resistance, coupled with the metabolic utilization of the herbicide as a carbon and nitrogen source in a nitrogen-fixing cyanobacterium. In: *49th National Symp. Recent Adv. Genet. Pl. Breed. Res. India, (Abst.)* Indian Society of Genetics and Plant Breeding, Banaras Hindu University, India, p. VIII–4: 0–6.

Vaishampayan, A. and Mishra, A.K., 1990. Cyanobacteria: A gene bank for the pesticide-resistance markers. In: *Proc. III World Cong. Environ. Hazards of Pesticides,* Cairo, Egypt, pp. 1–51.

Vaishampayan, A., Mishra, A.K. and Singh, V.P., 1992a. Uptake of monuron in a cyanobacterial mutant. *Biomed. Let.*, 47: 39–46.

Vaishampayan, A., Reddy, Y.R., Singh, B.D. and Singh, R.M., 1992b. Reduced phosphorus requirement of a mutant *Azolla-Anabaena* symbiotic nitrogen-fixing complex. *J. Exp. Bot.*, 43: 851–856.

Vaishampayan, A. and Prasad, A. B., 1980. Biological effects of the fungicide blitox on *Nostoc linckia* and *Nostoc muscorum*. In: *Proc. Golden J. Symp. Nat. Acad. Sci.,* Allahabad, India, pp. 66–166.

Vaishampayan, A. and Prasad, A.B., 1981a. A pesticide-resistant mutant of the nitrogen-fixing blue-green alga *Nostoc muscorum. Experientia,* 37: 1285–1286.

Vaishampayan, A. and Prasad, A.B., 1981b. Mutagenic effects of two pesticides in the blue-green alga *Nostoc muscorum*. In: *Proc. III Internat. Cong. Environ.*, Mutagens. Tokyo, Japan, 2: 38–118.

Vaishampayan, A. and Prasad, A.B., 1981c. Isolation and characterization of pesticide-resistant mutants of the blue-green alga *Nostoc muscorum*. In: *Proc. III Internat. Cong. Environ.*, Mutagens, Tokyo, Japan, 3: 18–19.

Vaishampayan, A. and Prasad, A.B., 1982. Blitox-resistant mutants of the nitrogen-fixing blue-green algae *Nostoc linckia* and *Nostoc muscorum. Environ. Exp. Bot.*, 22: 427–435.

Vaishampayan, A., and Prasad, A.B., 1984a. Inter-strain transfer of a pestiide-resistance marker in the nitrogen-fixing cyanobacterium *Nostoc muscorum. Molec. Gen. Genet.*, 193: 195–197.

Vaishampayan, A. and Prasad, A.B., 1984b. Pesticides in relation to the nitrogen-fixing apparatus in cyanobacteria. In: *Recent Trends in Botanical Researches,* (Ed.) R.P. Sinha. R.P. Roy Commemoration Fund, Patna University, Patna, pp. 193–202.

Vaishampayan, A. and Sahay, A., 1984a. Isolation of a cyanobacterial mutant resistant to virus N–II. In: *Proc. VIII Internat. Biophys. Cong.*, Bristol, U.K., pp. 78–121.

Vaishampayan, A. and Sahay, A., 1984b. Changes in cell morphology as a result of UV-irradiation in a cyanobacterium *Nostoc muscorum. Proc. Internat, Cell Biol. Cong.* Tokyo, Japan, pp. 430–487.

Vaishampayan, A. and Sahay, A. 1984c. A mutant of the cyanobacterium *Nostoc muscorum* resistant to virus N–II.. In: *Proc. VI Internat. Cong. Virol.* Sendai, Japan, pp. 48–64.

Vaishampayan, A. and Singh, H.N., 1981a. Mutation in the blue-green alga *Nostoc muscorum*. I. Methylamine-resistant mutants. *Biochem. Physiol. Pflanzen*, 176: 625–630.

Vaishampayan, A. and Singh, H.N., 1981b. Mutation in the blue-green alga *Nostoc muscorum*. II. L-methionine-DL-sulfoximine-resistant mutants. *Biochem. Physiol. Pflanzen*, 176: 631–637.

Vaishampayan, A., Singh, H.R. and Singh, H.N., 1978. Biological effects of rice-field herbicide *stam f-34* on various strains of the nitrogen-fixing blue-green alga *Nostoc muscorum*. *Biochem. Physiol. Pflanzen*, 173: 410–419.

Vaishampayan, A. and Singh, U.P., 1995. Somaclonal mutagenesis of *Azolla-Anabaena* symbiotic nitrogen-fixing complex for use as an economic bio-N fertilizer with rice culture. In: *Proc. Internat. Symp. Strategies for Exploiting Biotechnology Potentials*, (Eds.) V.I. Chopra and G.M. Reddy. Osmania University, Hyderabad, India (In Press).

Vasil, I.K., Vasil, V., White, D.W.R. and Berg, R.H., 1980. Plant tissue culture in plant improvement. In: *Developments in Plant Biology*, (Ed.) F. Sala. Biomedical Press, Amsterdam, pp. 215–226.

Venkataraman, G.S., 1962. Studies on nitrogen fixation by blue-green algae. III. Nitrogen fixation by *Anabaena azollae*. *Ind. J. Agric. Sci.*, 32: 22–24.

Venkataraman, G.S., 1981. Blue-green algae: A possible remedy to nitrogen scarcity. *Curr. Sci.*, 50: 253– 256.

Von Neumann, D., Ackerman, M. and Jacob, F., 1970. Zur Feinstruktur der endophytischen Cyanophyceen von *Gunnera chilensis* Lam. *Biochem. Physol. Pflanzen* 161: 483–492.

Vose, P.B., 1983. Developments in non-legume nitrogen-fixing systems. *Canad. J. Microbiol.*, 29: 837–847.

Watanabe, I., 1982. *Azolla-Anabaena* symbiosis: Its physiology and use in tropical agriculture. In: *Microbiology of Tropical Soil and Plant Productivity*, (Eds.) Y.R. Dommergues and H.G. Diem. Martinus Nijhoff/W. Junk Publ., The Hague, Boston-London, pp. 169–185.

Watanabe, I. and Kiyohara, T., 1973. Symbiotic blue-green algae in lichens, liverworts and cycads. In: *Studies on Micro-algae and Photosynthetic Bacteria*. Japanese Society of Plant Physiologists, Tokyo, Japan, pp. 189–202.

Watanabe, I., Espinas, C.R., Berja, N.S. and Alimagno, B.V., 1977. Utilization of the *Azolla-Anabaena* complex as a nitrogen fertilizer for rice. International Rice Research Institute Paper Series 11: 1–15.

Wolk, C.P., 1973. Physiology and cytological chemistry of blue-green algae. *Bacteriol. Rev.* 37: 73–79.

Wolk, C.P., Vonshank, A., Kehoe, P. and Elhai, J., 1984. Construction of shuttle vectors capable of conjugative transfer from *Escherichia coli* to nitrogen fixing filamentous cyanobacteria. In: *Proc. Nat. Aced. Sci., USA*, 81: 1561–1567.

Yagodina, I.B., 1983. The dynamics of growth processes in associative culture of tobacco cells and cyanobacterium. In: 4th *All Union Conf. Plant Cell Culture and Biotechnology (Abst.)*, (Eds.) R.G. Butenko. Stiinca Kishinev, Russia, p. 182.

Zhu, C., 1982. Fine structure of blue-green algae and the cells lined along the endophyte cavity in the coralloid roots of *Cycas*. *Acta Bot. Sin.*, 24: 109–118.

Chapter 3

Cyanobacterial Symbioses: An Overview

S. N. Tripathi and Meera Pandey

Department of Botany, Banaras Hindu University, Varanasi – 221 005

Symbiosis is a state of affairs in which organisms live together for mutual benefit in such close relationship that each has lost at least some degree of physiological independence (Rai *et al.*, 2000). Cyanobacteria a group of algae possessing capacity to occupy a wide range of habitats are known to be earliest oxygen evolving photosynthetic diazotrophs to occupy the earth. Its fossils existing in sediments of the early Precambrian period are over three billion years old. Besides, their remarkable ability to adapt varying environmental conditions such as fresh to marine water, hot to cold climate, anaerobic to aerobic conditions, etc., cyanobacteria are able form associations, more precisely symbiotic, with a range of eukaryotic hosts including algae, fungi, bryophytes, pteridophytes, gymnosperms, angiosperms, sponges and protists. Cyanobacterial symbionts called as phaeosomes are recorded in number of marine dianoflgellates *Ornithocercus, Histoneis* and *Citharistes*. However, *Rhizosolenia* sp. is capable of growth even in absence of the cyanobiont. *Richelia intracellularis* has also been reported in diatoms of the genus *Hemiaulus* (Adams 2000). Lichens are widely occurring form of mycobiont and phycobiont symbiotic association. Nearly 8-15 per cent of the 15000-20000 species lichens are known to have symbiotic association with a cyanobacterium as phycobiont such as species of *Nostoc, Calothrix, Scytonema* and *Fisherella* (Peters *et al.*, 1986; Rai, 1990b; Hill, 1994). Hepatatceae (liverworts), Anthocerotae (hornworts), and Musci (mosses) of bryophytes show epiphytic or endophytic association with cyanobacteria, primarily *Nostoc. Hapalosiphon* sometimes also grows closely associated with *Sphagnum. Azolla* is well known member of

Peridophytes forming endophytic symbiotic association with cyanobacteria usually-referred to as *Anabaena azollae.* Among Gymnosperms, cycads are known to form symbiotic association with cyanobacteria *Nostoc/Anabaena.* Among all the angiosperms, only genus *Gunnera,* forms symbiotic association with cyanobacteria.

Keeping aside epiphytic association, cyanobacteria mainly located intercellularly with the host plant under symbiotic conditions. *Gunnera* is the only one of the plant-cyanobacteria symbioses in which the cyanobiont *(Nostoc)* is found intracellularly. Cyanobacterial symbionts known as phaeosomes (Taylor, 1990) are found in a variety of tropical marine dianoflagellates (Lucas, 1991; Gordon *et al.*, 1994; Janson *et al.*, 1995a). Heterotrophic dianoflagellates of genera *Orinthocercus, Histioneis* and *Citharistes* have been shown to contain cyanobacterial symbionts *Synechococcus* and *Synechocystis* (Gordon *et al.*, 1994). The cyanobacteria are located between upper and lower lists of the horizontal groove of cells of *Ornithocercus* and *Histoneis,* and within special chamber inside the cells of *Citharistes. Richelia intracellularis,* which fixes atmospheric nitrogen and carbon dioxide has been known as symbiont in the marine pinnate diatom *Rhizosolenia* sp. (Rai, 1990a). The apex of *Rhizosolenia clevei* contains cytoplasm only in the region occupied by the cyanobacterium, although the cyambiont is always external to the host cytoplasm, and its cell wall remains intact (Janson *at al.*, 1995b). *Richelia intracellularis* is also found in diatoms of the genus *Hemiaulus* (Rai, 1990a). *Rhopalodia gibba* and *Epithemia turgida* members of the family Epithemiaceae are known to possess unicellular cyanobacterial symbionts in their cytoplasm (Rai, 1990a; De Yoe *et al.*, 1992).

Occurrence of symbiotic or mutualistic associations between cyanobacteria and some animals such as sponges, ascidians, echiuroid worms and chironomid larvae are known. Nearly 40 genera of marine sponges are known to contain, inter/intracellularly cyanobacterial species. Mostly these cyanobacteria, such as *Aphanocapsa, Phormidium, Oscillatoria,* are non-heterocystous. However, sponges, such as *Theonella swinhoei* and *Siphonochalia tabernacula* are reported to have association with nitrogen fixing cyanobacteria. American Pacific coast ascidians and echiuroid worms contain numerous tunic pockets filled with a variety of algae and cyanobacteria mainly *Dermocarpa, Xenococcus* and *Synechcoccus* species (Lambert *et. al.*, 1996). *Oscillatoria, Spirulina* and *Anabaena* are also reported in the inter-tunic cavity of double-tunic ascidians *Pyura cancellata* and *P. carnea* (Lambert *et al.*, 1996). Mutualistic association has also been observed between larvae of chironomid midge *Cricotopus nostocicola* living inside colonies of *Nostoc parmelioides* (Sheath and Cole, 1992; Dodds *et al.*, 1995).

Characteristically *Nostoc* is the heterocystous cyanobacterium, which has almost always been found forming symbiotic association with other plants (algae, fungi, bryophytes, pteridophytes, gymnosperms and angiosperms) (Leizerovitch *et al.*, 1990; Zimmerman and Culley, 1991; Bergman *et al.*, 1992b; Lotti *et al.*, 1996). As it is stated earlier, the association is mostly intercellular than intracellular, excepting *Gunnera.* The ability to form hormogonia has been proposed for successful association of *Nostoc* as cyanobiont in other plants. Secondly, occurrences of *Nostoc* to a wide range of habitat, its potential to survive extreme conditions are treated for its better partner in the symbiotic association. However, cyanobacteria other than *Nostoc* are also observed to form symbiotic association. *Encepharatos*, a cycad, and *Phaeoceros*, a hornwort, has been found to possess *Calothrix* spp. as cyanobionts (Grobbelaar *et al.*, 1987). Much diversified cyanobionts, with filamentous, colonial and unicellular, are reported to form association with lichens (Rai, 1990b). Among the filamentous form, heterocystous cyanobacteria *Nostoc, Calothrix. Scytonema* and *Fischerella* are much common. Of the colonial and unicellular cyanobacteria, the most common are *Gloeocapsa, Glocothece* and *Synechocystis.*

Identification of cyanobacteria at the species or genera level, particularly as cyanobiont, occurrence of one or more than species in an association with the other symbiotic partner have been some of the acute, mostly unresolved, problems. At genera level most of the difficulties have been among *Nostoc, Anabaena* and *Calothrix.* Morphological variations of the cyanobiont in the coralloid root and under culture conditions, and restriction fragment polymorphism (RFLP) have shown that a variety of cyanobiont *(Nostoc/Anabaena)* or their strains can be involved, although, mostly, single *Nostoc/Anabaena* strain is predominant in a particular root (Grobbelaar *et al.*, 1987; Lindblad *et al.*, 1989). This observation is further supported by conclusions drawn from comparisons of lectin binding, electrophoretic protein profiles and zymograms of the cyanobiont from cycads at a single site (Zimmerman and Rosen, 1902). Possibility of infection of several *Nostoc* spp. in *Gunnera* exists (Zimmerman and Bergman, 1990, Soderback, 1992; Bergman *et al.*, 1992a), only one or few are present in a single plant (Bergman *et al.*, 1992b). while only one cyanobacterial strain was observed in a single corallod root (Costa *et al.*, 1999). A recent study by Zheng *et al.* (2002) has shown that many genetically different *Nosloc* strains were present in a single coralloid root. Likewise, uncertainess of taxonomic positions of the cyanobiont of fungal partners highly persists. However, presence of a single cyanobiont in a plant is favoured by the study of PCR amplification of tRNA^{Leu} (UAA) introns from the *Nostoc* symbionts of lichens *Nemphroma arcticum, Peltigera aphthosa, P. membranacea* and *P. canina* (Paulsrud and Lindblad, 1998). A high degree of specificity was observed in the cyanobiont of the lichens collected from different localities (Paulsr *et al.*, 1998). Moreover, possibilities of infection of same cyanobiont in different plants have also been anticipated (Mollenhauer, 1992; Kluge *et al.*, 1994).

Controversy about the cyanobiont of *Azolla* has been much more acute, though, conventionally, it has been referred as *Anabaena azollae.* Strasburger (1873) described *Nostoc* as the cyanobiont of *Azolla,* and later, he renamed it as *Anabaena azollae* (Strasburger, 1884). Based on *nif genes* (Meeks *et al.*, 1988) and superoxide dismutase isoenzymes (Canini *et al.*, 1992a), cyanobiont was identified as genus *Nostoc.* This view was further supported by Kim *et al.* (1997) by applying Polymerase Chain Reaction (PCR) analysis of the *Azolla* cyanobiont. On the other hands, fatty acid analyses were not in accordance with either *Nostoc* or *Anabaena* (Caudales *et al.*, 1992). Also. immunological (Arad *et al.*, 1985) and lectin haemagglutination studies (McCowen *et al.*, 1987, Zimmerman *et al.*, 1989), RFLP analyses (Franche and Cohen-Bazire, 1985, 1987; Meeks *et al.*, 1988; Plazinski *el al.*, 1988; Gebhardt and Nierzwicki-Baur, 1991), and enzymatic comparisons (Zimmerman *et al.*, 1989) have confirmed that the readily culturable cyanobionts of *Azolla* are not the same as the freshly-isolated symbiont.

In the process of development of symbiotic association between the cyanobacterial and plant partners, the ability of the cyanobacterial partner to develop both heterocysts and hormogonia has been described as essential characteristic factor (Campbell and Meeks, 1989; Meeks, 1990; Johnsson and Bergman, 1994; Bergman *et al.*, 1996; Meeks, 1998). The ability of cyanobacteria to fix nitrogen and carbon has also been proposed as reasons for development of symbiotic association with cyanobacteria. Certainly, all hormogonia forming and heterocysts bearing cyanobacterial forms are not able to form symbiotic associations (Meeks, 1990; Johansson and Bergman, 1994;. Bergman *et al.*, 1996). Similarly, carbon and nitrogen fixation ability is much predominant in many cyanobacteria, but they fail to have symbiotic relations with other plants or animals. It has been found that plant hosts produce extracellular signals that trigger hormogonia formation in specific cyanobacteria. *Anthoceros punctatus* release a low molecular mass, heat-labile product that stimulates hormogonia formation in N*ostoc* strains (Campbell and Meeks, 1989). This phenomenon is not applicable to the hornwort growing in the culture medium containing NH_4^+, means that the compound is produced when the plant, is deprived of nitrogen. Rasmussen *et al.* (1994) have also shown presence of hormogonia-inducing signal in the acidic mucilage of the *Gunnera* stem glands, but in any other parts of the plant. The active component has been suggested as a heat-labile less than 12 KDa protein. Characteristically, arabinogalactan proteins (AGPs), components of the mucilage, are suggested to influence differentiation of compatible cyanobacteria prior to infection of the plant (Rasmussen *et al.*, 1996). Phenolic compounds found in the vicinity of *Gunnera* symbiotic tissue are also potential signals in this system, with possible roles in attraction, defense or the transcriptional activation of cyanobacterial genes involved in symbiosis (Bonnett, 1990; Bergman *et al.*, 1992a). Flavonoid-responsive genes showing homology to legumes inducible genes in rhizobia are also reported in the cyanobacteria infecting *Azolla* (Plazinski *et al.*, 1991) and *Gunnera* (Rasmussen *et al.*, 1996). Signals from cyanobacterium to *Gunnera* have also described (Bergman *et al.*, 1996). Besides, signals involvement in cyanobacteria specific symbioses, recognition processes is also reported in the infection of host by cyanobacteria. The host plants, lichens, bryophytes and *Azolla*, are known to produce lectins that recognize specific sugar residues on the cyanobacterial cell surface (Braun-Howland and Nierzwicki-Bauer, 1990; Kardish *et al.*, 1991; Bergman *et al.*, 1992, Honegger, 1993).

Though, there is specificity between the two partners, mode of infection and thereafter organization of the symbiotic structures are found varying depending on the two partners. The auricles, spherical structures on the ventral side of thallus of *Blasia*, contain cyanobacterial symbiont, cyanobionts of *Anthoceros* and *Phaeoceros* possess cyanobionts within their thick thalli. Microaerobic conditions are reported in the cavity of *Anthoceros*, which may be responsible for presence of defective envelope of the heterocysts of the cyanobiont (Campbell and Meeks, 1992). Infolding of the epidermis of dorsal leaf lobes of *Azolla* has been suggested for forming a cavity, which occupied with the cyanobiont (Peters *et al.*, 1986; Peters and Meeks, 1989; Braun-Howland and Nierzwicki-Bauer, 1990). The cavity is extracellular and initially open to outside, becoming close as the leaf matures (Uheda, 1986, Braun-Howland and Nierzwicki-Baur, 1990; Uheda and Kitoh, 1991). The cyanobiont remain associated with *Azolla* throughout its life cycle and continuity of symbiosis is maintained through megasporocarps. The cyanobiont persistent in the apical meristem is transferred to newly developing leaves (Peters and Meeks, 1989; Braun-Howland and Nierzwicki-Bauer, 1990; Peters and Perkins, 1993).

Coralloid roots of cycads are more often found below soil surface. However, in older plants they are observed on soil surfaces also (Lindblad *et al.*, 1985; Lindblad and Bergman, 1990; Ahern and staff, 1994). This is proposed that a break in dermal layer of the roots of cycads allow the cyanobiont to enter inside the roots and gradually reaches to the cortex through a channel (Lindblad and Bergman, 1990; Staff and Ahern, 1993; Ahern and Staff, 1994). Invasion by cyanobacteria stimulates irreversible changes in the precoralloids, transforming them into typical coralloid roots (Ahern and Staff 1994). At the base of each petiole, cyanobacterial invasion take place in the mucus secreting glands and ultimately results in *Gunnera-Nostoc* association (Towata, 1985; Peters *et al.*, 1986; Bonnet 1990; Bergman *et al.*, 1992b). Invasion takes place only when the cyanobacterium penetrates the host cell walls and becomes intracellular and ultimately results in formation of glands.

Establishment of symbiotic association between cyanobacterium and the host plant generally result in a distinct modification in the structural and functional behaviour of mainly the cyanobacterial partner than the host plant. In most of the cases, cell division, hormogonia formation, morphology, and physiological and biochemical phenomena pertaining to carbon and nitrogen metabolism of the cyanobacterium have been found considerably altered. The growth rate of the cyanobacterium of the symbiotically associated strains has been recorded considerably low than its free-living strains. It has been suggested that the rate of cell division of cyanobiont is linked with the host partner, which varies depending on the part of the host plant also (Peters and Meeks,1989; Hill, 1989; Braun-Howland and Nierzwicki-Bauer, 1990). In *Azolla,* the rate of cell division is slower in leaf cavities than at the at the shoot apex where host cell growth is rapid (Hill, 1989). It is suggested that the primary control is extended on cell division of the symbiont. resulting in reduced demand for nutrients, the excess becoming available for use by the host (Hill, 1989). This has recently been observed that a viable symbiotic colony could not be established if, immediately upon infection, the cyanobacteria were induced to form hormogonia once more, though, hormogonia are required for establishment of infection of the cyanobacterium in the host plant. On the other hands, hormogonia do not form heterocysts, a key factor in the development of symbiotic relationship between cyanobiont with its host partner. Therefore, there must be a means for repressing hormogonia formation (Cohen and Meeks, 1997; Meeks, 1998). Evidence for such a system has been demonstrated recently by analysis of a transposon-induced mutant of *Nostoc punctiforme* that is 50-fold infective than the wild-type in *Anthoceros* (Cohen *et al.*, 1994). Two genes, *hrmA* and *hrmU,* were identified as part of an operon, and expression of both was induced by an aqueous extract of *Anthoceros* tissue, but not by the hormogonia inducing factor (HIF) of the hornwort. The aqueous extract contains a hormogonia formation in wild-type *N. punctiforme,* and it seems that the mutant is more infective because it continues to respond to HIF, producing hormogonia and extending it infective state. Based on these observations, it is said that the gene products of the *hrmU* operon block hormogonium formation perhaps by production of an inhibitor or by catabolism of an activator (Cohen and Meek, 1997).

Cyanobionts are generally observed to undergo morphological modifications during or after establishment of symbiosis with their host partners. In liverworts and hornwort cyanobiont vegetative cells become enlarged and spherical with weak cell-to-cell junctions (Meeks, 1990). Filamentous cyanobionts *Scytonema* and *Calothrix* of lichens are most often transformed into unicellular forms (Rai, 1990), whereas, *Nostoc* cells are found enlarged (Ahmadjian, 1993). The cyanobiont of *Azolla,* having enlarged vegetative cells, possesses a variable cell size and shape between filaments (Peters, 1991). The *Nostoc* cyanobiont of *Cycas revoluta* increases in cell size on symbiosis establishment (Sharma *et al.*, 1992), as does the *Nostoc* symbiont of *Geosiphon pyriforme* after phagocytosis by the fungal host (Kluge *et al.*, 1992, 1994, Mollenhauer *et al.*, 1996). In general, heterocyst frequency compared to their free living strains has been observed to be very high in cyanobionts. The low level of CO_2 fixing capacity of the cynobionts has been attributed to enhancement in the high frequency of heterocysts also (Adams, 2000). However, heterocyst frequency is lower in the new symbiont than the older ones. In the newest *Azolla* leaf cavities, close to the apical meristem of each stem, the heterocyst frequency is almost zero, but increases as leaves age, until senescence begins and nitrogenase activity declines (Kaplan *et al.*, 1986; Peters *et al.*, 1986; Grilli Caiola *et al.*, 1989; Canini *et al.*, 1990; Bergman *et al.*, 1992b). Heterocyst frequency is high in the older (basal) leaf cavities, but nitrogenase activity is lowered (Canini *et al.*, 1990). Similarly, in cycad coralloid roots and also in *Gunnera* the frequency of heterocyst is lowest at the root tip and growing tips and highest in the basal, older parts, where high frequencies of two, three or even four adjacent heterocysts are found (Lindblad *et al.*,1985a, b; Peters, 1991; Soderback *et al.*, 1990; Bergman *et al.*, 1985; Stock and Silvester, 1994). Heterocyst differentiation has been proposed to be regulated by various factor in th free-living heterocystous cyanobacteria. Differentiation of heterocyst in the symbiotic cyanobacteria is not well understood. In free-living cyanobacteria the stimulus to develop heterocysts is depletion of nitrogen in the medium (Wolk *et al.*, 1994). In symbiosis, starvation of nitrogen has not been observed; heterocyst number has been found increased. In cycads *Zamia* and *Cycas* heterocyst frequency is highest in the basal region, yet nitrogenase activity is highest at the coralloid root tip (Lindblad and Bergman, 1990). It has been proposed that host plant regulates development of heterocyst (Adams, 2000); though, the true significance of the elevated heterocyst frequencies found in many cyanobacterial symbioses is difficult to assess.

The primary function of the cyanobiont in the plant-cyanobacteria symbioses is to make combined nitrogen available to the host plant. Almost all the known cyanobionts are capable to fix nitrogen (Rai, 1990). The enzyme nitrogenase involved in nitrogen fixation is confined to heterocysts if not all cyanobacterial symbioses (Bergman *et al.*, 1986; Braun-Howland *et al.*, 1988; Rai *et al.*, 1989; Soderback *et al.*, 1990). Heterocysts provide the necessary anaerobic conditions for nitrogenase. In other forms, such as aerobic leaf cavities of *Azolla,* where chances of creation of aerobic condition prevail (Grill Caiola *et al.*, 1989; Mutuskin and Kolesnikov, 1991), some additional protective mechanism such as involvement of superoxide dismutase

(SOD) are reported to remove superoxide radical from the cellular atmosphere. SOD has been observed highest in the leaves that show maximum nitrogenase activity (Canini *et al.*, 1992b). In contrast, the symbiotic colonies in the cycad *Macrozamia riedlei* are thought to be microaerobic, and freshly isolated symbionts exhibit increased sensitivity to oxygen (Lindblad *et al.*, 1991b), possible because of the incomplete development of heterocyst wall layers in absence of oxygen (Rippka and Stanier, 1978).

In most of the cases of cyanobacteria-plant symbioses, host plants are much more dependent on their cyanobiont partner for supply of nitrogen. In even the largest *Gunnera* plants, with leaves two meter across, this is achieved by a cyanobiont mass that may be as little as 1 per cent of the total plant weight (Osborne *et al.*, 1992). This has been attributed to the little capacity to use alternative sources of combined nitrogen by *Gunnera*, and consequently be almost entirely dependent on the cyanobionts for its nitrogen (Osborne, 1989; Osborne *et al.*, 1992). Another feature, which has been most commonly observed, is about occurrence of heterocyst and nitrogen fixing capacity by the cyanobiont. In younger colonies of the cyanobionts, heterocyst numbers are recorded less than the older colonies and according nitrogen fixation has been observed more in the older than newer ones. However, at the same times, heterocysts present in chains, *e.g.* in *Gunnera*, were found to fix nitrogen inefficiently, mainly because of poor transfer of photosynthate from vegetative cells (Bergman *et al.*, 1986, 1992; Soderback *et al.*, 1990).

Nitrogen fixed by the cyanobiont is released as ammonia to the host plant such as *Anthoceros, Blasia, Azolla, Gunnera* (Meeks *et al.*, 1985a, b; Bergman *et al.*, 1992; Silvester *et al.*, 1996) and many lichens (Rai *et al.*, 1990). An unknown organic compound (most likely citrulline or glutamine) is proposed to be transferred in cycads after fixation of nitrogen. Only a small fraction of fixed nitrogen is generally retained by the cyanobiont. 5 per cent in some lichens (Rai *et al.*, 1990a; Ahmadjian, 1993), 10 per cent in *Gunnera* (Silvester *et al.*, 1996). 20 per cent in the hornwort *Anthoceros* (Meeks *et al.*, 1985a), and 60 per cent in *Azolla* (Meeks *et al.*, 1985b, 1988; Peters and Meeks, 1989) combined nitrogen fixed is retained by the cyanobiont. Some of the fixed nitrogen has been observed to be released in the form of ammonia, mainly due to less activity of glutamine synthetase of glutamine-synthetase-glutamate synthase (GS-GOGAT) pathway, which primary route of ammonia assimilation in cyanobacteria and in bryophytes and *Azolla* (Meeks, 1990; Bergman *et al.*, 1992), but not in lichens which use glutamate dehydrogenase (Rai, 1990b). The decrease in GS activity in *Azolla* results from a decrease in the amount of enzyme in the symbiont (Lee *at al.*, 1988; Peters and Meeks, 1989). Ten-fold decrease in the level of GS (glnA) mRNA has been observed (Nierzwicki Bauer and Haselkorn, 1986). Likewise, in lichens both GS and GOGAT activity is reduced by over 95 per cent due to a reduction in the amount of GS protein (Rai, 1990a). In *Anthoceros*, an undetermined post-trancriptional modification of the enzyme is thought to be responsible for the decreased GS activity, because the amount of GS protein in free-living and symbiotically associated *Nostoc* is little different (Joseph and Meeks, 1987; Lee *et al.*, 1988;, Rai *et al.*, 1989; Meeks, 1990). The amount of GS protein and its activity could not be correlated in the case of *Gunnera* as in the cyanobiont GS specific activity is 70 per cent of that in the free living state although the amount of protein remains unchanged (Bergman *et al.*, 1992b; Silvester *et al.*, 1996). In cycads, GS activity and protein both are almost unchanged (Lindblad and Bergman, 1986), which is in accordance with the release of organic nitrogen to the plant (Pate *et al.*, 1988).

The level of GS protein in the heterocysts and vegetative cells is found same in the case of cyanolichens (Hallbom *et al.*, 1986), bryophytes (Rai *et al.*, 1989) and *Azolla* (Bergman and Rai, 1989). However, GS protein cannot be detected in the heterocysts of the *Calothrix* symbiont present in the apical regions of the lichen *Lichina confinis* (Janson *et al.*, 1993). Whereas, in the heterocysts of free-living cyanobacteria in which heterocysts have approximately twice the GS activity and protein levels of vegetative cells (Bergman *et al.*, 1985; Rai *et al.* 1989; Renstrom-Kellner *et al.*, 1990; Bergman *et al.*, 1992b).

Transfer of the fixed nitrogen from the cyanobacterium to the host partner does not seem to lead to nitrogen starvation, as most of the nitrogen containing as well as storage bodies of the cyanobiont remain functional and intact during the symbiotic conditions. Phycobiliproteins, which serve as both photosynthetic accessory pigments and nitrogen reserves in cyanobacteria, are still present in the cyanobionts of bryophytes (Rai *et al.*, 1989; Meeks, 1990), *Azolla* (Kaplan *et al.*, 1986; Braun-Howland and Nierzwicki-Bauer, 1990), cycads (Lindblad and Bergman, 1989) and *Gunnera* (Soderback and Bergman, 1992). Also, cyanophycin, a co-polymer of arginine and aspartic acid (Simon, 1987). working as a unique cyanobacterial nitrogen reserve, is also present in the cyanobionts of bryophytes (Rai *et al.*, 1989), *Gunnera* (Soderback *et al.*, 1990; Jager *et al.*, 1997), cycads (Lindblad *et al.*, 1985a) and *Azolla* (Bergman *et al.*, 1992b).

Ribulose-l,5-bisphosphate carboxylase/oxygenase is the primary carboxylating enzyme in cyanobacteria, part of which is contained in carboxysomes, polyhedral cellular inclusions mainly made up of Rubisco proteins (Codd and Marsden, 1984; Tabita, 1994). Light dependent O_2 evolution and CO_2 fixation in the cyanobionts of bryophytes (Meeks, 1990). cycads (Lindblad and Bergman, 1990; Lindblad *et al.*, 1991a) and *Gunnera* are either greatly reduced, or undetectable, immediately after their separation from the host plant. *In vitro*, carbon fixation in cycads is comparable to that of free-living isolates. This has been suggested that in these cases the cyanobiont grow photo-or chemoheterotrophically, receiving fixed carbon from its photosynthetic host, probably in the form of sucrose (Steinberg and Meeks, 1991; Bergman *et al.*, 1992; Soderback and Bergman, 1993), similarly in *Azolla* (Peters and Meeks, 1989; Braun-Howland and Nierzwicki-Bauer, 1990; Peters, 1991). The cyanobiont of *Azolla* does not show more than 5 per cent of the carbon fixation of the symbiosis (Kaplan and Peters, 1988; Peters and Meeks, 1989). The level of mRNA is at least five-fold lower than free-living *Anabaena* (Nierzwicki-Bauer and Haselkorn, 1986). Interestingly, just after isolation from the host, *Anabaena azollae* showed a similar rate of light dependent CO_2 fixation to that of the free-living organisms, whereas, *Nostoc* from *Anthoceros* has a rate of light dependent CO_2 fixation eight-fold lower than the same grown under free-living conditions (Steinberg and Meeks, 1989; Meeks, 1990). Both the partners have different level of Rubisco protein, and regulation of activity is suggested by post-translational modification of the enzyme (Steinberg and Meeks, 1989; Rai *et al.*,

1989; Meeks, 1990). In contrast, the freshly isolated cyanobiont of *Gunnera chilensis* shows Rubisco activity higher than the free-living *Nostoc,* even though the symbiotically associated symbiont showed lower CO_2-fixing capacity (Soderback and Bergman, 1993), which does not follow the post-translational modifications hypothesis. It seems that the photosynthetic machinery of the symbiont is largely intact, with chlorophyll-*a* (Soderback *et al.,* 1990). phycobiliproteins and Rubisco (Soderback and Bergman, 1992) being present at levels equivalent to the free living symbiont, and that a reduction in the photosynthetic electron transport is responsible for the reduced *in vivo* carbon fixing activity of the cyanobiont. A similar situation may arise in cycads such as *Cycas revoluta*, in which, despite the lack of *in vivo* carbon fixation, considerable Rubisco activity can be measured in crude extracts of the freshly isolated cyanobiont (Lindblad *et al.,* 1987; Lindblad and Bergman, 1990). Exceptionally, high rate of photosynthesis (oxygen evolution) has been observed in the immediately isolated cyanobiont of *Cycas circinalis*, possibly, because the coralloid roots being close to surface and thus they are exposed to light (Perraju *et al.,* 1986). The cyanobionts of both bipartite and tripartite lichens have shown photosynthetic carbon fixation and Rubisco enzyme activity detected by immunogold labeling (Bergman and Rai, 1989; Janson *et al.,* 1993). The photosynthates are transferred from the cyanobiont to their host partners in the form of glucose (Rai, 1990b; Feige and Jensen, 1992; Bonger *et al.,* 1993; Ahmadjian, 1993; Hill, 1994), which is later converted into mannitol (Ahmadjian, 1993). In bipartite lichens, mycobiont is entirely dependent on cyanobiont for carbon source.

The cyanobacteria, in free-living conditions, generally, are photoautotrophs possessing higher plant type oxygen evolving photosynthetic process. In the recent years some reviews (Adams, 2000; Rai *et al.,* 2000; Costa and Lindblad, 2002) have come up covering different aspects of cyanobacterial photosynthesis under symbioses, as it has been described above also, no clear view has been made whether the cyanobacterial partner under symbioses are functionally photosynthetic. This is much more relevant to the cyanobiont of cycads. In cycads, cyanobionts are part roots, which are supposed to be present below soil surface. Reaching of light to the cyanobiont has been always under question. In fact, based on available information it can be said that till date no substantive efforts have been made to investigate the phenomena of cyanobacterial symbioses especially photosynthesis pertaining to photochemical reactions in cycads.

Acknowledgement

Thanks are due to the Head, Department of Botany, Banaras Hindu University for providing facilities.

References

Adams, D.G., 2000. Symbiotic interactions. In: *The Ecology of Cyanobacteria,* (Eds.) B.A. Whitton and M. Potts. Kluwer Academic Publishers, Dodrecht, pp. 523–561.

Ahem, C.P. and Staff, I.A., 1994. Symbiosis in cycads: The origin and development of coralloid roots in *Marcrozamia communis* (Cycadaceae). *Am. J. Bot.,* 81: 1559–1570.

Ahmadjian, V., 1993. *The Lichen Symbiosis.* John Wiley and Sons, New York.

Arad, H., Keysari A., Tel-or, E. and Kobiler, D., 1985. A comparison between cell antigens in different isolates of *Anabaena azollae*. *Symbiosis*, 1: 195–204.

Bergaman, B., Lindblad, P., Pettersson, A., Renstrom, E. and Tiberg, E., 1985. Immunogold localization of glutamine synthetase in a nitrogen-fixing cyanobacterium, *Anabaena cylindrica*. *Planta*, 166: 329–334.

Bergman, B., Lindblad, P. and Rai, A.N., 1986. Nitrogenase in free-living and symbiotic cyanobacteria immunoelectronmicroscopic localization. *FEMS Microbiol. Lett.*, 35: 75–78.

Bergman, B. and Rai, A.N., 1989. The *Nostoc-Nephroma symbiosis*: Localization, distribution pattern and levels of key proteins involved in nitrogen and carbon metabolism of the cyanobiont. *Physiol. Plant.*, 77: 216–224.

Bergman, B., Johansson, C. and Soderback, E., 1992a. Tansley Review No. 42: The *Nostoc-Gunnera* symbiosis. *New Phytol.*, 122: 379–400.

Bergman, B., Rai, A.N., Johansson, C. and Soderback, E., 1992b. Cyanobacterial-plant symbioses. *Symbiosis*, 14: 61–81.

Bergman, B., Matveyev, A. and Rasmussen, U., 1996. Chemical signaling in cyanobacterial-plant symbiosis. *Trends Plants Sci.*, 1: 191–197.

Bogner, E., Wastlhuber, R., Schlegl, I. and Loos, E., 1993. Glycogen, amylase and alpha-glucosidase as possible components in the glucose-release system of the cyanobiont of *Peltigera horizontalis*-partial-purification and characterization. *Symbiosis*, 14: 487–489.

Bonnett, H.T., 1990. The *Nostioc-Gunnera* association. In: *CRC Handbook of Symbiotic Cyanobacteria*, (Ed.) A.N. Rai. CRC Press, Boca Raton, Florida, pp 161–171.

Braun-Howland, E.B. and Nierzwicki-Bauer, S.A., 1990. *Azolla-Anabaena* symbiosis biochemistry, physiology, ultrastructure, and molecular biology. In: *CRC Handbook of Symbiotic Cyanobacteria*, (Ed.) A.N. Rai. CRC Press. Boca Raton, Florida, pp 65–117.

Braun-Howland, E.B., Lindblad, P., Nierzwicki-Bauer, S.A. and Bergman, B., 1988. Dinitrogenase reductase (Fe-protein) of nitrogenase in the cyanobacterial symbionts of three *Azolla* species: Localization and sequence of appearance during heterocyst differentiation. *Planta*, 176: 319–332

Campbell, E.L. and Meeks, J.C., 1989. Characteristics of hormogonia formation by symbiotic *Nostoc* spp. in response to the presence *Anthoceros punctatus* or its extracellular product. *Appl. Environ. Microbiol.*, 55: 125–131.

Campbell, E.L. and Meeks, J.C., 1992. Evidence for plant-mediated regulation of nitrogenase expression in the *Anthoceros-Nostoc* symbiotic association. *J. Gen. Microbiol.*, 138: 473–480.

Canini, A., Grilli Caiola, M., Civitareale, P. and Galiazzo, F., 1992a. Superoxide dismutase in symbiotic, free living and wild *Anabaena* and *Nostoc* (Nostocales Cyanophyta). *Phycologia*, 31: 225–230.

Canini, A., Bergman, B., Civitareale, P., Rotilio, G. and Grilli Caiola, M., 1992b. Localization of iron superoxide dismutase in the cyanobiont of *azolla filiculoides* Lam. *Protoplasma,* 169: 1–8.

Canini, A., Grilli Caiola, M. and Mascini, M., 1990. Ammonium content, nitrogenase acitivity and heterocyst frequency within the leaf cavities of *Azolla filiculoides* Lam. *FEMS Microbiol. Lett.,* 71: 205–210.

Caudales, R., Wells, J.M. and Antoine, A.D., 1992. Cellular fatty acid composition of symbiotic cyanobacteria isolated from the aquatic fern *Azolla. J. Gen. Microbiol.,* 138: 1489–1494.

Cohen, M.F. and Meeks, J.C., 1997. A hormogonium regulating locus, hrmUA, of the cyanobacterium *Nostoc punctiforme strain* ATCC, 29133 and its response to an extract of a symbiotic plant partner *Anthoceros punctatus. Mol. Plant-Microbe Interactions,* 10: 280–289.

Cohen, M.F., Wallis, J.G., Campbell, E.L. and Meeks, J.C., 1994. Transposon mutagenesis of *Nostoc* sp. *strain* ATCC 29133, a filamentous cyanobacterium with multiple cellular differentiation alternative. *Microbiol.,* 140: 3233–3240.

Costa, J.L., Paulsrud, P. and Lindblad, P., 1999. Cyanobiont diversity within coralloid roots of selected cycad species. *FEMS Micriobiol. Ecol.,* 28: 85–91.

Costa, J.L and Lindblad, P., 2002. Cyanobacteria in symbiosis with cycads. In: *Cyanobacteria in Symbiosis,* (Ed.) Rai *et al.* Kluwer Academic Publishers, pp. 195–205.

De Yoe, H.R., Lowe, R.L. and Marks, J.C., 1992. Effects of nitrogen and phosphorus on the endosymbiont load of *Rhyopalodia gibba* and *Epithemia turgida* (Bacillariophycaea). *J. Phycol.,* 28: 773–777.

Dodds, K., Gudder, D.A. and Mollenhauer, D., 1995. The ecology of *Nostoc. J. Phycol.,* 31: 2–18.

Feige, G.B and Jensen, M., 1992. Basic carbon and nitrogen metabolism of lichens. In: *Algae and Symbioses: Plants, Animals, Fungi, Viruses, Interactions Explored,* (Ed.) W. Reisser. Biopress Ltd., Bristol, p. 277–299.

Franche, C. and Cohen-Bazire, G., 1985. The structural *nif* genes of four symbiotic *Anabaena azollae* show a highly conserved physical arrangement. *Plant Sci.,* 39: 125–131.

Gebhardt, J.S. and Nierzwicki-Bauer S., 1991. Identification of common cyanobacterial symbiont associated with *Azolla* spp. Through molecular and morphological characterization of free living and symbiotic cyanobactena. *Appl. Environ. Microbiol.,* 57: 2141–2146.

Grilli Caiola, M., Canini, A. and Moscone, D., 1989. Oxygen concentration nitrogenous activity and heterocyst from the leaf cavities of *Azolla fliculoides Lam.* FEMS *Lett.,* 59: 283–288.

Grobbelaar, N., Scott, W.E., Harttingh, W. and Marshall, J., 1987. The identification of the coralloid root endophytes of the Southern African cycads and the ability of the isolates to fix dinitrogen. *South African J. Bot.,* 53: 111–118.

Gordon, N., Angel, D.L., Neori, A., Kress, N. and Kimor, B., 1994. Heterotrophic dinoflagellates with symbiotic cyanobacteria and nitrogen limitation in gulf of Aqaba. *Marine Ecology Progress Series*, 107: 83–88.

Hailborn, L., Bergman, B. and Rai, A.N., 1986. Immunogold localization of glutamine synthetase in the cyanobionts of the lichens *Peltigera aphthosa* and *Peltigera canina* Lichen. *Physiol. Biochem.*, 1: 27–34.

Honegger, R., 1993. Tansley Review No. 60, Development biology of lichens. *New Phytol.*, 125: 659–677.

Hill, D.J., 1989. The control of the cell cycle in microbial symbionts. *New Phytol.*, 112: 175–184.

Hill, D.J., 1994. *The Nature of the Symbiotic Relationship in Lichens*. Endeavour, New Series, 18: 96–103.

Jager, K.M., Johansson, C., Kunz, U. and Lehmann, H., 1997. Sub-cellular element analysis of a cyanobacterium (*Nostoc* sp.) in symbiosis with *Gunnera manicata* by ESI and EELS. *Bot. Acta*, 110: 151–157.

Janson, S., Carpenter, E.J. and Bergman, B., 1995a. Immunolabeling of Phycoerythrin, ribulose 1,5 bisphosphate carboxylase/oxygenase and nitrogenase in the unicellular cyanobionts of *Ornithocercus* spp. (Dinophyceae). *Phycologia*, 34: 171–176.

Janson, S., Rai, A.N. and Bergman, B., 1995b. Intracellular cyanobiont *Richelia intracellularis*: Ultrastructure and immunolocalization of phycoerythrin, nitrogenase, Rubisco and glutamine synthetase. *Marine Biol.*, 124: 1–8.

Janson, S., Rai, A.N. and Bergman, B., 1993. The marine lichen *Lichina confinis* (O.F. Mull) c. Ag.: Ultra structure and localization of nitrogenase, glutamine synihetase, phycoerythrin and ribulose 1, 5 bisphosphate carboxylase/oxygenase in the cyanobiont. *New Phytol.*, 124: 149–160.

Johansson, C. and Bergman, B., 1994. Reconstruction of the symbiosis of *Gunnera manicata* Linden: cyanobacterial specificity. *New Phytol* 126: 643–652.

Joseph, C.M and Meeks, J.C., 1987. Regulation of expression of glutamine synthetase in a symbiotic *Nostoc* strain associated with *Anthoceros punctatus*. *J. Bacteriol.*, 169: 2471–2475.

Kaplah, D. and Petsers, G.A., 1988. Interaction of carbon metabolism in the *Azolla-Anabaena* symbiosis. *Symbiosis*, 6: 53–68.

Kaplan, D., Calvert, H.E. and Peters, G.A., 1986. The *Azolla-Anahaena azollae* relationship XII, Nitrogenase activity and phycobiliproteins of the endophyte as a function of leaf age and cell type. *Plant Physiol.*, 80: 884–890.

Kardish, N., Silverstein, L., Fleminger, G. and Galun, M., 1991. Lectin from the lichen *Nephroma leavigatum*. Ach. localization and function. *Symbiosis*, 11: 47–62.

Kim, J.J.H., Krawerzyk, K., Lorentz, W.P. and Zimmerman, W.J., 1997. Fingerprinting cyanobionts and hosts of the *Azolla* symbiosis by DNA amplification. *World J. Microbiol. Biotecthnol.*, 13: 97–101.

Kluge, M., Mollenhauer, D. and Mollenhauer, R., 1994. *Geosiphon pyriforme* (Kutzing) F-V Wettstein: A promising system for studying endocyanoses. *Prog. Bot.*, 55: 130–141.

Lambert, G., Lambert, C.C. and Waaland, J.R., 1996. Algal symbionts in the tunics of six New Zealand Ascidians (Chordata, Ascidiacea). *Invertebrate Biol.*, 115: 67–78.

Lee, K.Y., Joseph, C.M. and Meeks, J.C., 1988. Glutamine synthetase specific activity and protein concentration in symbiotic *Anabaena* associated with *Azolla caroliniana. Antonie van Leeuwenhoek*, 54: 345–355.

Leizerovitch, I., Kardish, N. and Galun, M., 1990. Comparison between eight symbiotic, cultured *Nostoc* isolates and a free-living *Nostoc* by recombinant DNA. *Symbiosis*, 8: 75–85.

Lindblad, P. and Bergman, B., 1986. Glutamine synthetase: Activity and localization in cyanobacteria of the cycads *Cycas revoluta* and *Zamia skinneri. Planta*, 169: 1–7.

Lindblad, P. and Bergman, B., 1989. Occurrence and localization of phycoerythrin in symbiotic *Nostoc of Cycas revoluta* and in the free-living isolate *Nostoc* 7422. *Plant Physiol.*, 89: 783–785.

Lindblad, P. and Bergman, B., 1990. The cycad-cyanobacterial symbiosis. In: *CRC Handbook of Symbiotic Cyanobacteria*, (Ed.) A.N. Rai. CRC Press, Boca Raton, Florida, pp 137–159.

Lindblad, P., Atkins, C.A. and Pate. J.S., 1991a. Nitrogen fixation by freshly isolated *Nostoc* from coralloid root of the Cycad Macro *zamia riedlei* (*Fisch.* Ex Gaud.) Gardn. *Plant Physiol.*, 95: 753–759.

Lindblad, P., Arkins, C.A. and Pate, J.S., 1991b. N_2-fixation by freshly isolated *Nostoc* from coralloid roots of the cycad *(Macrozamia nediei)* Fisch.ex. (Gaud) Gardn. *Plant Physiol.*, 95: 753–759.

Lindblad, P., Hailbom, L. and Bergam, B., 1985a. The cyanobacterium-*Zamia* symbiosis: C_2H_2 reduction and heterocyst frequency. *Symbiosis*, 1: 19–28.

Lindblad, P., Hailbom, L. and Bergam, B., 1985b. The cyanobacterium-*Zamia* symbiosis: C_2H_2 reduction and heterocyst frequency. *Symbiosis*, 1: 19–28.

Lindblad, P., Rai, A.N. and Bergman, B., 1987. The *Cycas revoluta-Nostoc* symbiosis: Enzyme activities of nitrogen and carbon metabolism in the cyanobiont. *J. Gen. Microbiol.*, 133: 1695–1699.

Lotti, F., Giovannetti, L., Margheri, M.C., Ventura, S. and Materassi, R., 1996. Diversity of DNA methylation pattern and total DNA restriction pattern in symbiotic *Nostoc. World J. Microbiol. and Biotechnol.*, 12: 38–42.

Lucas, I.A.N., 1991. Symbionts of the tropical *Dinophysiales* (Dinophyceae). *Ophelia*, 33: 213–224.

McCowne, S.M., MacArthur, L. and Gates, J.E., 1987. *Azolla* fern lectins that specifically recognize endosymbiotic cyanobacteria. *Curr. Microbiol.*, 14: 329–333.

Meeks, J.C., Enderlin, C.S., Joseph, C.M., Chapman, J.S. and Lollar, M.W.L., 1985a. Fixation of [^{13}N]N_2 and transfer of fixed nitrogen in the *Anthoceros-Nostoc* Symbiotic association. *Planta*, 164: 406–414.

Meeks, J.C., Joseph, C.M. and Haselkom, R., 1988. Organization of the *nif* genes in cyanobacteria in symbiotic association with *Azolla* and *Anthocerous, Arch. Microbiol.*, 150: 61–71.

Meeks, J.C., Steinberg, N., Joseph, C.M., Enderlin, C.S., Jorgensen, P.A. and Peters, G.A., 1985b. Assimilation of exogenous and dinitrogen-derived $^{13}NH_4^4$ by *Anabaena azollae* separated from *Azolla caroliniana. Wild. Arch. Microbiol.*, 142: 229–233.

Meeks, J.C., 1988. Symbiotic associations. In: *Methods in Enzymology*, (Eds.) L. Packer and A.N. Glazer. Academic Press Inc., London, 167: 113–121.

Meeks, J.C., 1990. Cyanobacterial-bryophyte associations. In: *CRC Handbook of Symbiotic Cyanobacteria*, (Ed.) A.N. Rai. CRC Press, Boca Raton Florida, pp. 43–63.

Meeks, J.C., 1998. Symbiosis between nitrogen fixing cyanobacteria and plants. *Bioscience*, 48: 2266–2276.

Mollenhauer, D., 1992. *Geosiphon Pyriforme.* In: *Algae and Synbioses: Plants, Animals, Fungi, Viruses Interactions Explored*, (Ed.) W. Reisser. Biopress Ltd., Bristel, pp. 339–351.

Mollenhauer, D., Mollenhauer, R. and Kluge, M., 1996. Studies on initiation and development of the partner association in *Geosiphom pyriforme* (Kutz.) v. Wettstein. A unique endosymbiotic system of a fungus (Glomales) and the cyanobacterium *Nostoc punciforme* (Kutz.) Hariot. *Protoplasma*, 193: 3–9.

Mutuskin, A.A. and Kolesnikopv, PA., 1991. Symbiotic *Azolla-anabena* association: some photosynthetic characteristics, nitrogenase activity and oxidative systems. *Photosynthetica*, 25: 639–694.

Nierzwicki-Bauer, S.A. and Haselkorn, R., 1986. Differences in mRNA levels in *Anabaena* living freely or in symbiotic association with *Azolla. EMBO J.*, 5: 29–35.

Osborne, B.A., 1989. Comparisons of photosynthesis and productivity of *Gunnera tinctoria* (Molina) Mirbel with and without the phycobiont *Nostoc punctiforme* L. *Plant Cell and Environ.*, 12: 941–946.

Osborne, B.A., Cullen, A., Jones, P.W. and Campbell, G.J., 1992. Use of nitrogen by the *Nostoc-Gunnera tinctoria* (Molina) Mirbel symbiosis. *New Phytol.*, 120: 481–487.

Pate, J.S., Lindblad, P. and Atkins, C.A., 1988. Pathways of assimilation and transfer of fixed nitrogen in coralloid roots of cycad-*Alosfocsymbioses*. *Planta*, 176: 461–471.

Paulsrud, P., Rikkinen, J. and Lindblad, P., 1998. Cyanobiont specificity in some *Nostoc*-containing lichens and in a *Peltigera aphthosa* photosymbiodeme. *New Phytol.*, 139: 517–524.

Perraju, B.T.W., Rai, A.N., Kumar, A.P. and Singh, H.N., 1986. *Cycas circinalis–Anabaena cycadeae* symbiosis: photosynthesis and the enzymes of nitrogen and hydrogen metabolism in symbiotic and cultured *Anabaena cycadeae*. *Symbiosis*, 1: 239–250.

Peters, G.A., 1991. *Azolla* and other plant cyanobacteria symbioses: Aspects of form and function. *Plant Soil*, 137: 25–36.

Peters, G.A. and Meeks, J.C., 1989. The *Azolla–Anabaena* symbiosis: Basic biology. *Annu. Rev. Plant Physiol.*, 40: 193–210.

Peters, G.A. and Perkins, S.K., 1993. The *Azolla–Anabaena* symbiosis endophyte continuity in the *Azolla* life cycle is facilitated by epidermal trichomes. II. Re-establishment of the symbiosis following gametogenesis and embryogenesis. *New Phytol.*, 123: 65–76.

Peters, G.A., Toia, Jr. R.E., Calvert, H.E. and Marsh, B.H., 1986. Lichens to *Gunnera* with emphasis on *Azolla*. *Plant and Soil*, 90: 17–34.

Plazinski, J., Croft, L., Taylor, R., Zheng, Q., Rolfe, B.G. and Gunning, B.E.S., 1991. Indigenous plasmids in *Anabaena azollae:* Their taxonomic distribution and existence of regions of homology with symbiotic genes of *Rhizobium*. *Can. J. Microbiol.*, 37: 171–181.

Rai, A.N., 1990a. Cyanobacteria in symbiosis. In: *CRC Handbook of Symbiotic Cyanobacteria*, (Ed.) A.N. Rai. CRC Press, Boca Raton, Florida, pp. 1–7.

Rai, A.N., 1990b. Cyanobacterial-fungal symbioses: The cyanolichens. In: *CRC Handbook of Symbiotic Cyanobacteria*, (Ed.) A.N. Rai. CRC Press, Boca Raton, Florida, pp. 9–41.

Rai, A.N., Borthakur, M., Singh, S. and Bergman, B., 1989. *Anthoceros–Nostoc* symbiosis: Immuno-electron microscopic localization of nitrogenase, glutamine synthetase, phycoerythrin and ribulose-1,5 bisphosphate carboxylase/oxyenase in the cyanobiont and the cultured (free living) isolate *Nostoc* 78. *J. Gen. Microbiol.*, 135: 385–395.

Rasmussen, U., Johansson, C. and Bergman, B., 1994. Early communication in the *Gunnera–Nostoc* symbiosis: Plant induced cell differentiation and protein synthesis in cyanobacterium. *Mol. Plant-Microbe Interaction*, 7: 696–702

Rasmussen, U., Johansson, C., Renglin, A., Petersson, C. and Bergman, B., 1996. A molecular characterization of the *Gunnera–Nostoc* symbiosis: comparison with *Rhizobium*- and *Agrobacterium*-plant interactions. *New Phytol.*, 133: 391–398.

Renstrom-Kellner, E., Rai, A.N. and Bergman, B., 1990. Correlation between nitrogenase and glutamine synthetase expression in the cyanobacterium *Anabaena cylindrical*. *Physiol. Plant.*, 80: 12–19.

Rippka, R. and Stanier, R.Y., 1978. The effects of anaerobiosis on nitrogenase synthesis and heterocyst development by nostocacean cyanobacteria. *J. Gen. Microbiol.*, 105: 83–94.

Sharma, A., Mishra, D.P. and Kumar, A., 1992. Characterization of a symbiotic, heterocystous. N_2-fixing cyanobacterium from *Cycas* coralloid roots. *World J. Microbiol. Biotechnol.*, 8: 529–531.

Sheath, R.G. and Cole, K.M., 1992. Biogeography of stream macroalgae in North America. *J. Phycol.*, 28: 448–460.

Silvester, W.B., Parsons, R. and Watt, P.W., 1996. Direct measurement of release and assimilation of ammonia in the *Gunnera Nostoc* symbiosis. *New Phytol.*, 132: 617–625.

Simon, R.D., 1987. Inclusion bodies in the cyanobacteria. In: *The Cyanobacteria*, (Eds.) P. Fay and C. Van Baalen, pp. 199–225.

Soderback, E. and Bergman, B., 1992. The *Nostoc-Gunnera magellanica* symbiosis: Phycobilproteins, carboxysomes and Rubisco in the cyanobiont. *Physiol. Plant.*, 84: 425–432.

Soderback, E. and Bergman, B., 1993. The *Nostoc-Gunnera* symbiosis: Carbon fixation translocation. *Physiol. Plant.*, 89: 125–132.

Soderback, E., Lindblad, P. and Bergman, B., 1990. Developmental patterns related to nitrogen fixation in the *Nostoc-Gunnera magellanica* Lam. *Symbiosis Planta*, 182: 355–362.

Steinberg, N.A. and Meeks, J.C., 1989. Photosynthetic CO_2 fixation and ribulose biophosphate carboxylase/oxygenase activity of a *Nostoc* sp. Strain UCD 7801 in symbiotic association with *Anthoceros punctatus*. *J. Bacteriol.*, 171: 6227–6233.

Steinberg, N.A. and Meeks, J.C., 1991. Physiological sources of reductant for nitrogen fixation activity in *Nostoc* sp. strain UCD 7801 in symbiotic association with *Anthoceros puncatatus*. *J. Bacteriol.*, 73: 7324–7329.

Staff, I.A. and Ahern, C.P., 1993. Symbiosis in cycads with special reference to *Macrozamia communis*. In: *The Biology, Structure and Systematics of the Cycadales Proceedings of CYCAD 90: The Second International Conference on Cycad Biology*, (Eds.) D.W. Stevenson and J.K. Norstog. Palman Cycad Societies of Australia Ltd., Queensland, Australia, pp 200–210.

Stock, P.A. and Silvester, W.B., 1994. Phloem transport of recently-fixed nitrogen in the *Gunnera–Nostoc* symbiosis. *New Phytol.*, 126: 259–266.

Strasburger, E., 1873. *Uber Azolla*. Herman Dabis Verlag, Jena, pp. 76–86.

Strasburger, E., 1884. Das Botanische Practicum. Gustav Fischer, Jena and Leipzig.

Tabita, F.R., 1994. The biochemistry and molecular regulation of carbon dioxide metabolism in cyanobacteria. In: *The Molecular Biology of Cyanobacteria*, (Ed.) D.A. Bryant. Kluwer Academic Publishers, Dordrecht, pp 437–467.

Taylor, F.J.R., 1990. Symbiosis in marine protozoa. In: *Ecology of Marine Protozoa*, (Ed.) G.M. Capriulo. Oxford University Press, Oxford, pp. 323–340.

Towata, E.M., 1985. Mucilage glands and cyanobacterial colonization in *Gunnera kaalensis* (Haloragaceae). *Bot Gaz.*, 146: 56–62.

Uheda, E., 1986. Isolation of empty packets from Anabaena-free *Azolla*. *Plant Cell Physiol.*, 27: 1187–1190.

Uheda, E. and Kitoh, S., 1991. Electron microscopic observations of the envelopes of isolated algal packet of *Azolla*. *Can. J. Bot.*, 69: 1418–1419.

Wolk, C.P., Ernst, A. and Elhai, J., 1994. Heterocyst metabolism and development. In: *The Molecular Biology of Cyanobacteria*, (Ed.) D.A. Bryant. Kluwer Academic Publishers, Dordrecht, pp. 769–823.

Zheng, W.W., Song, T., Bao, X., Bergman, B. and Rasmussen, U. High cyanobacterial diversity in coralloid roots of *Cycas*. *FEMS Microbiol. Ecol.* (In press).

Zimmerman, W.J. and Bergman, B., 1990. The *Gunnera* symbiosis: DNA restriction fragment length polymorphism and protein comparison of *Nostoc* symbionts. *Microb. Ecol.*, 19: 291–302.

Zimmerman, W.J. and Rosen, B.H., 1992. Cyanobiont diversity within and among cycads of one field site. *Can. J. Microbiol.*, 38: 1324–1328.

Zimmerman, W.J., Rosen, B.H. and Lumpkin, T.A., 1989. Enzymatic, lectin, and morphological characterization and classification of presumptive cyanobionts from *Azolla* Lam. *New Phyiol.*, 113: 497–503.

Chapter 4

Algal Biodiversity in Ghazipur, U.P.

Harendra Nath Das*

Department of Botany, Government Girl's P.G. College, Ghazipur – 233 001, U.P.

Introduction

Besides its well established role in agriculture by fixing atmospheric nitrogen, soil reclamation, in medicine, food, sewage disposal, and commercial production of agar-agar, carageeneen, alginates and iodine etc. algae are contributing significantly in modern biotechnical research, bioremediation and waste water recycling. Recently it has been found that algae have a major role in nanotechnology. The algae have now become enterprising on account of its commercial uses, medicinal value, food and feed implications and its use as biofertilizer. The mass cultivation of algae is generating the employment opportunity for rural men and women and is playing a major role in rural development (Sahoo, 2006). Day *et al.* (2005) in their project study have stated that microalgae are one of the most biologically important elements of worldwide ecology and could be the source of diverse new products and medicines. They also emphasized the need to conserve the algal diversity with its main objective to optimize the methodologies of its conservation. Owing to these applied aspects of algae, basic research with reference to algai diversity can not be ignored, as algae are likely to be exploited more and more in the coming future.

The degree of algal diversity is exhibited in their habitats, sizes and level of organization, life histories, photosynthetic pigments and metabolism. Horiguchi

* E-mail: drhndas@gmail.com

(2006) outlined the relationship between algal diversity and chloroplast acquisition through endosymbiosis. They are ubiquitous and occur in a wide range of habitats as oceans, rivers, lakes, soils, rocks, hot springs, snow etc.

Norton *et al.* (1996) assessed the relevance of algae to the current debate on global biodiversity. According to them algal diversity has been considered at the levels of richness of species and of higher taxonomic ranks and as the variety of habitats algae dominate and their functional importance in the process they mediate. The geographical distribution of algae indicates substantial differences in species richness and degree of endemism in different regions. Coasts with the greatest diversity of algae occur in both temperate and tropical seas.

Assessing thoroughly the potential vulnerability of algae and algal communities and the possible consequences of their depletion or extinction, they also emphasized the urgent need to inventory the world's algae, conservation of species in culture collections and genetic stocks in seed banks as well as to preserve natural communities by protecting sites of phycological importance.

Algal diversity studies have been made by various authors in India and abroad in different ways. In India, lakes, rivers and other freshwaters support a large diversity of biota representing almost all taxonomic groups. Algae in open waters represent the floristic diversity and macrophytes dominate the wetlands (Prasad *et al.* 2002). According to them it is difficult to analyze the algal diversity in India with reference to different habitats, endemicity to India, as well as the changes that occur due to anthropogenic disturbances.

The majority of freshwater algae are coming under Chlorophyta (green algae), Cyanophyta (blue green algae) and Bacillariophyceae. Cyanobacteria and Chlorophyta comprise two composite groups of photosynthetic organisms adapted to a broad range of environmental conditions. The green algae, on the other hand spanning a wider range of cellular organization than in any other algal 'phyla', exhibit a diverse spectrum of growth responses to critical environmental variables. A good number of literatures are available on significant works made abroad related to algal diversity with objectives for making inventory as well as to identify their habitats, phycological spots and for assessment of water quality and pollution loads their in (Costelloe, 2005; Kullberg, 1968; Munari, 1999; Olquin, 2004; Tsarenko, 2002).

The diversity of habitats, climatic conditions, environmental conditions and geographical locations in India provides a good home to algal flora. If the world algal diversity is assessed, India will be one among the best algal hot-spots of the world. Literatures on such studies are accumulating from different parts of the country. Epiphytic algal diversity have been studied by Ratha *et al.* (2003) reporting significantly a greater number of blue green algae with a very marked degree of adaptation to the conditions of an aquatic environment. Algal diversity with physico-chemical factors of two ponds at Faizabad were studied by Dwivedi and Pandey (2002) reporting seventy species of algae, Cyanophyceae dominating the summer seasons. Shashtri and Pendse (2001) reported algal diversity of Dahikhuta reservoir reporting nineteen algal taxa. Khare and Suseela (2004) and Suseela (2005) studied the freshwater algal diversity of the water spring of piner area and of Garuda Tal near Nainital respectively. Sampathkumar (2004) studied the importance of algal diversity as an indicator of pollution in freshwater. Mishra *et al.* (2002) recorded seven taxa of green algae from Basti and Dwivedi *et al.* (2004) reported twenty four taxa of desmids from Western Uttar Pradesh.

Extensive survey works on algal floristic and diversity studies have been made in different cities and areas of Uttar Pradesh since early years of twentieth century. Most of the studies were restricted to major cities and areas like Varanasi (Bharadwaja, 1935; Rao, 1936, 1937; Singh, 1939, 1942; Venkatraman, 1957 etc.) Allahabad (Chadha and Pandey,1977,1978; Pandey and Pandey, 1980, 1980a, 1980b; Singh *et al.*. 1981; Pandey and Pandey,1982; Ramakant and Pandey, 1985 etc.), Lucknow (Prasad and Mehrotra, 1976; Prasad and Vijaykumari, 1976; Prasad and Mehrotra, 1980; Fatma, 1985, 1990; Tiwari *et al.*, 2001; Mishra *et al.*, 2001a, 2001b, 2002 etc.) and Rohilkhand division, Bareilly (Chaturvedi and Pandey, 1976; Ramjee and Bhatnagar, 1978; Chaturvedi *et al.*,1979; Pandey and Chaturvedi, 1979; Chaturvedi and Pandey, 1980; Chaturvedi and Bhatnagar, 1982; Pandey, 1982; Chaturvedi, 1985; Pandey and Gangwar, 1986; Singh and Pandey, 1986; Chaturvedi *et al.*, 1987; Shukla *et al.*, 1988; Habib and Chaturvedi, 1993; Habib, 2002 etc). There are very few reports from other towns/areas like Kanpur (Pandey and Tripathi, 1988; Shukla and Shukla, 1988), Agra (Sengar and Sharma, 1982; Tiwari and Chauhan, 2001; Tiwari *et al.*, 2001), Sultanpur (Khan, 1985), Gorakhpur (Sahai *et al.*,1985), Saharanpur (Pal, 1977), Aligarh (Pundhir *et al.*, 1994), Rampur (Singh, 2001), Mahoba (Habib, 2001) and from Ballia and Ghazipur (Pandey, 1965, 1966; Gupta, 1984). These areas are less investigated so far as algal diversity is concerned and needs more attention. The earlier studies made in Ghazipur is restricted to rice field soil algae only. The present investigation, therefore, is aimed to bring on record both the crop field algae as well as those inhabiting water bodies.

Experimental Area and Methodology

The district Ghazipur, a part of the eastern Uttar Pradesh lies between the parallels of 25°19′ and 25°54′ North latitude and 83°4′ and 83°58′ east longitude, 67.50m above the sea level. The town Ghazipur is situated at 25°35′N latitude and 83°35′E longitudes. It is a part of Mid Gangetic plain garlanded by the Ganges, Karmnasa and Gomati (Plate I). The climate of Ghazipur is typical monsoon type with temperature ranging between 5°C and 17°C during winter and between 30°C and 40°C during summers. The annual rainfall ranges between 800mm and 1200mm. Relative humidity are found ranging between 70 per cent and 25 per cent. The political boundary of Ghazipur is covered by the districts Ballia in the north east, Maunath Bhanjan in the north, Azamgarh and Jaunpur in the west and adjoining districts of Bihar in the south east.

The climatic conditions of Ghazipur together with its geographical situation, as it is traversed and garlanded by a number of rivers, provide a favourable condition and habitat for reproduction and luxuriant growth of various algae. The present work, therefore, has been undertaken with the objective to bring on record the algal diversity from different locality and habitats of Ghazipur.

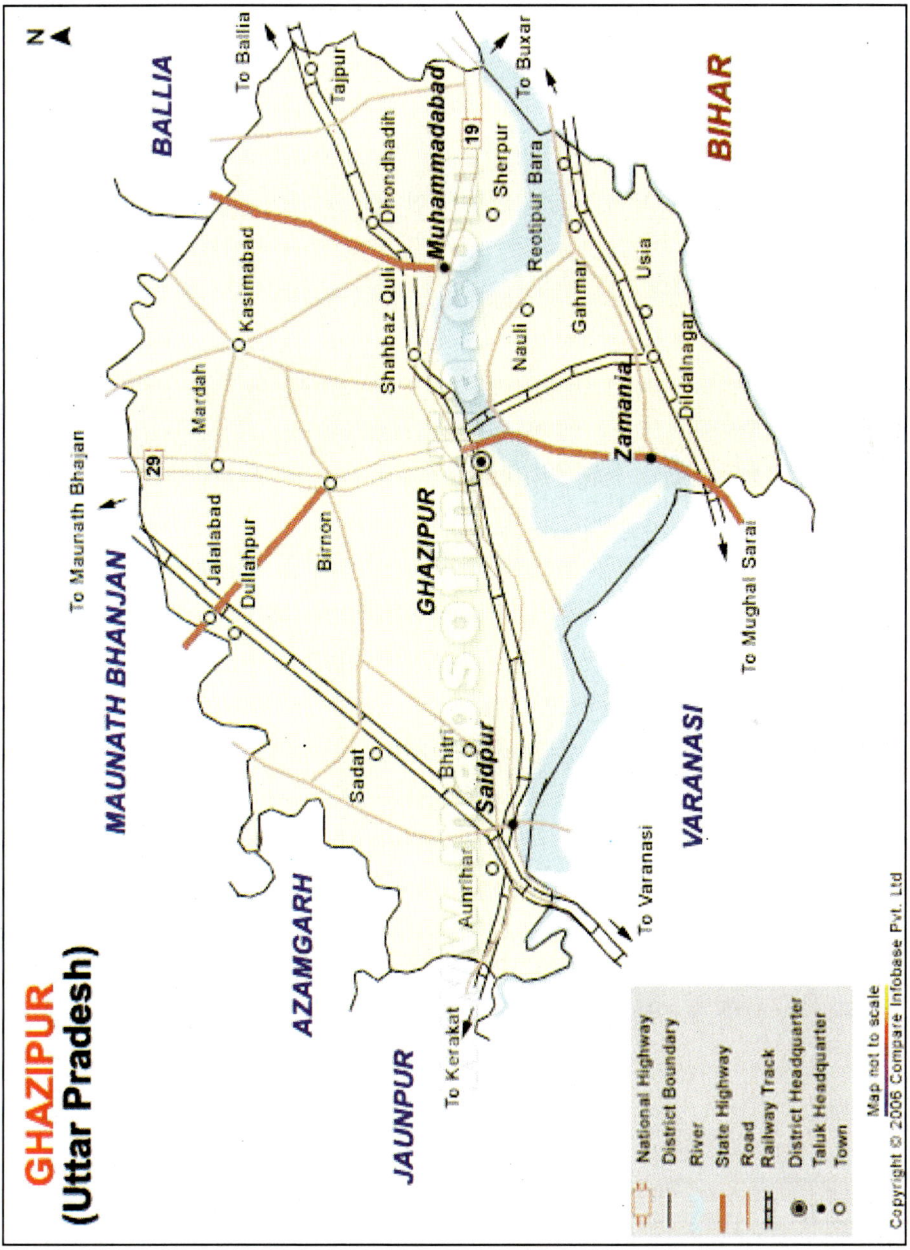

Plate I

***Source:* www.mapsofindia.com**

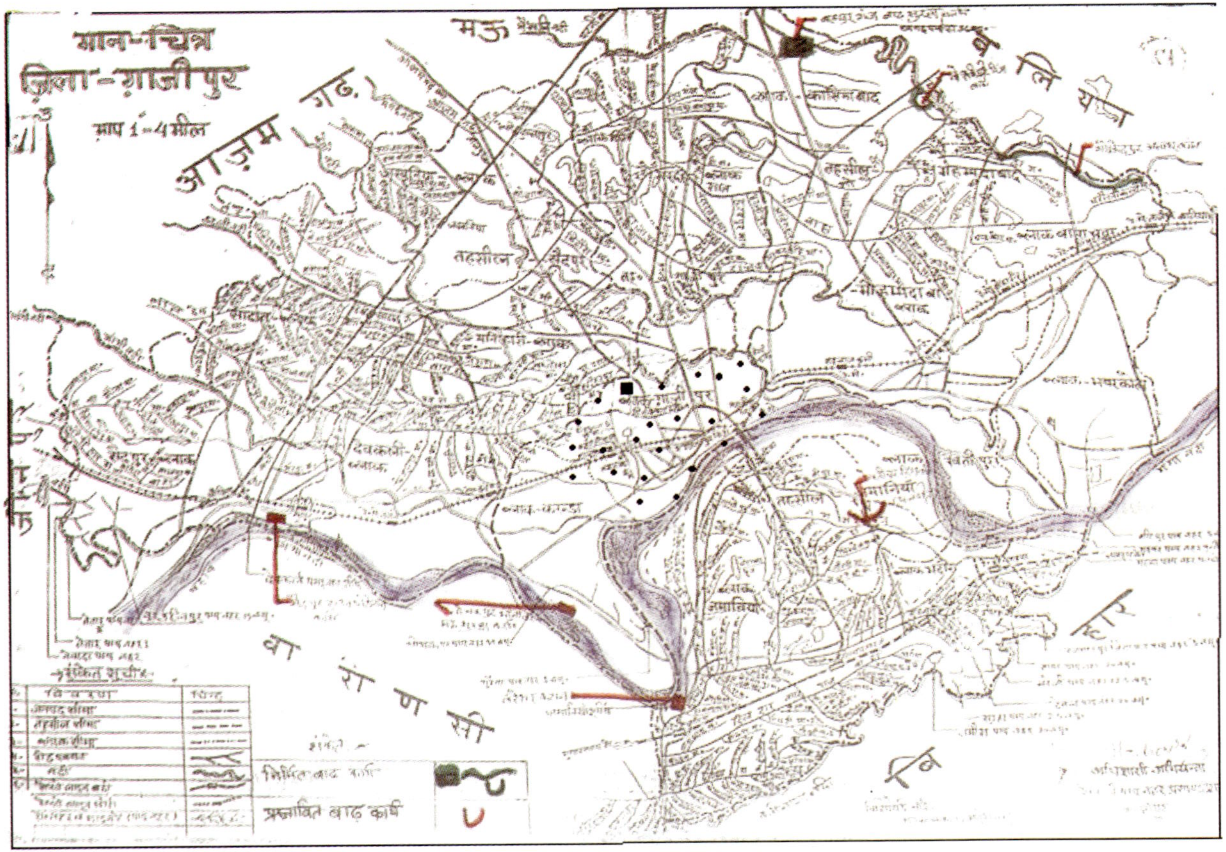

Plate II: Area Undertaken for Study

• Nawkapura.• Dadarighat • Koyalaghat • Steamerghat • Collectorghat • Tulasipur • Saklenabad • Rauza • Prakashnagar • Mauroad • Phullanpur • Gorabazar • Maharajganj • Bhutahiatar Ghat • Jangipur • Buzurga • Chawaniline.

Source: N.I.C., District Unit, Ghazipur. www.ghazipur.nic.in

The algae were collected by random sampling from temporary and permanent water bodies like the river Ganges, ponds, puddles sewage and temporary water loggings; and from the cultivated fields. Areas covered in the study are shown in the map (Plate II). In addition to the visual collection of sample, soils were also incubated in Bold's Basal medium under laboratory conditions illuminated with fluorescent light for the algal emergence. Samples collected were brought to laboratory, preserved in 4 per cent formaldehyde and were identified on the basis of morpho-taxonomic characters as per standard text and literatures cited else. Sketches were drawn with the help of Camera Lucida and photographs were snapped where ever possible. For the seasonal variation study, a pond was selected, from where the samples were collected seasonally and were analyzed for physico-chemical and biological variations.

Algal Inventory of Ghazipur

A complete systematic list of algae reported from Ghazipur including those investigated in present study is presented (Table 4.1) together with reporting authors' reference. Taxa reported in present study have been marked with asterisks (*). Arrangement of taxa is as per the classification given by Smith (1938).

Taxonomic Enumeration of Investigated Taxa

An inventory of one hundred and forty nine taxa from Ghazipur has been presented out of which eighty one algal taxa have been recorded in the present study. The taxa were identified with the help of Prescott (1951), related monographs *viz.* Desikachary (1959), Gonzalves (1981), Iyengar and Desikachary (1981), Pal *et al.* (1962), Philipose (1967), Ramnathan (1964) and Randhawa (1959) and standard literatures published by various authors. The arrangement of taxa is as per Desikachary (loc. cit.) for blue green algae and Prescott (loc. cit.) for other algae following the classification adopted by Smith (1933). Algae encountered in the present investigation (asterisk marked in the inventory) are being taxonomically described here briefly.

Division: CHLOROPHYTA

Class: CHLOROPHYCEAE

Volvocales: (Chlamydomonadaceae)

1. *Chlamydomonas angulosa** Dill, 1895 Prescott, 1962(revised ed.), p.70, Pl. 1. Figure 3. Pl. XI. Figure 11

Cells broadly ovoid, truncated anteriorly with a prominent papilla. Contractile vacuole 2, below the flagella. Chloroplast a massive, parietal cup shaped with a large angular pyrenoid in the base. Cells 13-16μ in diameter, 18-22μ long.

Collected from cultivated field Paddy field soil, Ghazipur (Pandey, 1966).

Ulotrichales: (Ulotrichaceae)

2. *Ulothrix tenerrima** Kuetzing, 1843 Prescott, 1962 (revised ed.), p.96, Pl. 6. Figure 12. Pl. VIII. Figure 6. Pl. XI. Figure 1.

Filaments free floating or attached in a short sections, composed of short cylindrical cells with constrictions at the cross walls. Chloroplast as an irregularly

Table 4. 1

Division	*Class/Orders*	*Name of the Taxa*	*References*
CHLOROPHYTA	CHLOROPHYCEAE		
	Volvocales:	1. *Chlamydomonas angulosa** Dill, 1895	Pandey, 1966
	(Chlamydomonadaceae)	2. *Chlamydomonas gloeogama* Korschikoff.	
	Ulotrichales:	3. *Ulothrix tenerrima** Kuetzing, 1843	
	(Ulotrichaceae)		
	Chaetophorales:	4. *Stigeoclonium lubricum** (Dillw.) Kuetzing, 1845.	
	(Chaetophoraceae)		
	Cladophorales:	5. *Cladophora glomerata** (L.) Kuetzing, 1845.	
	(Cladophoraceae)	6. *Rhizoclonium hookeri** Kuetzing, 1849.	
		7. *Rhizoclonium fontanum** Kuetzing, 1843.	
	Oedogoniales:	8. *Oedogonium microgonium** Prescott, 1944.	
	(Oedogoniaceae)	9. *Oedogonium cantonense** var. indicum Gonz. & Jain, 1968.	
	Chlorococcales:	10. *Chlorochytrium paradoxum* (Klebs)G. S. West, 1916.	Pandey, 1966
	(Endosphaeraceae)		
	(Chlorococcaceae)	11. *Chlorococcum humicola** (Naegeli) Rabenhorst, 1868	Pandey, 1966
	(Hydrodictyaceae)	12. *Hydrodictyon reticulatum** (L) Lagerheim, 1883.	Pandey, 1966
		13. *Pediastrum simplex** (Meyen) Lemmermann, 1897	
		14. *Pediastrum araneosum** (Racib.) G. M. Smith, 1916.	
		15. *Pediastrum biradiatum**Meyen, 1829	
	(Oocystaceae)	16. *Chlorella vulgaris** Beyerinck, 1890	
		17. *Tetraedron minimum** (A. Br) Hensgirg, 1888a.	
		18. *Tetraedron regulare** var. *minus* (Reinsch)De Toni, 1889.	
		19. *Dimorphococcus lunatus** A. Braun, 1885.	

Contd...

Table 4.1–Contd...

Division	Class/Orders	Name of the Taxa	References
	Zygnematales:	20. *Spirogyra maravillosa** Transeau, 1938.	
	(Zygnemataceae)	21. *Spirogyra gratiana** Transeau, 1938	
		22. *Spirogyra azygospora** Singh, 1938	
		23. *Spirogyra singularis** Nordstedt, 1880, Transeau, 1951.	Pandey, 1966
		24. *Spirogyra crenulata* Singh.	Pandey, 1966
	(Desmidiaceae)	25. *Cosmarium tetraophthalmum** Breb	
		26. *Closterium lanceolatum** Kutz.	
		27. *Closterium* sp.	
	(Mesotaeniaceae)	28. *Cylindrocystis* sp*Kaufman, 1914	
	CHAROPHYCEAE	29. *Nitella hyaline** (De Cond.) Agardh.	
	Charales:		
	(Characeae)		
CHRYSOPHYTA	BACILLARIOPHYCEAE	30. *Nitzschia* sp* Hassal.	
		31. *Pinnularia viridis** (Nitzsch) Her.	
		32. *Eunotia* *	
		33. *Pinnularia* sp*	
		34. *Fragillaria* sp*	
		35. *Pleurosigma spenceri** W. Sm.	
		36. *Navicula* sp.	

Contd...

Table 4.1–Contd...

Division	*Class/Orders*	*Name of the Taxa*	*References*
EUGLENOPHYTA	EUGLENOPHYCEAE		
	Euglenales:		
	(Euglenaceae)	37. *Phacus acuminatus**Stokes, 1885a	
		38. *Phacus curvicauda** Swirenko, 1915a	
		39. *Phacus orbicularis** Huebner, 1886	
		40. *Phacus suecicus** Lemm. in Pascher & Lemmermann, 1913	
CHLOROMO-NADOPHYTA	CHLOROMONADINEAE (Chloromonadinaceae)	41. *Gonyostomum* sp*	
PYROPHYTA	DINOPHYCEAE		
	Peridiniales:		
	(Ceratiacea)	42. *Ceratium sp.* *	
CYANOPHYTA	CYANOPHYCEAE		
	Chroococcales:	43. *Microcystis protocystis* Crow 1923	Pandey, 1965; Gupta, 1984.
	Chroococcaceae)	44. *Microcystis aeruginosa** Kuetzing, 1846.	Pandey, 1965
		45. *Microcystis flos-aquae** (Wittr.) Kirchner 1898.	Pandey, 1965;Gupta, 1984.
		46. *Microcystis pulverea* (Wood) Forti, 1907	Pandey, 1965;Gupta, 1984.
		47. *Chroococcus westii* (W. West) Boye-Petersen, 1923	Gupta, 1984
		48. *Chroococcus minutus** (Kuetzing) Naegeli, 1849	Pandey, 1965;Gupta, 1984.
		49. *Chroococcus montanus** Hansgirg, 1888.	Pandey, 1965;Gupta, 1984.
		50. *Chroococcus pallidus* Naegeli 1849	Pandey, 1965;Gupta, 1984.
		51. *Gloeocapsa decorticans** (A. Br.) Richter. 1925.	Pandey, 1965;Gupta, 1984.
		52. *Gloeocapsa crepidinum* (Thuret) 1876.	Gupta, 1984.

Contd...

Table 4.1–Contd...

Division	*Class/Orders*	*Name of the Taxa*	*References*
		53. *Gloeothece rupestris**(Lyngb.) Bornet, 1892.	Pandey, 1965;Gupta, 1984.
		54. *Gloeothece samoensis* var. *major** Wille, 1915.	Gupta, 1984.
		55. *Aphanocapsa pulchra* (Kuetzing) Rabenh. 1865.	Pandey, 1965;Gupta, 1984.
		56. *Aphanocapsa banaresensis* Bharadwaja, 1935.	Pandey, 1965;Gupta, 1984.
		57. *Aphanocapsa biformis** A. Br. 1865	Gupta, 1984.
		58. *Aphanocapsa grevillei* (Hass.) Rabenh. 1865.	Gupta, 1984.
		59. *Aphanocapsa montana** Cramer, 1862	Pandey, 1965;Gupta, 1984.
		60. *Aphanothece stagnina* (Spreng.) A. Br. , 1865.	Pandey, 1965;Gupta, 1984.
		61. *Aphanothece saxicola* Nag., 1849	Gupta, 1984.
		62. *Aphanothece pallida** (Kutz.) Rabenh. 1865.	Pandey, 1965;Gupta, 1984.
		63. *Aphanothece naegelii** Wartm. 1865.	Gupta, 1984
		64. *Aphanothece microscopica* Nag. 1849.	Pandey, 1965
		65. *Aphanothece microspora** (Menegh) Rabenh. 1863.	Pandey, 1965;Gupta, 1984.
		66. *Synechococcus cedrorum* Sauvageau, 1892.	Gupta, 1984.
		67. *Merismopedia tenuissima** Lemmermann, 1898.	Pandey, 1965
		68. *Chlorogloea fritschii* Mitra, 1950.	Pandey, 1965
	Chamaesiphonales (Chamaesiphonaceae)	69. *Chamaesiphon minutus* (Rostafa.) Lemm.	Pandey, 1965; Gupta, 1984.
	Nostocales:	70. *Arthrospira gomontiana** Setchell, 1895	Pandey, 1965
	(Oscillatoriaceae)	71. *Oscillatoria obscura* Bruhlet Biswas, 1922	Gupta, 1984.
		72. *Oscillatoria curviceps** Ag. ex Gomont, 1824	Pandey, 1965;Gupta, 1984.
		73. *Oscillatoria curviceps* var *angusta** Ghose, 1927b	Pandey, 1965;Gupta, 1984.

Contd...

Table 4.1–Contd...

Division	*Class/Orders*	*Name of the Taxa*	*References*
		74. *Oscillatoria princeps** Vaucher ex Gomont, 1803	Pandey, 1965
		75. *Oscillatoria anguina* (Bory Gomont)1892.	Pandey, 1965;Gupta, 1984.
		76. *Oscillatoria proboscidea* Gomont , 1892.	Pandey, 1965;Gupta, 1984.
		77. *Oscillatoria laete-virens* (Crounan) Gomont, 1892	Pandey, 1965;Gupta, 1984.
		78. *Oscillatoria irrigua**(Kutz.) Gomont, 1892.	Pandey, 1965;Gupta, 1984.
		79. *Oscillatoria formosa* Bory, 1827.	Pandey, 1965;Gupta, 1984.
		80. *Oscillatoria acuta** Bruhl et Biswas, 1922.	Pandey, 1965;Gupta, 1984.
		81. *Phormidium angustissimum* W. et G. S. West, 1897	Pandey, 1965
		82. *Phormidium fragile* (Meneghini) Gomont. 1892.	Gupta, 1984.
		83. *Phormidium mucicola* Hub Pestalozzi et Naumann, 1929.	Pandey, 1965;Gupta, 1984.
		84. *Phormidium africanum* Lemm. 1911.	Gupta, 1984.
		85. *Phormidium foveolarum* (Mont) Gomont, 1892.	Gupta, 1984.
		86. *Phormidium molle* (Kuetzing) Gomont, 1892.	Gupta, 1984.
		87. *Phormidium dimorphum** Lemm. 1908.	Pandey, 1965
		88. *Phormidium truncicola** Ghose1924.	Pandey, 1965;Gupta, 1984.
		89. *Phormidium tenue** (Menegh) Gomont, 1892.	Pandey, 1965
		90. *Phormidium rubroterricola* Gardner, 1927	Pandey, 1965;Gupta, 1984.
		91. *Phormidium luridum* (Kuetzing) Gomont, 1892.	Pandey, 1965
		92. *Phormidium valderianum** (Delp.) Gomont, 1892.	Pandey, 1965;Gupta, 1984.
		93. *Phormidium corium** (Ag.) Gomont, 1890.	Pandey, 1965.
		94. *Phormidium ceylanicum** Wille 1915.	Gupta, 1984.
		95. *Phormidium autumnale* (Ag.)Gomont. 1892.	Gupta, 1984.

Contd...

Table 4.1–Contd...

Division	*Class/Orders*	*Name of the Taxa*	*References*
		96. *Lyngbya epiphytica** Hieron, 1898.	Pandey, 1965.
		97. *Lyngbya spirulinoides** Gomont, 1890.	Gupta, 1984.
		98. *Lyngbya lagerheimii** (Mob.) Gomont, 1890.	Gupta, 1984.
		99. *Lyngbya limnetica* Lemm. 1898	Pandey, 1965.
		100. *Lyngbya birgei** Smith, G. M., 1916.	Gupta, 1984.
		101. *Lyngbya hieronymusii** Lemm. , 1905.	Pandey, 1965.
		102. *Lyngbya rubida* Fremy. 1929.	Pandey, 1965;Gupta, 1984.
		103. *Lyngbya ceylanica* Wille, 1914	Pandey, 1965.
		104. *Lyngbya ceylanica* Wille var. *major* Pandey et Mitra, 1965b.	Gupta, 1984.
		105. *Lyngbya dendrobia** Bruhl et Biswas, 1923.	Gupta, 1984.
		106. *Lyngbya aestuarii* Liebm. ex. Gomont. 1841	Pandey, 1965;Gupta, 1984.
		107. *Lyngbya truncicola** Ghose var terrestris Pandey et Mitra, 1965b.	Gupta, 1984.
		108. Lyngbya aerugineocoerulea (Kuetzing) Gomont, 1892.	Gupta, 1984.
		109. *Lyngbya putealis** Mont. ex. Gomont , 1840.	Pandey, 1965;Gupta, 1984.
		110. *Lyngbya martensiana* Meneghvar *ghazipurensis* Pandey et Mitra, 1965b.	Gupta, 1984.
		111. *Microcoleus chthonoplastes** Thuret ex Gmont, 1875.	Pandey, 1965;Gupta, 1984.
		112. *Microcoleus acutissimus* Gardner, 1927	Gupta, 1984.
		113. *Microcoleus lacustris** (Rabenh.) Farlow, 1877.	Pandey, 1965.
		114. *Microcoleus sociatus* West et West, 1897.	Pandey, 1965.

Contd...

Table 4.1–Contd...

Division	*Class/Orders*	*Name of the Taxa*	*References*
	(Nostocaceae)	115. *Anabaenopsis circularis* (G. S. West) Wolosz et Miller, 1923.	Gupta, 1984
		116. *Anabaenopsis ambigua* Pandey et Mitra, 1962.	Pandey, 1965;Gupta, 1984.
		117. *Anabaenopsis raciborskii** Wolosz, 1913	Pandey, 1965;Gupta, 1984.
		118. *Cylindrospermum gorakhpurense* Singh, R. N. 1939	Gupta, 1984.
		119. *Cylindrospermum licheniforme* Kuetzing ex Born et Flah, 1847.	Gupta, 1984.
		120. *Nostoc punctiforme*(Kuetz.) Hariot, 1891.	Pandey, 1965.
		121. *Nostoc paludosum* Kuetzing ex Born et Flah, 1850.	Pandey, 1965;Gupta, 1984.
		122. *Nostoc paludosum* Kuetzing var major Mitra, 1951.	Pandey, 1965;Gupta, 1984.
		123. *Nostoc linckia* (Roth.) Bornet ex Born et Flah, 1880.	Pandey, 1965;Gupta, 1984.
		124. *Nostoc humifusum* Carmichael ex Born et Flah, 1833.	Pandey, 1965.
		125. *Nostoc muscorum** Ag. ex Born. et Flah, 1812.	Pandey, 1965;Gupta, 1984.
		126. *Nostoc commune** Vaucher ex Born et Flah, 1803.	Pandey, 1965;Gupta, 1984.
		127. *Nostoc sphaericum* Vacher ex Born et Flah, 1803.	
		128. *Anabaena fertilissima** Rao, C. B. 1937b.	
		129. *Anabaena ambigua** Rao, C. B. 1937.	
		130. *Anabaena naviculoides* Fritsch 1949.	
		131. *Anabaena variabilis** Kuetzing ex Born et Flah, 1843.	
		132. *Aulosira fertilissima* Ghose, 1924	

Contd...

Table 4.1–Contd...

Division	Class/Orders	Name of the Taxa	References
	(Scytonemataceae)	133. *Plectonema nostocorum* Bornet ex Gomont, 1887.	
		134. *Plectonema nostocorum* Bornet forma *major* n. forma, Pandey, 1965.	
		135. *Scytonematopsis ghazipurensis* Pandey et Mitra, 1965c.	
		136. *Scytonema ocellatum* Lyngbye ex Born et Flah, 1819.	
		137. *Scytonema hofmanni* Ag. ex Born et Flah, 1817.	
		138. *Tolypothrix nodosa* Bharadwaja, 1934.	
		139. *Tolypothrix tenuis* (Kuetzing) Johs. Schmidt em. 1899.	
	(Microchaetaceae)	140. *Microchaete tenera* Thuret ex Born et Flah., 1875.	
		141. *Microchaete ghazipurensis* Pandey et Mitra, 1965b.	
	(Rivulariaceae)	142. *Calothrix javanica* de Wilde, 1897.	
		143. *Calothrix braunii* (A. Br.) Bornet et Flah, 1886.	
		144. *Calothrix membranacea* Schmidle, 1901.	
		145. *Calothrix anomala* Mitra, 1947	
	(Stigonemataceae)	146. *Hapalosiphon fontinalis* (Ag.) Bornet, 1889.	
		147. *Fischerella muscicola* (Thuret) Gomont, 1895.	
RHODOPHYTA	RHODOPHYCEAE	148. *Compsopogon aeruginosus** (J. Ag.) Kuetzing	
	(Bangiales)	149. *Compsopogon aeruginosus* J. Ag. Kuetzing Var *catenatum*	

folded plate, about half the length of the cell with one pyrenoid. Cells 8-11μ in diameter, 12.5-15.7μ long.

Collected from road side canal, Mahuabagh, Ghazipur. Reported from Bareilly (Chaturvedi and Pandey, 1976)

Chaetophorales: (Chaetophoraceae)

3. *Stigeoclonium lubricum** (Dillw.) Kuetzing, 1845. Prescott, 1962 (revised ed.), p. 115, Pl. 11. Figure 1, 2. Pl. XIII. Figure 6.

Filaments elongate and robust, branches opposite, developed from barrel shaped axial cells; ending in a blunt point. Cells of main filament up to 17μ in diameter, 22-28μ long; branch cells 6-7μ in diameter. Prostrate portion of the thallus well developed.

Collected from college campus. Reported earlier from Bareilly (Shukla *et al.*, 1988)

Cladophorales: (Cladophoraceae)

4. *Cladophora glomerata** (L.) Kuetzing, 1845. Prescott, 1962 (revised ed.), p.138, Pl. 20. Figures 8,9; Pl. 21, Figures 1,2. Pl. VI. Figure 4-6.

Attached forming dark green fluffy thalli in flowing water, filaments successively and regularly branched, the branches crowded in the upper limits. Apical cell slightly attenuated and bluntly pointed. Main axis 75-100μ in diameter, 6-7 times the diameter in length; cells in the branches 30-51μ in diameter, 115-250μ in length.

Found attached to the cemented bricks under flowing water at Dadarighat of the river Ganges in Ghazipur. Reported earlier from Rohilkhand division (Pandey and Chaturvedi,1979), Gomati river, Sultanpur (Khan, 1985), Kanpur (Tiwari *et al.*, 2001)

5. *Rhizoclonium hookeri** Kuetzing, 1849. Prescott, 1962 (revised ed.), p.143, Pl. 23. Figure 4-7. Pl. IV. Figure 6, Pl. V. Figure 1.

Filaments crisp, freely branching, developed from long, cylindrical and irregularly inflated cells. Cells of main axis are 60-80μ in diameter and 850μ in length. Cells of the branches are of the same diameter as those of main filament.

Collected from a drain canal, Mahuabagh, Ghazipur.

6. *Rhizoclonium fontanum** Kuetzing, 1843. Prescott, 1962 (revised ed.), p.142, Pl. 23. Figure 2. Pl. VI. Figure 1-3.

Filaments very coarse, crooked or straight, branched. Cells cylindrical with uneven lateral wall, 30μ in diameter, 300μ long.

Collected from bank of the river Ganga, near Koyalaghat, Ghazipur.

Oedogoniales: (Oedogoniaceae)

7. *Oedogonium microgonium** Prescott, 1944. Prescott, 1962 (revised ed.), p.184, Pl. 36. Figures 11-14; Gonzalves, p.524, Figure 9.430 Pl. XII. Figure 5.

Vegetative cells distinctly capitate, 8-9μ in diameter, 15-23μ long; basal cell elongate, 20μ long. Oogonia solitary, little wider than vegetative cell, globose, operculate; 12.5μ in diameter, 11μ long.

Collected from the side of a pond.

8. *Oedogonium cantonense* var. indicum* Gonz. and Jain, 1968. Gonzalves, p.523, Figure 9.430 Pl. XII. Figure 9.

Vegetative cells capitellate, 17μ in diameter, 56-84μ long; basal cell elongate, 20-24μ in diameter, 40-45μ long; oogonium single, subglobose, 51μ in diameter, 56μ long.

Collected from water logged cultivated fields.

Chlorococcales: (Chlorococcaceae)

9. *Chlorococcum humicola** (Naegeli) Rabenhorst,1868 Prescott, 1962 (revised ed.), p. 212, Pl. 45. Figure 1; Philipose,p. 73, Figure 3. Pl.XI. Figure 10.

Cells spherical, solitary or in small clumps, variable in size; cells 14-17μ in diameter.

Obtained from drain canal. Earlier reported from paddy field soil of Ghazipur (Pandey, 1966), Kitham lake Agra (Tiwari *et al.*, 2001), Mahoba (Habib and Chaturvedi, 2001).

Chlorococcales: (Hydrodictyaceae)

10. Hydrodictyon reticulatum* (L) Lagerheim, 1883. Prescott, 1962 (revised ed.), p.219, Pl. 47. Figure 1; Philipose,p.134, Figure 48. Pl.VII. Figure 1-3.

Thallus macroscopic, composed of cylindrical cells adjoined at their ends to form a cylindrical net, 5-7 sided meshes, chloroplast parietal plate or reticulum with many pyrenoids, cells multinucleate. Cells of the meshes are 0.31-0.36mm long and 80-100μ in diameter, widely distributed, an indicator species of hard water.

Habitat: East side of Dadarighat of the river Ganges at Ghazipur; also collected from a pond near Bhutahiatar Ghazipur. Massive growth of the taxa was observed near Dadarighat and a few near Koyalaghat and collectorghat. Reported earlier from Nakatia river, Bareilly (Chaturvedi and Pandey, 1976), Gomati river, Sultanpur (Khan, 1985), Kitham lake Agra (Tiwari *et al.*, 2001).

11. *Pediastrum simplex** (Meyen) Lemmermann1897. Prescott, 1962 (revised ed.), p.227, Pl. 50. Figure 2; Philipose,p.113, Figures 36a-c. Pl.V. Figure 2., Pl. XIII. Figure 12.

Colonies circular to oval, of 4-8-16-32 or more cells. Inner side of marginal cells nearly straight, outer side produced into a gradually tapering process, sides concave. Inner cells polygonal. Intercellular spaces when present, very small and few in number. Cells 15-20μ long, 7.5-12.5μ broad.

Collected from temporary waterlogging near Mau Road and from a permanent water body near Rauza. Reported from Suraha 'tals' U.P. (Singh, 1959), Agra (Sengar and Sharma, 1982), Ganga river, Kanpur (Gupta and Shukla, 1994), Kitham lake Agra (Tiwari *et al.*, 2001).

12. *Pediastrum araneosum** (Racib.) G.M.Smith, 1916. Prescott, 1962 (revised ed.), p.221, Pl. 47. Figure 4; Philipose,p.120, Figure 42b. Pl. XIII. Figure 11.

Colonies circular to oval, of 8-16-32-64 or more, closely arranged cells without perforations or with minute perforations. Inner cells 5-6 sided, transversely elongated and with the outer side slightly concave. Marginal cells closely arranged, deeply emarginate and with the two lobes slightly concave on the back. Lobes short and pointed or rounded at the tip, sometimes with two indentations. Cells 15-32μ in diameter

Suraha 'tals' U.P. (Singh,1959). Collected from water logged field.

13. *Pediastrum biradiatum** Meyen, 1829. Prescott, 1962(revised ed.), p.222, Pl. 47. Figures 5,6; Philipose,p.127, Figure 44a. Pl. XIII. Figure 1.

Colonies 4-8-16-32-64 celled with medium sized perforations. Marginal cells in contact at the base only, and provided with two lobes formed by an incision reaching the middle of the cell. Lobes dilated and incised at the apex. Inner cells with lobes which are not dilated or incised. Cells 9-22μ broad, 15-30μ long. Colonies 32-celled, 125μ in diameter.

Collected as plankton from river water.

Chlorococcales: (Oocystaceae)

14. *Chlorella vulgaris** Beyerinck, 1890. Prescott, 1962(revised ed.), p.237, Pl. 53. Figure 13; Philipose,p.173, Figure 82. Pl.XII. Figure 4.

Cells spherical, scattered among other algae or free living, chloroplast a parietal cup, with a pyrenoid sometimes indistinct; cells 7-9μ in diameter.

In paddy field soil culture from U.P. (Singh, 1939; Mitra, 1951, Pandey, 1966 from Ballia and Ghazipur) Kanpur (Tiwari *et al.*, 2001), Kitham lake, Agra (Tiwari *et al.*, 2001).

15. *Tetraedron minimum** (A.Braun) Hensgirg, 1888a. Prescott, 1962 (revised ed.), p.267, Pl. 60. Figures 12-15; Philipose,p.138, Figures 53a-c. Pl.XI. Figure 12.

Cells small and quadrangular with the sides concave and angles rounded. Cell wall smooth. Cells 6-25μ in diameter.

Ubiquitous with other alga. Collected from water logged abandoned field, Buzurga road.

16. *Tetraedron regulare* var. minus* (Reinsch) De Toni,1889. Prescott, 1962 (revised ed.), p.269, Pl. 61. Figures 4-7; Philipose,p.146, Figure 60. Pl. XI. Figure 7.

Cells tetragonal with the sides slightly convex and with a long massive spine from each corner. Cells, 31-33μ in diameter with spines.

Collected from pond water. Reported earlier from Mahoba (Habib and Chaturvedi, 2001).

17. *Dimorphococcus lunatus** A. Braun, 1885. Prescott, 1962 (revised ed.), p.252, Pl. 55. Figure 8; Philipose,p.205, Figure 115. Pl.X. Figure 10.

Colonies irregular, cells in groups of four and arranged alternately in a zig-zag fashion. Outer cells of each group reniform or somewhat crescent shaped, inner cells elongate-ovoid to ellipsoid. Ends of cells rounded. Chloroplast a parietal plate nearly covering the entire cell wall in mature cells. Cells 7-8μ broad, 12-15μ long. Colonies up to 100μ in diameter.

Characteristically found in soft waters. Collected from pond water as Plankton. Reported from Mahoba (Habib and Chaturvedi, 2001).

Zygnematales: (Zygnemataceae)

18. *Spirogyra maravillosa** Transeau,1938. Prescott, 1962 (revised ed.), p.252, Pl. 55. Figure 8. Pl. XI. Figure 16.

Vegetative cells 28.40x255μ, with plane end walls; 2 to 3 chloroplasts making 2 to 5 turns in the cell. Conjugation unknown; aplanospores broadly ellipsoid, 28-36 x 43-60μ; median wall smooth, yellow brown, sporangia enlarged or slightly inflated.

Collected from road side water loggings.

19. *Spirogyra gratiana** Transeau,1938. Prescott, 1962(revised ed.), p.315, Pl. 74. Figure 9. Pl. XI. Figure 15.

Vegetative cells 34x340μ, with replicate end walls; usually 3 chloroplasts, sometimes 2 or 4. Zygospores ellipsoid, 35-47 x 108-223μ, aplanospores infrequent, smaller, 38-40 x47-72μ

River side water from collector ghat.

20. *Spirogyra azygospora** Singh, 1938. Prescott, 1962 (revised ed.), p.252, Pl. 55. Figure 8.

Vegetative cells 85-90 x 270-300μ with plane end walls; chloroplast 5. Conjugation unknown; reproduction by aplanospores which are compressed globose, 71-77 x 60-67μ median wall thick, brown and smooth.

Collected from road side ditch. Reported from Gorakhpur, U.P.

21. *Spirogyra singularis** Nordstedt, 1880, Transeau, 1951. Prescott, 1962(revised ed.), p.320. Pl. VIII. Figure 1 and 3

Vegetative cells 22-28x102-115μ; end walls plane; chloroplast 1,making 3 to 7 turns. Conjugation scalariform, zygospores ellipsoid, 27-36 x 46-70μ, median spores wall yellow, smooth.

Collected from road side ditch. Reported from Banares, U.P., Saharanpur, U.P. (Pal, 1977); Sultanpur, U.P. (Khan, 1985)

Zygnematales: (Desmidiaceae)

22. *Cosmarium tetraophthalmum** Breb Fritsch, p.339, Figure 106A-B. Pl. XIII. Figure 2.

Free floating, freshwater desmids, flattened, wall consists of two halves, deep median constriction, pores are grouped symmetrically around the bases of the spines, warts etc. of the outer layer of the wall. Chloroplast one or two, axile in each semi cell. Cells 40μ in diameter.

Collected as plankton from pond water and river water.

23. *Closterium lanceolatum** Kutz. Fritsch p.343, Figure 108A. Pl. IX. Figure 2.

Vegetative cell attenuated, pores are usually arranged in longitudinal rows in the narrow grooves between the fine ridges on the membrane. The cells perform pendulum like movements about one fixed extremity. Single nucleus in the region of isthmus and axile chloroplast with a row of pyrenoids. Chloroplast in each semi cell bears longitudinal ridges undulated or anastomose. Cells 15μ broad and 90μ long.

Collected from mud side of a pond. Reported as planktonic form in a ditch, Izatnagar (Pandey *et al.*, 1987).

Zygnematales: (Mesotaeniaceae)

24. *Cylindrocystis*sp* Kaufman,1914. Fritsch p.311. Pl. XI. Figure 3.

Cells rod shaped or oblong, without median constriction, cell wall smooth, cell with a pair of stellate axile chloroplasts each with a massive central pyrenoid. Cell 42μ long and 15.56μ broad.

Collected from field soil water.

Charales: (Characeae)

25. *Nitella hyaline** (De Cond.) Agardh. Pl. VIII. Figure 4 and 5.

Monoecious, internodes about 2-4 times the length of the branchlets, primary branchlets 7-8, 1-3 times furcated; primary rays ½–3/5 of the whole branchlets; rays at first furcation 6-10, of which few remain unbranched, at the second 4-6; dactyls of equal size uniformly 2-celled; the lower cell tapering towards the base of the upper cell; upper cell normally conical; accessory branchlets–the upper ones furcated only once or simple, the lower ones 1-2 times furcated; antheridia and oogonia are born at the furcation of all the branchlets; generally found at the first furcation of the primary branchlets; antheridia 272-320μ in diam.; oogonia solitary 354-480μ long; 288-320μ broad; spiral cell showing 8-9 convolutions; coronula 28-36μ high, oospore dark brown, showing 6-7 broadly flanged ridges, 330-335μ long, 272-280μ broad.

Occasionally collected from a ditch near railway line, Tulasipur, Ghazipur. The taxa in question might have been migrated from other places as they are not found only once. Reported from Banaras, Gonda, Oudh U.P. and from Moradabad (Ramjee and Bhatnagar, 1978).

Division: CHRYSOPHYTA

Class: BACILLARIOPHYCEAE

26. *Nitzschia sp** Hassal Fritsch, p.571. Pl. XII. Figure 3.

Keel of the valve placed diagonally with evenly spaced carinal dots, valves 31.40μ long and 6.28μ broad. Collected from pond water. Earlier reported from Rohilkhand division, Bareilly, U.P. (Chaturvedi, 1985)

27. *Pinnularia viridis** (Nitzsch) Ehr. Pl. XII. Figure 6.

Valves 56.52μ long and 15.7μ broad, transverse striations medianly radial, polarly convergent, 8-9 striations in 10μ crossed by a wide longitudinal band. Collected from pond water. Earlier reported from Rohilkhand division, Bareilly, U.P.(Chaturvedi, 1985).

28. *Pinnularia sp** Pl. XI. Figure 14.

Valves 59.66μ long and 12.56μ broad, transverse striations medianly radial, 8-9 striations in 10μ Reported from Allahabad (Pandey and Pandey, 1980).

1. ***Anabaena ambigua*** **(x 625)**

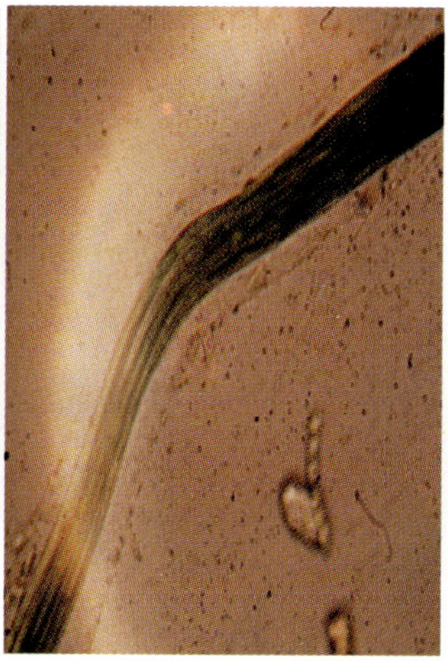

2. ***Microcoleus lacustris*** **(x 200)**

3. ***Lyngbya truncicola*** **(x 600)**

4. ***A.ambigua*** **(x 600)**

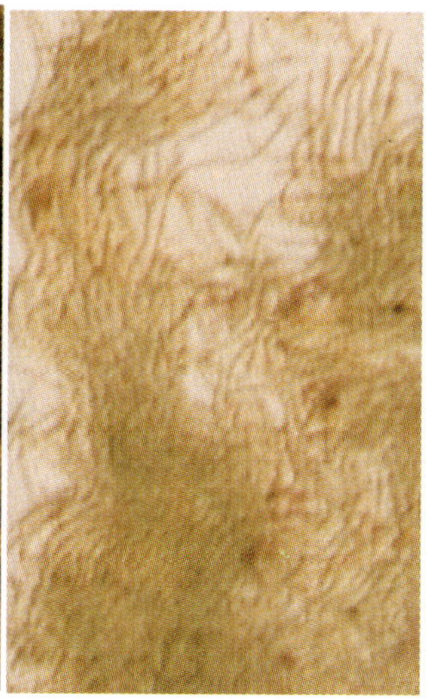

5. ***A. ambigua*** **(x156)**

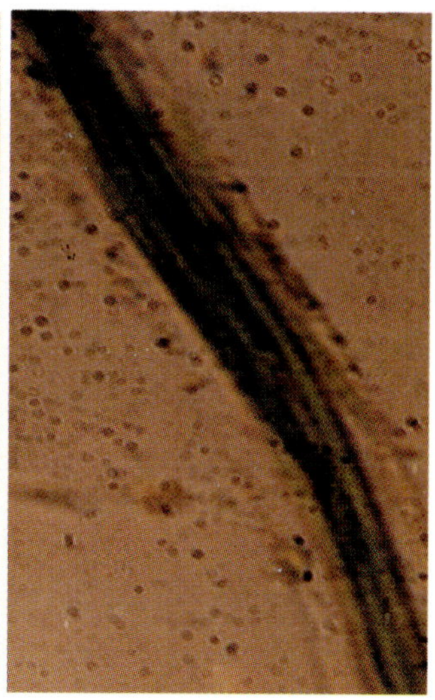

6. ***M.lacustris*** **(x 600)**

Plate III

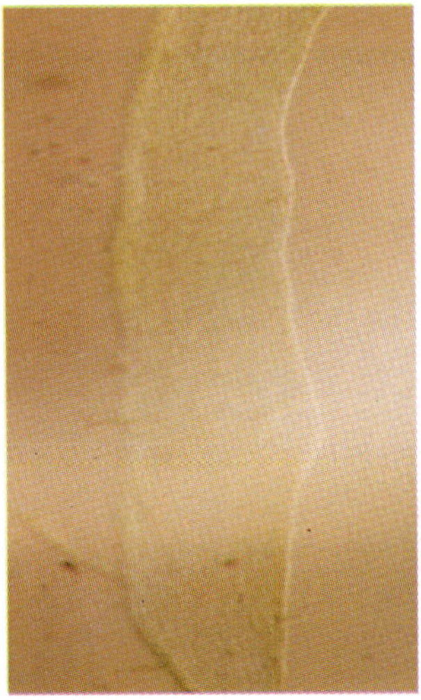

1. Unidentified (x 600)

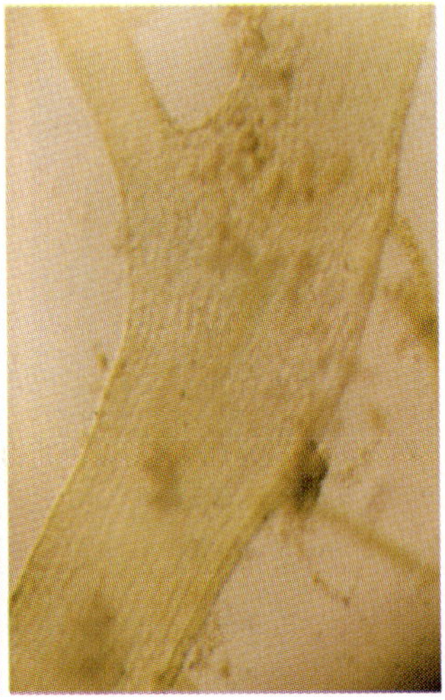

2. Unidentified (x 600)

3. Unidentified (x 600)

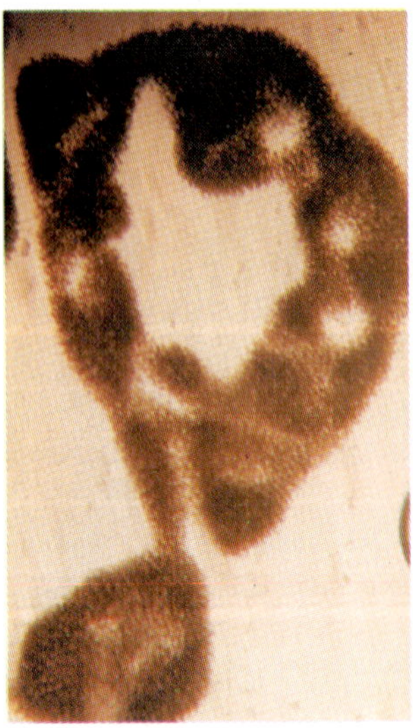

4. *Microcystis aeruginosa* (x125)

5. Unidentified (x 125)

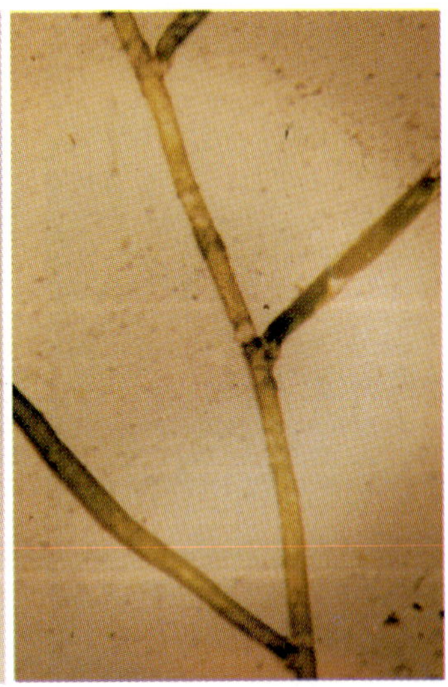

6. *Rhizoclonium hookeri* (x 150)

Plate IV

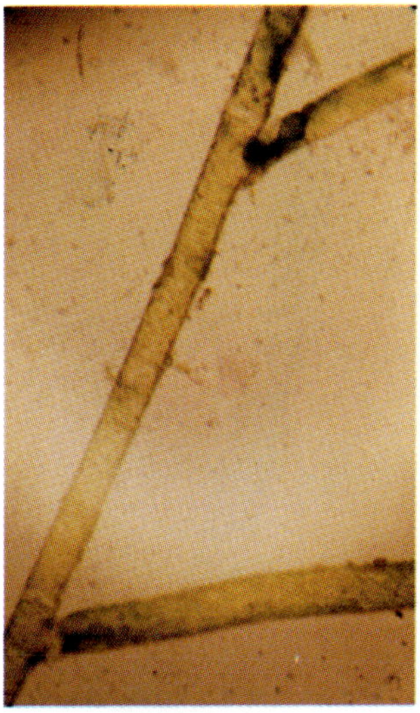

1. ***Rhizoclonium hookeri*** **(x100)**

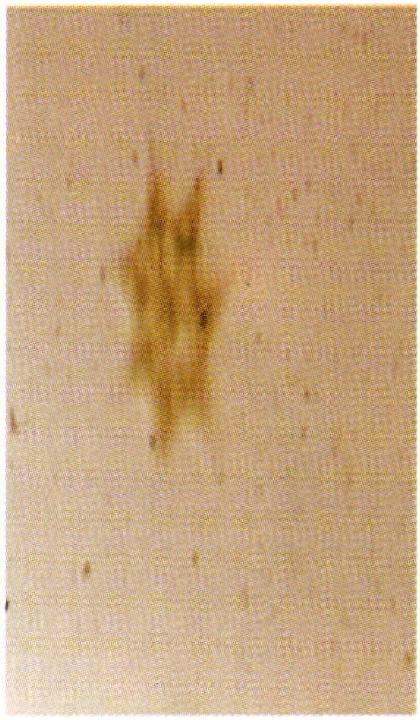

2. ***Pediastrum simplex*** **(x 750)**

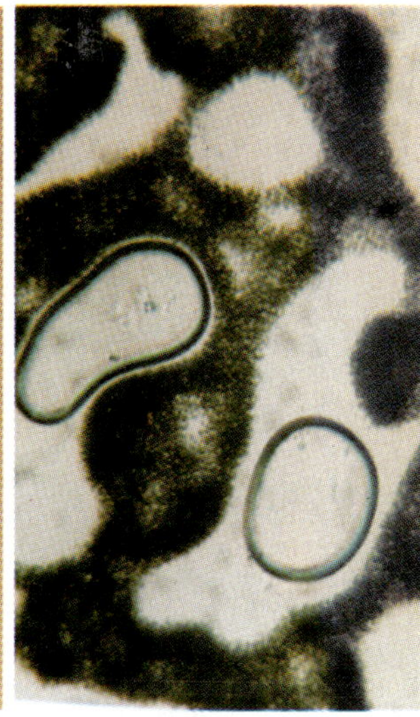

3. ***Microcystis aeruginosa*** **(x 50)**

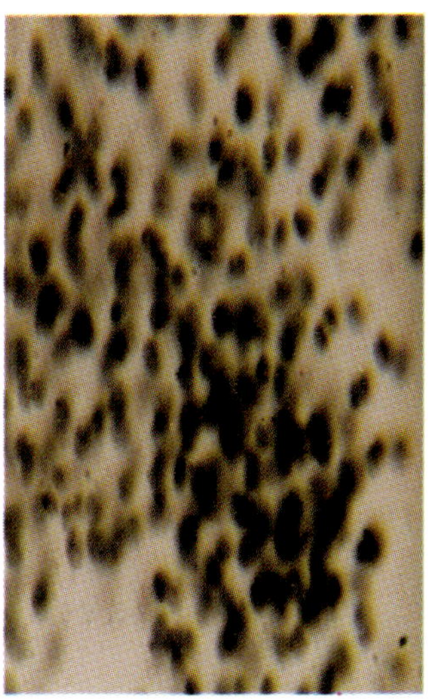

4. ***M. aeruginosa*** **(x 500)**

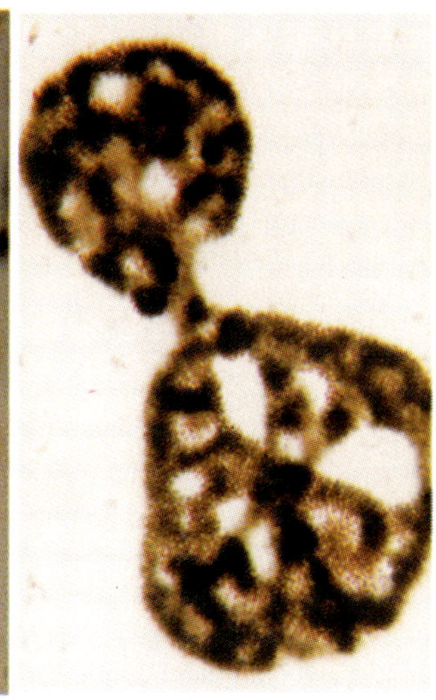

5. ***M. aeruginosa*** **(x100)**

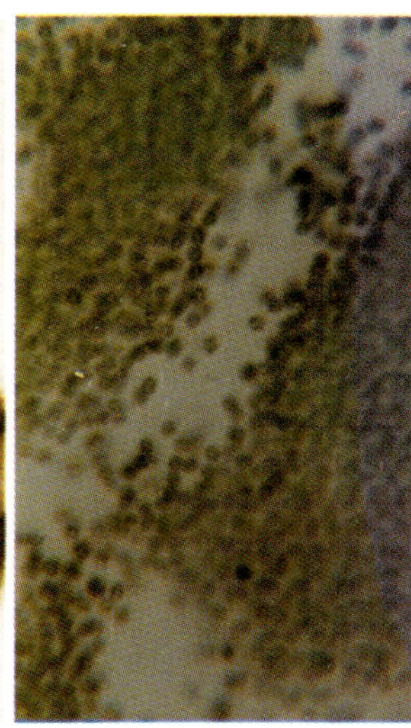

6. ***M. aeruginosa*** **(x 375)**

Plate V

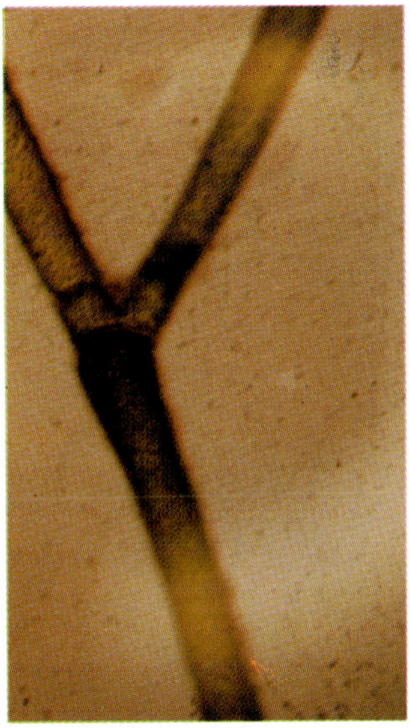

1. ***Rhizoclonium fontanum*** **(x 330)**

2. *R. fontanum* (x 130)

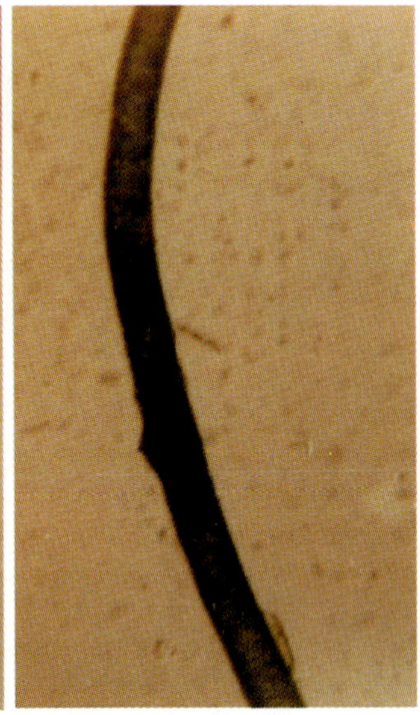

3. *R. fontanum* (x 330)

4. *Cladophora glomerata*

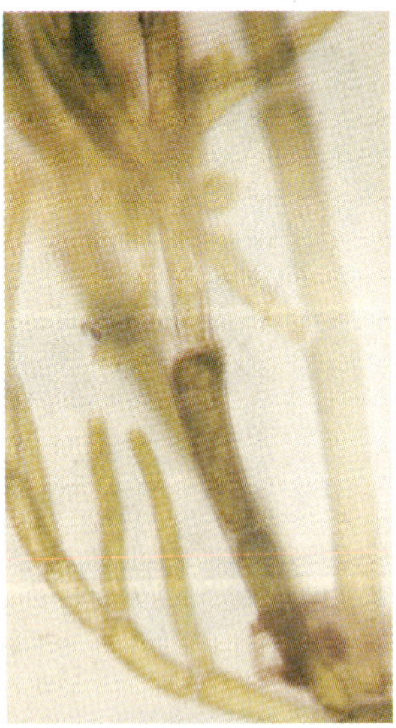

5. *C. glomerata* (x125)

6. *C. glomerata* (x 50)

Plate VI

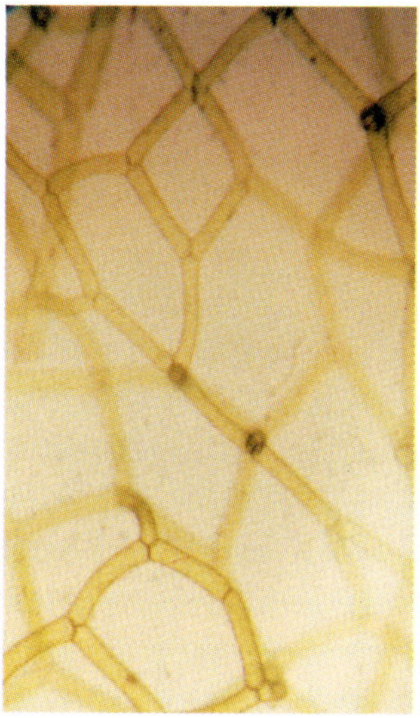

1. ***Hydrodictyon reticulatum*** **(x 50)**

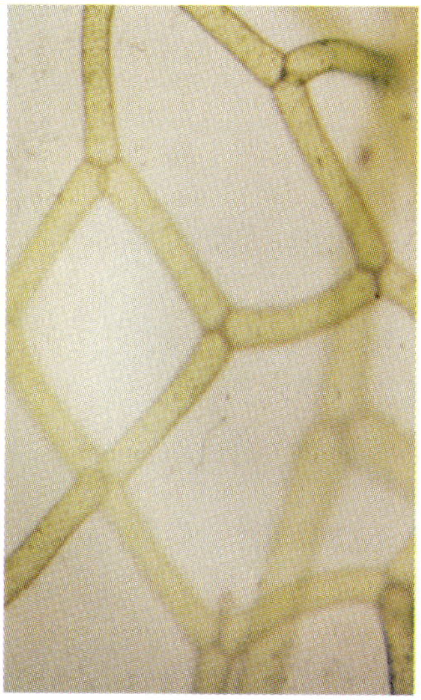

2. ***H. reticulatum*** **(x 100)**

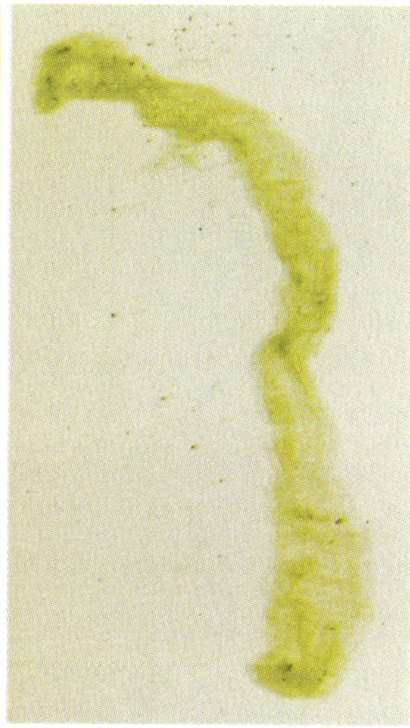

3. ***H. reticulatum*** **(Water net)**

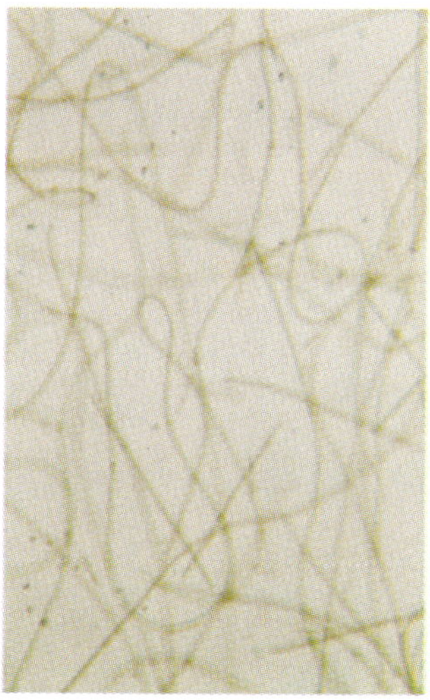

4. ***Anabaena variabilis*** **(x 125)**

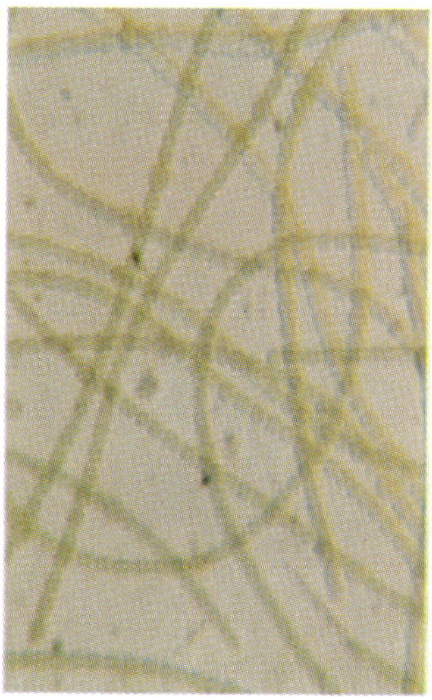

5. ***A. variabilis*** **(x 500)**

6. ***Oscillatoria curviceps*** **(x 450)**

Plate VII

1. *Spirogyra singularis* (x 100)

2. *S. singularis* (x100)

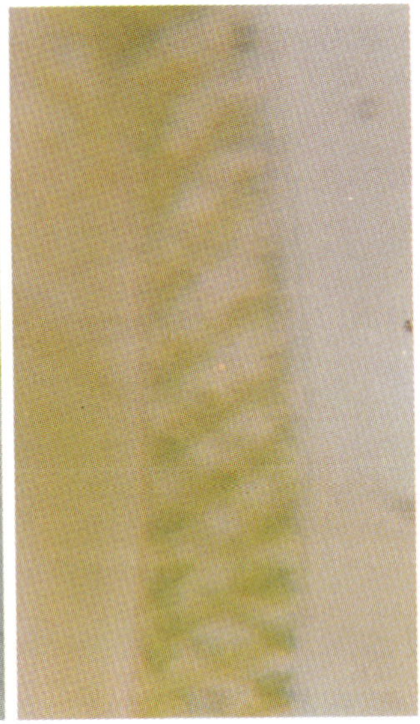

3. *S. singularis* (x 600)

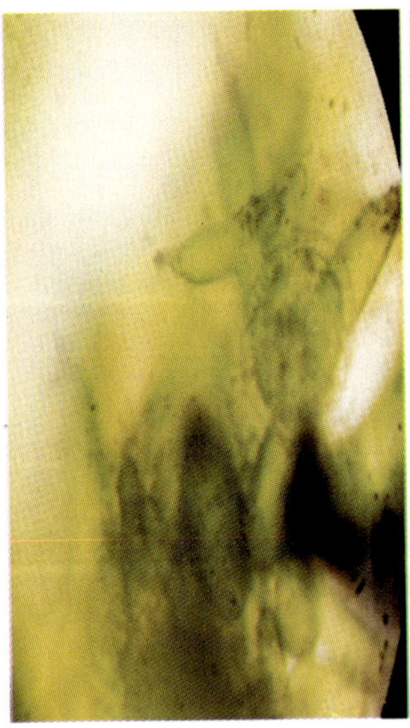

4. *Nitella hyaline* (x 100)

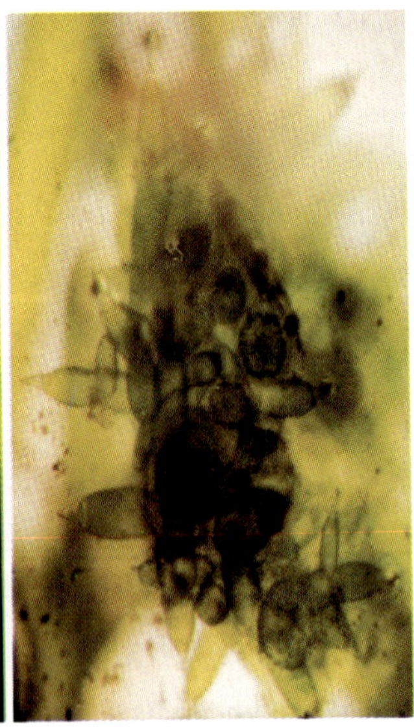

5. *Nitella hyaline* (x 50)

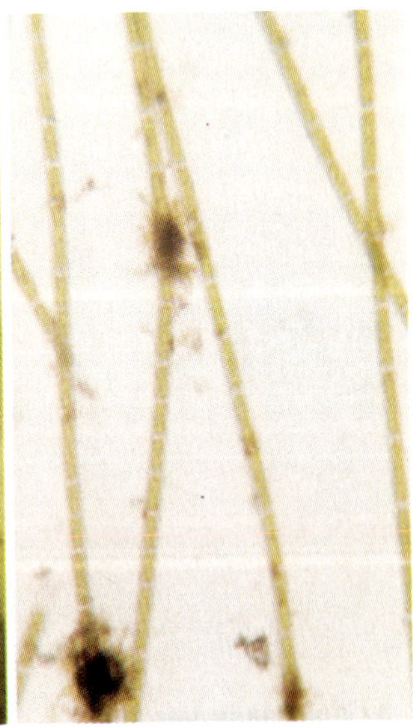

6. *Ulothrix tenerrima* (x 150)

Plate VIII

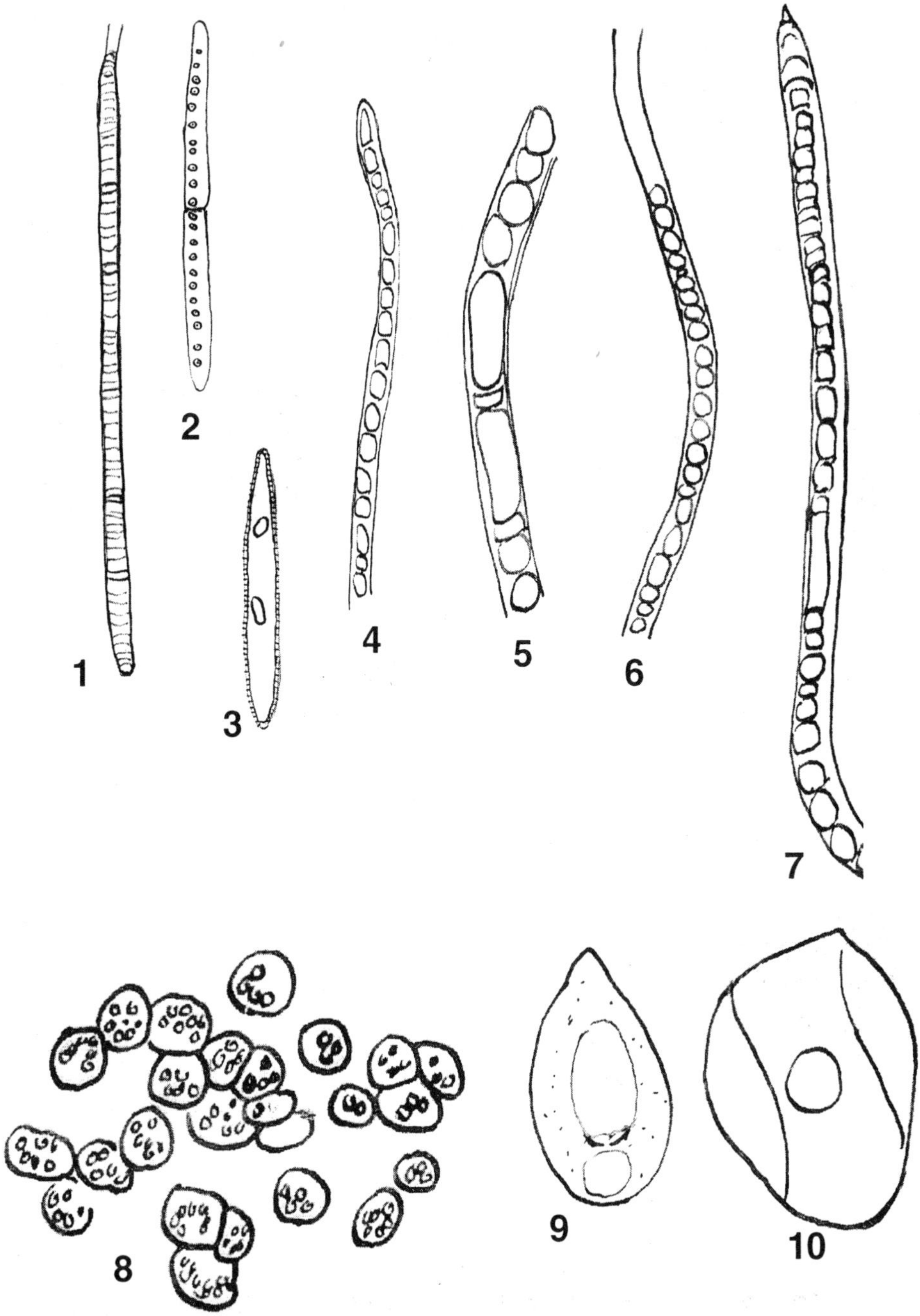

Plate IX: Figures 1-10

1. *Lyngbya epiphytica* (x 650). 2. *Closterium lanceolatum* (x 700).
3. *Fragillaria sp* (x 750). 4. *Phormidium dimorphism* (x 750).
5-7. *Stichosiphon sansibaricus*. (?) 8. *Microcystis flos-aquae* (x 2000).
9. *Phacus acuminatus* (x 1000). 10. *P. suecicusss* (x 1300).

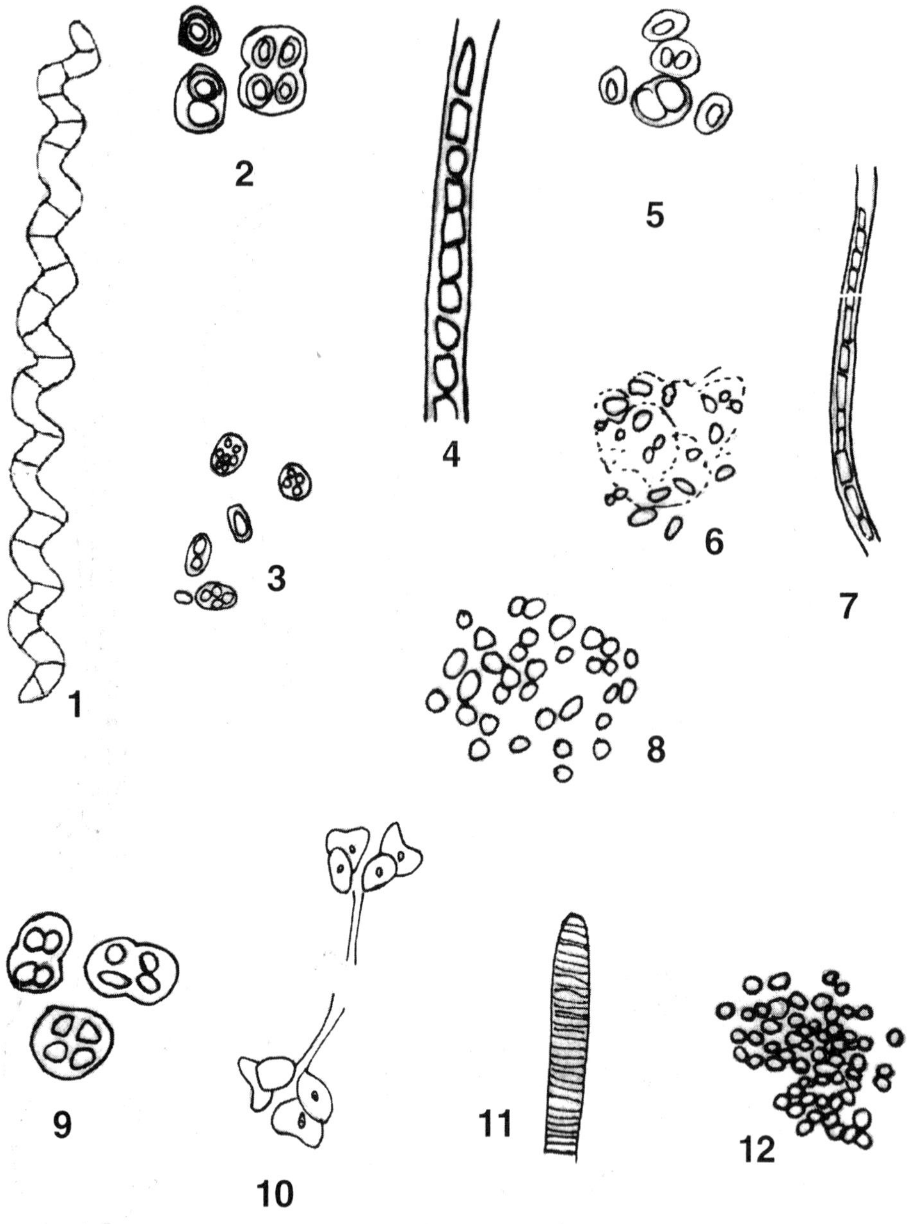

Plate X: Figures 1-12

1. *Arthrospira gomontiana*. (x 1000) 2. *Gloeocapsa decorticans*. (x 500) 3. *Aphanothece microspora*. (x 400) 4. *Phormidium dimorphum*. (x 1000) 5. *Gloeothece rupestris*. (x 150) 6. *Aphanothece pallida*. (x 650) 7. *Phormidium valderianum*. (x 750) 8. *Aphanocapsa montana*. (x 1000) 9. *Chroococcus minutus*. (x 750) 10. *Dimorphococcus lunatus*. (x 750) 11. *Oscillatoria princeps*. (x 350) 12. *Microcystis flos-aquae*. (x 400).

29. *Eunotia sp** Pl.XI. Figure 6.

Rectangular frustules united by valve faces, undulated margin, 28.26μ long and 6.28μ broad. Reported from Allahabad (Pandey and Pandey, 1980).

30. *Fragillaria sp** Pl. IX. Figure 3.

Striae separated by a narrow linear smooth area, devoid of markings, frustules quadrate in girdle view and fusiform in valve view, the poles narrowed from an enlarged central region. Valves 86μ long and 6.28μ broad. Reported from Allahabad (Pandey and Pandey, 1980). Collected from abandoned water body and rice field.

31. *Pleurosigma spenceri** W. Sm. Pl. XIV. Figure 1.

Two large plate like chromatophores connected by a median bridge, valve wall is occupied by chambers opening interior and exterior by narrow pores, raphe occupying the apical axis without keel. Valve 78.5μ long and 7.85μ broad.

32. *Navicula sp.* Pl. XI. Figure 13.

Valve broadly lanceolate, narrowly rostrate to capitate ends, 65.71μ long and 26.69μ broad. Reported from Allahabad (Pandey and Pandey, 1980).

Division: EUGLENOPHYTA

Class: EUGLENOPHYCEAE

Euglenales: (Euglenaceae)

33. *Phacus acuminatus**Stokes,1885a Prescott, 1962 (revised ed.), p.396, Pl. 88. Figure 4. Pl. IX. Figure 9.

Cells sub-orbicular in outline, broadly rounded posteriorly with a short, blunt apiculation; periplast longitudinally striated, paramylon bodies 1-2 ringlike discs; cells 30μ in diameter, 35μ long.

Collected from rice field soil.

34. *Phacus curvicauda** Swirenko,1915a Prescott, 1962 (revised ed.), p.399, Pl. 87. Figure 14., Pl.88, Figure 21. Pl. XIII. Figure 7.

Cells broadly ovoid to suborbicular in outline, slightly spiral in the posterior part, which is extended into a caudus that curves obliquely to the left, anterior end broadly rounded, periplast longitudinally finely striated, paramylon bodies 2, large discs; chloroplasts numerous, ovoid bodies; cells 24-26μ in diameter, 30-35μ long.

Collected from pond water.

35. *Phacus orbicularis** Huebner,1886. Prescott, 1962(revised ed.), p.401, Pl. 87. Figure 10. Pl.XIII. Figure 8.

Cells orbicular in outline, with a short caudus curved to the right; broadly rounded anteriorly; periplast finely striated longitudinally; flagellum as long as the body; paramylon bodies 2 disc-shaped plates; cells 40μ in diameter, 60μlong.

Collected from temporary Water logging.

36. *Phacus suecicus** Lemmermann in Pascher and Lemmermann,1913 Prescott, 1962 (revised ed.), p.403, Pl. 88. Figures 2,3. Pl. IX. Figure 10.

Cells broadly ellipsoid or ovate, but not quite symmetrical, with a relatively long, sharp caudus which is slightly deflected, truncate or slightly retuse at the anterior end but with a prominent, median, collar-like papilla through which the flagellum extends; flagellum as long as the cell body; periplast longitudinally striated,chloroplast numerous discs, paramylon bodies 2, cells(14)-19-22μ in diameter, 34-36μlong, 6-11μ thick.

Division: CHLOROMONADOPHYTA

CHLOROMONADINEAE (Chloromonadinaceae)

37. *Gonyostomum*sp.* Prescott, 1962 (revised ed.), p.422, Pl. 99. Figures 11,12. Pl. XIV. Figure 2.

Cells motile, dorsiventrally flattened, obovate-lanceolate, showing a short caudus, dorsal surface convex, ventral surface with a longitudinal furrow extending posteriorly from an opening which leads into a colourless cavity in the anterior region, flagella 2, usually longer than the body, chloroplasts numerous ovoid discs, crowded at the periphery, short radiating, rodlike trichocysts also at the periphery. Cells 22μ in diameter, 62μ long; nucleus central.

Collected from pond water.

Division: PYROPHYTA

Class: DINOPHYCEAE

Peridiniales: (Ceratiaceae)

38. *Ceratium*sp.* Prescott, 1962 (revised ed.), p.437, Pl. 5592. Figures 4,5. Pl. XV. Figure 5.

Cells broadly or narrowly fusiform in outline, very much flattened dorsiventrally; epitheca with sharply converging margins from just above the transverse furrow, then narrowed gradually to form a long horn, transverse furrow relatively narrow, horns usually 2, cells 100-400μ long.

Very common in hardwater lakes. Collected as plankton from pond water.

Division: CYANOPHYTA

Class: CYANOPHYCEAE

Chroococcales: (Chroococcaceae)

39. *Microcystis aeruginosa** Kuetzing, 1846. Desikachary (1959) p. 93, Pl. 17, Figures 1, 2, 6; and Pl. 18, Figure 10. Pl. IV. Figure 4, Pl. V. Figure 3-6.

Colonies when young round or slightly larger than broad, solid, when old becoming clathrate, with distinct hyaline colonial mucilage, cells 4-5μ in diameter, spherical, generally with gas vacuole.

Generally in plankton forming water-bloom in stagnant waters. Collected from Saklenabad pond forming bloom. Reported from Banaras (Bharadwaja, 1935), ricefield, Ghazipur (Pandey, 1965), Prasad and Mehrotra (1980), from Bareilly pond (Pandey, 1982), Kanpur (Shukla and Shukla, 1988), from the river Ganga, Kanpur (Gupta and Shukla, 1994).

40. *Microcystis flos-aquae** (Wittr.) Kirchner 1898. Desikachary(1959) p.94, Pl.17, Figure 11; and Pl. 18, Figure 11. Pl. IX. Figure 8., Pl. X. Figure 12.

Colonies roughly spherical, or elongate, not clathrate, with indistinct colonial mucilage; cells 2-5μ in diameter, spherical with gas vacuole.

Collected from paddy field soil of Ghazipur, reported earlier from Banaras (Bharadwaja, 1935), Ghazipur(Pandey,1965 and Gupta, 1984).

41. *Chroococcus minutus** (Kuetzing) Naegeli, 1849 Desikachary(1959) p.103, Pl.24, Figures 4; and Pl. 26, Figure 4.15. Pl. X. Figure 9

Cells spherical or oblong, single or in groups of 2-4, light blue-green, with sheath 6-15μ diam. And without sheath 4-5μ diam., colonies 14 x 20μ; sheath not lamellated, colorless.

Collected from paddy field soil of Jangipur,Ghazipur. Reported from Banaras (Rao,C.B.1937; Singh,1939); Allahabad (Mitra,1951); Ballia and Ghazipur (Pandey,1965, Gupta,1984).

42.*Chroococcus montanus**Hansgirg,1888. Desikachary(1959) p.108, Pl.26, Figures 12. Pl. XIII. Figure 9.

Thallus slimy,gelatinous, brownish or blackish brown; cells 4.5-9.4μ diam., single or 2-4, sometimes more in irregular groups of 19-25.

Collected from moist crop field soil, Ghazipur as well as from garden soil. Reported from Varanasi (Rao, C.B.,1937b), Prasad and Mehrotra(1980).

43. *Gloeocapsa decorticans** (A.Br.) Richter. 1925. Desikachary(1959) p.114, Pl.24, Figures 9. Pl. X. Figure 2.

Cells spherical or oval, blue-green, single or in groups of 2-4 together; cells 11 x 14μ and without sheath 7 x 9μ; in two celled stage with sheath 16 x 19μ, without sheath 11μ long, sheath colorless, thick and distinctly lamellated.

From culture of garden soil,Allahabad(Mitra,1951), collected from paddy field soil of Ghazipur, also reported earlier by Pandey, 1965; Gupta,1984.

44. *Gloeothece rupestris** (Lyngb.) Bornet,1892. Desikachary(1959) p.127, Pl.25, Figures 4. Pl. X. Figure 5.

Cells ellipsoidal to cylindrical, blue-green, without envelop 3-7μ broad, 6-8μ long, with envelop 8-12 μ broad, 2-4 or 8 together in oval to subglobose colonies, colonies 25-41μ diam., envelopes colorless.

From cultured soil, Ghazipur, also reported by Gupta,1984.

45.*Gloeothece samoensis var. major** Wille,1915. Desikachary(1959) p.128, Pl.23, Figures 3. Pl. XIII. Figure 3.

Cells ellipsoidal, without sheath 4-6μ broad, 8-10μ long, cells yellowish or bluish green, mostly 2-4 in a common envelope, envelope colorless, unlamellated.

Obtained from garden soil culture. Reported from Rohilkhand division (Pandey, 1982).

46. *Aphanocapsa biformis** A.Br. 1865. Desikachary(1959) p.134, Pl.21, Figures 3,4. Pl. XIII. Figure 4.

Thallus olive green, gelatinous, cells 4-7µ diam., spherical, with an envelope, loosely arranged, 2 to 4 together in a common mucilaginous envelope.

Collected from garden soil. Reported from U.P.(Prasad and Mehrotra, 1980).

47. *Aphanocapsa montana** Cramer,1862. Desikachary(1959) p.135, Pl.20,, Figure 8. Pl. X. Figure 8

Thallus gelatinous, yellow-green or blue-green, of no definite shape; cells 2.5-4µ diam., single or in pairs; mucilage colorless, diffluent.

Collected from garden soil. Earlier reported by Pandey,1965; and Gupta, 1984 from Ghazipur.

48. *Aphanothece pallida** (Kutz.)Rabenh.1865. Desikachary(1959) p.140, Pl.22, Figure 3. Pl. X Figure 6.

Thallus gelatinous blue-green upto 4-6mm diam., cells oblong,elliptical, 3-5µ broad and 4-6µ long, sheath distinct and lamellated.

Banaras(Rao,C.B.1936), Ghazipur, rice field soil(Pandey,1965; Gupta, 1984)

49. *Aphanothece naegelii** Wartm.1865. Desikachary(1959) p.141, Pl.22, Figures 7. Pl. XIII. Figure 10.

Thallus gelatinous, yellow-brown or olive-green; cell oval or spherical, 3.5-4.5 x 6.5-8µ in size, blue-green,sheath diffluent.

Collected from the river Ganga.Reported from moist soil and culture,Varanasi (Rao,C.B.1936), Ghazipur (Pandey, 1965; Gupta, 1984), Prasad and Mehrotra(1980) from U.P., Rohilkhand (Pandey,1982).

50. *Aphanothece microspora** (Menegh) Rabenh. 1863. Prescott,1962,p.468,Pl.105, Figures 7-9.; Pandey,1965,Nova Hedwigia, p.183,Pl.2, Figure 4, Pl.10, Figure 6. Pl. X. Figure 3.

Thallus of irregular outline, cells embedded in a common transparent mucilage; cells cylindrical,usually slightly curved, with or without individual sheath; cells found in 2-8 in a common sheath; cells 2.5-6µ broad and up to 11µ long, content homogenous, light blue-green.

Paddy field soil Ghazipur (Pandey, 1965).

51. *Merismopedia tenuissima** Lemmermann,1898. Desikachary(1959) p.154, Pl.29, Figure 7; and Pl. 30, Figure 8,9. Pl. XIII. Figure 5.

Cells pale blue-green, closely packed in colonies of 16-100 cells, sub-spherical, 1.5-2µ broad, sometimes individual cells with distinct mucilaginous envelope.

Collected from temporary water loggings. Reported from Banaras (Rao,C.B.1938) Kanpur (Shukla and Shukla, 1988, Gupta and Shukla, 1994).

Nostocales: (Oscillatoriaceae)

52. *Arthrospira gomontiana** Setchell,1895 Desikachary (1959) p.192 Pl. X Figure 1.

Free swimming, trichome 2.5-3μ broad, attenuated at the apices, regularly spirally coiled, spirals 6μ broad and 16-18μ distant; cells 4-5μ long, pale blue-green, cross wall commonly granulated.

Collected from rice field water. Prasad and Mehrotra (1980) reported the taxa from paddy fields, U.P.

53. *Oscillatoria curviceps**Ag.ex Gomont,1824 Desikachary(1959) p.209, Pl.38, Figure 2. Pl. XI. Figure 4., Pl.VII. Figure 6.

Thallus light or dark blue-green; trichomes more or less straight, bent at the end or spirally coiled, not attenuated, not constricted at the cross walls, 12μ broad, 2-3μ long, cross walls sometimes granulated; end cells flat rounded,not capitate.

Collected from sugarcane fields. Reported from Kanpur (Pandey and Tripathi,1988; Tiwari *et al.*, 2001).

54. *Oscillatoria curviceps var angusta** Ghose,1927b Desikachary(1959) p.210, Pl.39, Figure 12. Pl. XI. Figure 8.

Trichomes 5-6μ broad.

Collected from sugarcane field.

55. *Oscillatoria princeps** Vaucher ex Gomont,1803 Desikachary(1959) p.210, Pl.37, Figure 1,10,11,13,14. Pl. X. Figure 11.

Trichomes blue-green, mostly forming a thallus, straight, not constricted at the cross-walls, 16-20 μ broad, slightly attenuated and bent; cells 3-4μ long; end cells flatly rounded, slightly capitate.

Cosmopolitan in distribution. Collected from paddy field soil. Reported from Ghazipur (Gupta, 1984), Kanpur (Tiwari *et al.*, 2001).

56. *Oscillatoria irrigua** (Kutz.)Gomont,1892. Desikachary(1959) p.224, Pl.42, Figures 7,9. Pl. XI Figure 9.

Thallus blackish blue-green, straight, flexuous, 6-8μ broad, apex slightly attenuated, sub capitate; cells quadrate, 3-4μ long, septa not ordinarily granulated, apical cell convex.

Collected from road side mud. Reported from Kanpur (Pandey and Tripathi,1988)

57. *Oscillatoria acuta** Bruhl et Biswas, 1922. Desikachary(1959) p.240, Pl.39, Figures 5,8. Pl. XIV. Figure 3.

Trichomes hardy, brittle, not constricted at the cross walls, 7μ thick, 70-400 μ long,usually straight, narrow or acuminate towards the sub obtuse, non capitate apex, may be straight but more often abruptly bent aside; cells 4-4.7μ long.

Collected from garden soil and rice field. Reported from U.P. (Prasad and Mehrotra, 1980).

58. *Phormidium dimorphum** Lemm.1908. Desikachary(1959) p.256, Pl.54, Figure 8. Pl. IX., Figure 4. Pl. X. Figure 4.

Thallus membranous, broadly expanded, dark blue-green; trichome straight or bent, distinctly constricted at the cross-walls, 3-5 µ broad, blue-green, attenuated; sheath firm, colorless, thin, not lamellated; cells mostly quadrate, 3-6µ long, not granulated at the cross walls, end cells long, conical or short rounded.

Collected from moist soil.

59. *Phormidium truncicola** Ghose1924. Desikachary(1959) p.258, Pl.59, Figure 9. Pl. XIV. Figure 5.

Thallus expanded, membranous, yellow-green or blue-green, sheath diffluent, fibrous; trichomes 6-8 µ broad, nearly parallel,distinctly constricted at the cross walls, cells as long as broad, attenuated at the apices.

Collected from paddy field soil.

60. *Phormidium tenue** (Menegh) Gomont, 1892. Desikachary(1959) p.259, Pl.43, Figures 13-15; and Pl. 44, Figure 7-9. Pl. XII. Figure 7.

Thallus pale blue-green,trichome straight, slightly constricted at the cross walls, attenuated at the ends, 1-2 µ broad, sheath thin,diffluent,cell 6-8µ long, septa not granulated, cross walls not commonly visible; end cell acute conical, calyptra absent.

Collected from rice field soil. Reported earlier from Ghazipur(Pandey, 1965; Gupta, 1984), Kanpur (Tiwari *et al.*, 2001)

61. *Phormidium valderianum** (Delp.) Gomont, 1892. Desikachary(1959) p.263, Pl.44, Figures 5. Pl. X. Figure 7.

Thallus lubricous, lamellated, outer surface dull green, trichomes flexuous, densely entangled, not constricted at the cross walls, ends not attenuated, not curved, 2-2.5 µ broad, sheath thin, firm,more or less diffluent; cells longer than broad, 4-7µ long, end cell rounded.

Collected from paddy fields, mud bottoms and water logged soils. Reported from Kanpur (Pandey and Tripathi, 1988).

62. *Phormidium corium** (Ag.)Gomont, 1890. Desikachary(1959) p.269, Pl.44, Figures 10-11. Pl. XII. Figure 13.

Thallus membranous, brownish green, filaments long, densely entangled; sheath thin, trichome not constricted at the cross walls, ends straight, shortly attenuated, 3-4.5 µ broad, cells quadrate, 4-6µ long, not granulated at the cross walls; end cell obtuse conical.

Collected from temporary water logging. Reported from culture of rice field soils and uncultivated field soil near Allahabad(Mitra,1951), Kanpur (Shukla and Shukla, 1978), U.P.(Prasad and Mehrotra,1980).

63. *Phormidium ceylanicum** Wille1915. Desikachary(1959) p.272, Pl.43, Figures 18,19. Pl. XII. Figure 10.

Thallus dull green, flat, sheath diffluent, in a homogenous mucilage; trichome straight, slightly constricted at the cross walls, ends attenuated, cells 7-9μ broad, 5-8μ long.

Collected from moist soil.

64. *Lyngbya epiphytica** Hieron,1898. Desikachary(1959) p.284, Pl.53, Figure 7. Pl. IX. Figure 1.

Filaments growing in a thick or loosely spiral manner on other algae, 2-4μ broad; sheath thin, colorless; cells 1-2μ long, not constricted at the cross walls and not granulated; end cells hemispherical, not attenuated.

Epiphytic on other algae in rice field soil.

65. *Lyngbya spirulinoides** Gomont,1890. Desikachary(1959) p.289, Pl.52, Figures 4,5. Pl. XII Figure 2.

Thallus olive green; filaments entangled, regularly or loosely spirally coiled, distance between two spirals 73-108 μ; sheath thin, colorless, lamellated, mucilaginous, not constricted at the cross walls, cells 12-14μ broad; 3-5μ long, cross walls sometimes granulated; end cell rounded.

Collected from road side ditch and rice field soil.

66. *Lyngbya lagerheimii** (Mob.) Gomont1890. Desikachary (1959) p.290, Pl.48, Figure 6; and Pl. 53, Figure 2. Pl. XIV. Figure 4.

Filaments single, or entangled with one another, irregularly spirally coiled, sheath thin colorless, trichome about 2 μ broad; cells 1.2-3 μ long not constricted at the cross walls, pale blue-green, end cell rounded, not attenuated.

Reported earlier from rice field soil at Ghazipur (Pandey,1965), U.P.(Prasad and Mehrotra, 1980).

67. *Lyngbya birgei** Smith,G.M.,1916. Desikachary (1959) p.296, Pl.50, Figures 7,8. Pl. XI Figure 5

Filaments straight, seldom coiled, free floating, 25-30μ broad; sheath firm, colorless, seldom lamellated, 1.5μ thick; trichome not constricted at the cross walls, 18-20 μ broad, ends rounded, not attenuated, cells shorter than broad, 2-2.5 μ long, sometimes with gas vacuoles.

Collected from the moist soil near the tap. Gupta (1956) reported a form free floating in filter beds at Allahabad, U.P.(Prasad and Mehrotra, 1980).

68. *Lyngbya hieronymusii** Lemm.,1905. Desikachary(1959) p.297, Pl.48, Figure 4. Pl. XI. Figure 2., Pl.XII. Figure 11.

Filament single, free floating, straight or slightly bent, 15μ broad, sheath firm, homogenous, colorless, cells 11-13 μ broad, 2.5-4 μ long, not constricted at the cross walls, granulated, with gas vacuole, not attenuated, end cell broadly rounded.

Collected from road side mud. Reported from rice fields and ponds at Allahabad(Mitra,A.K.1951; Gupta,1956),paddy field soil of Ghazipur(Pandey,1965; Gupta,1984), Kanpur (Shukla and Shukla, 1978), Rohilkhand (Chaturvedi and Pandey, 1976; Pandey and Chaturvedi, 1979).

69. *Lyngbya dendrobia** Bruhl et Biswas,1923. Desikachary(1959) p.302, Pl.50, Figures 3,10; and Pl. 55, Figure 2-4. Pl. XII. Figure 12.

Thallus compact, thin, filaments long, interwoven, cells with sheath 10-11 μ thick, sheath thin 1-1.5 μ thick, smooth, colorless; trichomes 9-10 μ broad, less constricted at the cross walls, cells 3-5μ long, uniformly and densely granular.

Collected from cultivated moist soil.On the barks of tree at Lucknow and from moist soil, Varanasi(Rao,C.B.1937b)

70. *Lyngbya putealis** Mont.ex.Gomont,1840. Desikachary(1959) p.317, Pl.52, Figures 12. Pl. XII. Figure 8.

Thallus expanded up to 1cm long, penicillate, dull blue-green, filaments nearly straight, sheath colorless, trichome 11μ broad, distinctly constricted at the cross-walls, cross walls sometimes granulated, not attenuated, not capitate, cells quadrate,4-7μ long; end cells rounded.

Collected from cultivated moist soil.Banaras(Rao,C.B., 1937b), U.P. (Prasad and Mehrotra, 1980).

71. *Lyngbya truncicola* Ghose *var terrestris** Pandey et Mitra, 1965. Desikachary(1959) p.308, Pl.51, Figures 4. Pl. III. Figure 3.

Thallus extensive, blue green; filaments straight, 8-12μ broad, sheath thin, hyaline, diffluent, trichome blue green, 6.5-7μ broad, not constricted at crosswalls, cells short 1.5-3μ long, apical cell rotund.

Collected from ricefield soil, Tulasipur. Reported from Ghazipur paddy field soil(Pandey, 1965).

72. *Microcoleus chthonoplastes** Thuret ex Gomont,1875. Desikachary(1959) p.3343, Pl.60, Figures 7-9. Pl. XII. Figure 1.

Filaments forming an expanded dirty to dark green lamellated thallus, coiled, sheath gelatinizing, uneven thick, many closely grouped trichomes; constricted at the cross walls; cells blue green, 3-7μ long; end cells pointed conical.

Collected from rice field soil. Reported from Banaras, Basti,Gorakhpur and Mirzapur, Siwait near Allahabad, and Ghazipur in Uttar Pradesh(Bharadwaja,1935, Rao,C.B., 1937b, Singh, R. N.,1939, Mitra,1951,Pandey, 1965; Gupta, 1984)

73. *Microcoleus lacustris** (Rabenh.) Farlow,1877. Desikachary(1959) p.345, Pl.60, Figures 4,5. Pl. III. Figure 2, 6.

Thallus blackish blue-green, filaments contorted, seldom branched; sheath colorless, slimy, many trichomes in each; trichomes distinctly constricted at the cross walls, 4-5 μ broad; cells cylindrical, 6-12 μ long, end cell more or less rounded, conical.

Collected from rice field.

Nostocales: (Nostocaceae)

74. *Anabaenopsis raciborskii** Wolosz, 1913. Desikachary(1959) p.358, Pl.63, Figures 6,7. Pl.XV. Figure 4.

Filament short, straight, seldom spirally coiled with 1-2 ½ spirals up to 200 µ long, not constricted at the cross walls, with a heterocyst at either end; cells cylindrical, 2.5-4 µ broad, without gas vacuole.

Collected from pond water.

75. *Nostoc muscorum** Ag. ex Born. Et Flah,1812. Desikachary(1959) p.385, Pl.70, Figure 2. Pl. XIV. Figure 10.

Thallus gelatinous, filaments densely entangled; sheath distinct only at the periphery of the thallus, trichome 3-5µ broad; cells short barrel-shaped, 4µ long; heterocysts nearly spherical, 6µ broad; spores oblong, many in series, 4-8 µ broad.

From paddy field soil, Ghazipur(Gupta,1984)

76. *Nostoc commune** Vaucher ex Born et Flah,1803. Desikachary(1959) p.387, Pl.68, Figure 3. Pl. XIV. Figure 6.

Thallus firm, gelatinous, later flattened, expanding, or leathery, irregularly torn, blue-green, filaments flexuous, entangled, sheath distinct at the periphery, thick, lamellated; trichome 4-6 µ broad,cells short barrel-shaped or nearly spherical, 7 µ broad.Spores as big as the vegetative cell.

Collected from cultivated field soil.Paddy field soil, Ghazipur(Gupta,1984).

77. *Anabaena fertilissima** Rao, C.B.1937. Desikachary(1959) p.398, Pl.74, Figure 1. Pl. XIV. Figure 7.

Trichome single, straight or bent, with rounded end cells, 4-5µ broad, at the apex 4 µ broad; cells barrel-shaped, 4-5µ long; heterocysts almost spherical, 7.8µ broad; spores in long chains, 6-8µ broad and 3.2-8.8 µ long.

Collected from rice field soil.Banaras (Rao,C.B.1937b),Ghazipur (Gupta,1984).

78. *Anabaena ambigua** Rao, C.B.1937. Desikachary(1959) p.400, Pl.76, Figure 2. Pl. III. Figure 1, 4, 5.

Trichomes free or completely enclosed in a mucilaginous envelope or sheath; sheath usually firm, hyaline, with smooth or rough outline; 10-50 µ broad; trichomes bent, slightly tapering at the ends, end cells having rounded apices, cells barrel-shaped with deep constrictions at the joints; heterocysts at intervals, almost spherical, usually broader than the cells, 6.4-9 µ in diameter.

Attached to submerged plants in a pond, Banaras(Rao,C.B.1937a),Paddy fields in U.P., Allahabad,Ghazipur(Singh,R.N.1942, Mitra,1951; Gupta,1984)

79. *Anabaena variabilis** Kuetzing ex Born et Flah,1843. Desikachary(1959) p.410, Pl.71, Figure 5. Pl. VII. Figure 4 and 5.

Thallus gelatinous, dark green; trichome without any sheath, flexuous, 4-6 µ broad, slightly constricted at the cross walls, end cells conical, obtuse; cells barrel-shaped, sometimes with gas vacuoles, 2.5-6 µ long; heterocysts spherical or oval, 6 µ broad, up to 8 µ long.

Collected from water logged field. Reported from paddy field soil Ghazipur (Gupta, 1984), U.P.(Prasad and Mehrotra, 1980), Rohilkhand (Pandey, 1982).

Division: RHODOPHYTA

Class: RHODOPHYCEAE

Order: Bangiales

80. *Compsopogon aeruginosus** (J. Ag.) Kuetzing. Singh and Pandey(1986). Phycos;25: 84-87. Pl. XIV. Figure 8.,Pl. XV. Figure 3.

Thallus blue green, upto 1m length, attached to the substratum, main axis profusely branched at distal end, branches are of two types – long lateral and short laterals. Mature thallus 1-1.5 mm in diameter, older secondary branches with spines 600-650µ in diameter. Younger multiceriate branches 240-280 µ and uniseriate branches 50-60µ.

Collected from the river Ganga. Reported from Nakatia river near Bareilly district, U.P. (Singh and Pandey1986).

81.. *Compsopogon aeruginosus* J. Ag. Kuetzing *var catenatum*. Pl. XIV. Figure 9. Pl. XV. Figure 1 and 2.

Thallus bluish dark violet, erect filament nodulated with distinct wavy margins and with many short spines. Main axis 40-50µ broad and is multiseriate. Main axis and branches have central axial cell surrounded by corticating peripheral cells. Branches arise at 30° to 60°. Lateral branches alternate, 6-18µ long and 20-38µ broad. Spine like branches present on the main filament.

Collected from the river Ganga, Dadarighat, Ghazipur. Reported from Dorania river at Bareilly (Yadav and Pandey, 1980).

Seasonal Variation of Algal Diversity in a Pond

To study the seasonal variation in algal population a pond near Bhutahiatar, Ghazipur was selected. Selection of the pond was made keeping in view that they are surrounded by the fields actively used for cultivating major crops *viz.*, Rabi and Kharif supposed to fetch maximum diversity.

Surface water samples were collected from four corners of the pond quarterly in the months of January, April, July and October during the first week of respective months. Samples were collected by hauling about 100 L of water through a plankton net and brought to laboratory for phytoplankton studies. Visual collections of algae were also made. Complete physico-chemical and biological analysis of water is beyond the scope of the project, only the pH and temperature were recorded at the spot and identification of algae were made in the laboratory with the help of standard literatures cited earlier. Population density of countable algae was also recorded by Haemocytometer. Results are summarized in the Tables 4.2–4.4.

Table 4.2: Record of Temperature and pH of Water at the Time of Sample Collection

Parameters	*Months*			
	January	*April*	*July*	*October*
Temperature(°C)	13	30	32	26
pH	7.5	8.05	7.70	7.4

Table 4.3: Seasonal Variation in the Spectrum of Algae

Algal Taxa Recorded	*Months*			
	January	*April*	*July*	*October*
Chroococcus sp.	+	+	–	++
Merismopedia tenuissima	+	+	–	++
*Oscillatoria acuta**	+++	++	+	+++
Spirogyra sp*	++	++	+	++
Cosmariumsp.	+	++	–	++
Pediastrum sp.	++	+	+	++
Nitzschia sp.	+	+	+	+
Navicula sp.	+	+	+	+
Hydrodictyon sp.*	+	+	–	–

* Visually collected.

Table 4.4: Population Density of Countable Algae (Unit/L)

Name of the Taxa	*Months*			
	January	*April*	*July*	*October*
Chroococcus sp.	10526.31	13157.89	–	13157
Merismopedia tenuissima	8552.63	9210.52	–	10526.31
Cosmarium sp.	6578.94	8552.94	–	7894.73
Pediastrum sp.	9210.52	6578.94	5921.05	8552.63
Nitzschia sp.	5263.16	3289.47	3289.47	4605.26
Navicula sp.	5263.16	3947.36	3209.47	5921.05

Summary and Conclusion

Ghazipur a part of eastern Uttar Pradesh is located at 25° 35′N latitude and 83° 35′ E longitude. It is a part of mid Gangetic plain surrounded by the rivers Ganges, Karmnasa and Gomati. The climate is typically monsoon type with a temperature range of 5-17°C during winter and 30-40° C during summers. The climatic condition of Ghazipur provides a condusive environment for the growth and reproduction of algae. Since the locality of Ghazipur was unexplored for algal studies except a few (Pandey, 1965, 1966 and Gupta, 1984) restricted to the rice field algae only, the present study has been undertaken with the objective to bring on record the algal diversity of Ghazipur from different locality and habitat.

As per the objective of the project, an algal inventory of Ghazipur is prepared including the present and previous studies. The algal inventory of Ghazipur now consists a total of 148 algal taxa (Table 4.1) out of which eighty one have been recorded and identified in the present study. They are:

Chlamydomonas angulosa, Ulothrix tenerrima, Stigeoclonium lubricum, Cladophora glomerata, Rhizoclonium hookeri, Rhizoclonium fontanum, Oedogonium microgonium, Oedogonium cantonense var. *indicum, Chlorococcum humicola, Hydrodictyon reticulatum, Pediastrum simplex, Pediastrum araneosum, Pediastrum biradiatum, Chlorella vulgaris, Tetraedron minimum, Tetraedron regulare* var. *minus, Dimorphococcus lunatus, Spirogyra maravillosa, Spirogyra gratiana, Spirogyra azygospora, Spirogyra singularis, Cosmarium tetraophthalmum, Closterium lanceolatum, Cylindrocystis sp., Nitella hyaline, Nitzschia sp., Pinnularia viridis, Eunotia, Pinnularia sp., Fragillaria sp., Pleurosigma spenceri, Navicula sp., Phacus acuminatus, Phacus curvicauda, Phacus orbicularis, Phacus suecicus, Gonyostomum sp., Ceratium sp., Microcystis aeruginosa, Microcystis flos-aquae, Chroococcus montanus, Chroococcus minutus, Gloeocapsa decorticans, Gloeothece rupestris, Gloeothece samoensis* var. *major, Aphanocapsa biformis, Aphanocapsa Montana, Aphanothece pallida, Aphanothece naegelii, Aphanothece microspora, Merismopedia tenuissima, Arthrospira gomontiana, Oscillatoria curviceps, Oscillatoria curviceps* var. *angusta, Oscillatoria princes, Oscillatoria irrigua, Oscillatoria acuta, Phormidium dimorphum, Phormidium truncicola, Phormidium tenue, Phormidium valderianum, Phormidium corium, Phormidium ceylanicum, Lyngbya epiphytica, Lyngbya spirulinoides, Lyngbya lagerheimii, Lyngbya birgei, Lyngbya hieronymusii, Lyngbya dendrobia, Lyngbya truncicola Ghose* var. *terrestris, Lyngbya putealis, Microcoleus chthonoplastes, Microcoleus lacustris,. Anabaenopsis raciborskii, Nostoc muscorum, Nostoc commune, Anabaena fertilissima, Anabaena ambigua, Anabaena variabilis, Compsopogon aeruginosus, Compsopogon aeruginosus* var. *catenatum.*

Of these eighty one taxa, twenty five belong to various orders of division Chlorophyta, seven to Chrysophyta (Bacillariophyceae), four to Euglenophyta, one each to Chloromonadophyta (Chloromonadinaceae) and Pyrophyta (Dinophyceae), forty one to Cyanophyta and two to Rhodophyta.

Fifty nine out of the eighty one taxa recorded in the present study are being reported for the first time from Ghazipur, *viz. Chlamydomonas angulosa, Ulothrix tenerrima, Stigeoclonium lubricum, Cladophora glomerata, Rhizoclonium hookeri, Rhizoclonium fontanum, Oedogonium microgonium, Oedogonium cantonense* var. *indicum, Hydrodictyon reticulatum, Pediastrum simplex, Pediastrum araneosum, Pediastrum biradiatum, Tetraedron minimum, Tetraedron regulare* var. *minus, Dimorphococcus lunatus, Spirogyra maravillosa, Spirogyra gratiana, Spirogyra azygospora, Spirogyra singularis, Cosmarium tetraophthalmum, Closterium lanceolatum, Cylindrocystis sp., Nitella hyaline, Nitzschia sp., Pinnularia viridis, Eunotia, Pinnularia sp., Fragillaria sp., Pleurosigma spenceri, Navicula sp., Phacus acuminatus, Phacus curvicauda, Phacus orbicularis, Phacus suecicus, Gonyostomum sp., Ceratium sp., Chroococcus montanus, Gloeothece samoensis* var. *major, Aphanocapsa biformis, Merismopedia tenuissima, Arthrospira gomontiana, Oscillatoria curviceps, Oscillatoria curviceps* var. *angusta, Oscillatoria irrigua, Oscillatoria acuta, Phormidium dimorphum, Phormidium truncicola, Phormidium valderianum, Phormidium corium,, Phormidium ceylanicum Lyngbya epiphytica, Lyngbya spirulinoides, Lyngbya birgei, Lyngbya dendrobia, Lyngbya putealis, Microcoleus lacustris, Anabaenopsis raciborskii, Compsopogon aeruginosus, Compsopogon aeruginosus* var. *catenatum.*

Twenty five taxa *viz. Rhizoclonium hookeri, Rhizoclonium fontanum, Oedogonium microgonium, Oedogonium cantonense* var. *indicum, Pediastrum araneosum, Pediastrum biradiatum, Tetraedron minimum, Spirogyra maravillosa, Spirogyra gratiana, Cosmarium tetraophthalmum, Cylindrocystis sp., Pleurosigma spenceri, Phacus acuminatus, Phacus curvicauda, Phacus orbicularis, Phacus suecicus, Gonyostomum sp., Ceratium sp., Oscillatoria curviceps* var. *angusta, Phormidium dimorphum, Phormidium truncicola, Phormidium ceylanicum Lyngbya epiphytica, Lyngbya spirulinoides, Microcoleus lacustris* seems to have no earlier reports from the state, Uttar Pradesh.

The seasonal variation of algal diversity in a pond was also studied quarterly in a year (*i.e.* in the months of January, April, July, October) within the scope of the project. Temperature variation was recorded as lowest as 13°C in the month of January and a maximum of 32°C in the month of July. pH of the pond water was recorded optimum for the algal growth ranging between 7.4 and 8.05. Altogether nine algal taxa were recorded from the pond selected near Bhutahiatar, Ghazipur *viz. Chroococcus sp., Merismopedia tenuissima, Oscillatoria acuta, Spirogyra sp., Cosmarium sp. Pediastrum sp., Nitzschia sp., Navicula sp. and Hydrodictyon sp.* Of these, *Oscillatoria acuta, Spirogyra* sp., *Pediastrum* sp., *Nitzschia* and *Navicula* sp. were recorded in all the months of observation (Table 4.3) *Chroococcus* sp. witnessed the highest population diversity during the month of April and October about 13157 units/L of water sample (Table 4.4). Population of diatoms *Nitzschia* sp. and *Navicula* were recorded lowest ranging from 3209 to 3289 units/L in the month of July.

Interesting to note in the present study is that the massive growth of algal taxa, mostly the green algae in the river Ganga was observed near Dadarighat where discharge of Opium Factory(though properly treated) joins the river (Pl. XV. Figure 6,7 and 8). A detailed study is desirable in this direction which may reveal some interesting results. Algal populations are continuously being disturbed in various localities due to anthropogenic factor which is a major threat for their conservation. Their germplasm collection and preservation is highly recommended in view of the growing importance of algae in various field.

References

Bharadwaja, Y., 1935. The Myxophyceae of the United Provinces. India–I. *Proc. Indian Acad. Sci.*, B. 2: 95–107.

Chadha, A. and Pandey, D.C., 1977. Addition to the algal flora of Allahabad–I, Genus *Scenedesmus. Phykos*, 16(1 and 2): 69–75.

Chadha, A. and Pandey, D.C., 1978. Additions to the algal flora of Allahabad–II (Chlorophyta, Chlorococcales). *Phykos*, 17(1 and 2): 73–80.

Chaturvedi, U.K., 1985. Additions to algal flora of Rohilkhand Division, U.P., India: IX Diatoms from Bareilly District. *Phykos*, 24: 163–169.

Chaturvedi, U.K. and Pandey, U.C., 1976. A list of blue-green and green algae from Rohilkhand Division, U.P., India–IV. *Phykos*, 15(1 and 2): 127–131.

Chaturvedi, U.K., Pandey, U.C., Habib, I. and Shukla, H.M.1987. Desmids of Bareilly II. *Phykos*, 26: 95–102.

Chaturvedi, U.K., Bhatnagar, S.K. and Pandey, U.C., 1979. On *Batrachospermum* from Rohilkhand division, U.P., India. *Phykos*, 18(1 and 2): 53–56.

Chaturvedi, U.K. and Bhatnagar, S.K., 1982. *Compsopogon coeruleus* (Balbis) Mont. from Rohilkhand division, U.P., India. *Phykos*, 21: 91–92.

Chaturvedi, U.K. and Pandey, U.C., 1980. On a terrestrial *Oedogonium* from Rohilkhand division, U.P. *Phykos*, 19(1): 63–65.

Costelloe, J.F., Powling, J., Reid, J.R.W., Shiel, R.J. and Hudson, P., 2005. Algal diversity and assemblages in arid zone rivers of the lake Evrie Basin, Australia. *River Research and Applications*, 21(2–3): 337–349.

Day, J.G., Benson, E.E., Harding, K., Knowles, B., Idowu, M., Bremner, D., Santos, L., Santos, F., Friedl, T., Lorenz, M., Lukesova, A., Elaster, J., Lukavsky, J., Herdman, M., Rippka, R. and Hall, T., 2005. Cryopreservation and conservation of microalgae: The development of Pan-European Scientific and biotechnological resource (The COBRA Project). *Cryo. Letters*, 26(4): 231–238.

Desikachary, T.V., 1959. *Cyanophyta*. ICAR Publication, New Delhi.

Dwivedi, B.K. and Pandey, G.C., 2002. Physico-chemical factors and algal diversity of two ponds (Girija Kund and Maqubara pond), Faizabad, India. *Poll. Res.*, 21(3): 361–370.

Dwivedi, S., Mishra, P.K. and Suseela, M.R., 2004. Some Desmids from central and western Uttar Pradesh. *Phytotaxonomy*, 4: 64–73.

Fathi, A.A., Abdelzaher, H.M.A., Flower, R.J., Ramdani, M. and Kraiem, M.M., 2001. Phytoplankton communities of North African wetland lakes–The CASSARINA Project. *Aquatic Ecology*, 35(3–4): 303–318.

Fatma, T., 1985. Epiphytic algal flora of freshwater ponds of Lucknow–Chaetophorales. *Phykos*, 24: 173–174.

Fatma, T., 1990. Ecological studies on Chaetophorales of two ponds at Lucknow. *Phykos*, 29(1 and 2): 89–94.

Gonzalves, E.A., 1981. *Oedogoniales*. ICAR Publication, New Delhi.

Gupta, A.B., 1956. A contribution to the algal flora of the Allahabad district. *J. Res.*, D.A.V. College, Kanpur, 3(1): 76–81.

Gupta, R.K., 1984. On the occurrence of some additional blue-green algae in paddy fields of Ghazipur. *The Journal of Scientific Research*, B.H.U., 34(2): 161–167.

Gupta, P. and Shukla A.C., 1994. Bottom sand algae of river Ganga. *Phykos*, 33(1 and 2): 43–46.

Habib, I. and Chaturvedi, U.K., 1993. Contribution to Desmids of Rohilkhand division. *Phykos*, 32(1 and 2): 51–54.

Habib, I., 2002. Some Chlorococcales from Kumaun region (U.P.), India. *J. Indian Bot. Soc.*, 81(3–4): 195–198.

Habib, I. and Chaturvedi, U.K., 2001. A systematic account of Chlorococcales from Mahoba, India. *Phykos*, 40(1 and 2): 107–113.

Horiguchi, T., 2006. Algae and their chloroplasts with particular reference to the dinoflagellates. *Paleontological Research*, 10(4): 299–309.

Iyengar, M.O.P.and Desikachary, T.V., 1981. *Volvocales*. ICAR Publication, New Delhi.

Khan, M., 1985. Algal flora of Sultanpur (U.P.), India. *Phykos*, 24: 52–57.

Khare, R. and Suseela, M.R., 2004. Freshwater algal diversity of Nainital, Uttaranchal, India. *Phytotaxonomy*, 4: 19–24.

Kulberg, R.G., 1968. Algal diversity in several thermal spring effluents. *Ecology (Washington D.C.)*, 49(4): 751–755.

Mishra, P.K., Srivastav, A.K. and Jaiprakash, 2002. Morphotaxonomic survey on Ulotrichales and Cladophorales of North eastern U.P., India. *J. Indian Bot. Soc.*, 81(3–4): 345–350.

Mishra, P.K., Srivastav, J.P. and Anand, K., 2002. Filamentous green algae from Basti, Uttar Pradesh. *Phytotaxonomy*, 2: 130–134.

Mishra, U., Dixit, G. and Tiwari, G.L., 2001. Physiological Characterization of some Rivularian isolates from paddy fields of Uttar Pradesh. *Phykos*, 40(1 and 2): 65–69.

Mishra, U., Pabbi, S. and Singh, P.K., 2001. Cyanobacterial diversity in terai belt of Uttar Pradesh, India. I. Occurrence of some heterocystous forms. *Phykos*, 40(1 and 2): 89–93.

Mitra, A.K., 1951. The algal flora of certain Indian Soils. *Indian J. Agric. Sci.*, 21: 357–373.

Munari, C. and Vigna, M.S., 1999. Contribution to the knowledge of the phytoflora of Buenos Aires Cityponds. Revista-del-Museo, Argentino-de-Ciencias Naturales, Nueva-Serie., 1(1): 13–22.

Norton, T.A., Melkonian, M. and Andersen, R.A., 1996. Algal biodiversity. *Phycologia*, 35(4): 308–326.

Olquin, H.F., Puiq, A., Loez, C.R., Saliban, A., Topalian, M.L., Castane, P.M. and Rovedatti, M.G., 2004. An integration of water physicochemistry, algal bioassays, phytoplankton and zooplankton for ecotoxicological assessment in a highly polluted lowland river. *Water, Air and Soil Pollution*, 155(1–4): 355–381.

Pal, S., 1977. A preliminary list of Zygnemaceae of Saharanpur, U.P., India. *Phykos*, 16(1 and 2): 43–45.

Pal, B.P., Kundu, B.C., Sundarlingam, V.S. and Venkataraman, G.S., 1962. *Charophyta*. ICAR Publication, New Delhi.

Pandey, D.C., 1965. A study on the algae from paddy field soils of Ballia and Ghazipur Districts of Uttar Pradesh, India. Part IIA: Taxonomic considerations–Cyanophyceae. *Nova Hedwigia*, 10(1/2): 177–209.

Pandey, D.C., 1966. A study on thc algac from paddy field soils of Ballia and Ghazipur Districts of Uttar Pradesh, India. Part IIB: Taxonomic considerations–Chlorophyceae. *Nova Hedwigia*, 11: 75–88.

Pandey, D.C. and Mitra, A.K., 1965. Certain new Myxophyceae from the ricefield soils of India. *Nova Hedwigia*, 10(1/2): 85–86.

Pandey, S.N. and Tripathi, A.K., 1988. Studies on algae of Polluted ponds of Kanpur(India). VIII. Role of blue-green algae. *Phykos*, 27: 38–43.

Pandey, U.C., 1982. Additions to the algal flora of Rohilkhand division VI. Cyanophyceae. *Phykos*, 21: 137–140.

Pandey, U.C. and Chaturvedi, U.K., 1979. Algae of Rohilkhand division, U.P., India–V. *Phykos*, 18(1 and 2): 37–43.

Pandey, U.C. and Gangwar, F.C., 1986. Chlorococcales of Bareilly–I. *Phykos*, 25: 144–147.

Pandey, U.C. and Pandey, D.C., 1980. Addition to the algal flora of Allahabad.III, Desmids. *Phykos*, 19(1): 71–81.

Pandey, U.C. and Pandey, D.C., 1980. Addition to the algal flora of Allahabad.IV, Diatoms. *Phykos*, 19(2): 153–160.

Pandey, U.C. and Pandey, D.C., 1980. Addition to the algal flora of Allahabad.V, Desmids. *Phykos*, 19(2): 161–170.

Pandey, U.C. and Pandey, D.C., 1982. Additions to the algal flora of Allahabad VIII. Cyanophyta. *Phykos*, 21: 76–79.

Pandey, U.C., Habib, I., Gangwar, F.C. and Shukla, H.M., 1987. A contribution to our knowledge of Desmids from Bareilly. *Phykos*, 26: 86–94.

Philipose, M.T., 1967. *Chlorococcales*. ICAR Publication, New Delhi.

Prasad, B.N. and Mehrotra, R.K., 1976. *Calothrix lucknowense*: A new species from paddy field soil. *Phykos*, 15(1 and 2): 69–73.

Prasad, B.N. and Mehrotra, R.K., 1980. Blue-green algae of paddy fields of Uttar Pradesh. *Phykos*, 19(1): 121–128.

Prasad, B.N. and Kumari, Vijay, 1976. Observations on the cytology of Basicladia sp. *Phykos*, 15(1 and 2): 1–4.

Prasad, S.N., Ramchandra, T.V., Ahalya, N., Sengupta, T., Kumar, A., Tiwari, A.K., Vijayan, V.S. and Vijayan, L., 2002. Conservation of wetlands of India: A review. *Trop. Ecol.*, 43(1): 173–186.

Prescott, G.W., 1951. *Algae of the Western Great Lakes Area*. Cranbook Inst. of Science, Bloomfield Hills, Mich, 946pp.

Pundhir, H.S., Gautam, A. and Vidyavati, 1994. Some Charophytes from Aligarh District, Uttar Pradesh. *Phykos*, 33(1 and 2): 1–9.

Ramakant and Pandey, D.C., 1985. Observations on *Nitellopsis obtuse* (Desv. in Lois.) J. Groves. from Allahabad. *Phykos*, 24: 95–97.

Ramjee and Bhatnagar, S.K., 1978. Studies on Charophytes from Rohilkhand Division–I, Moradabad: Taxonomic enumeration and chromosome counts. *Phykos*, 17(1 and 2): 87–92.

Ramnathan, K.R., 1964. *Ulotrichales*. ICAR Publication, New Delhi.

Randhawa, M.S., 1959. *Zygnemaceae*. ICAR Publication, New Delhi.

Rao, C.B., 1936. The Myxophyceae of the United Provinces. India–II. *Proc. Indian Acad. Sci.*, B. 3(2): 165–174.

Rao, C.B., 1937a. A new species of Anabaena (*Anabaena ambigua sp.nov*). *Proc. Indian Acad. Sci.*, B. 5(3): 101–108.

Rao, C.B., 1937b. The Myxophyceae of the United Provinces. India–III. *Proc. Indian Acad. Sci.*, B. 6: 339–375.

Rao, C.B., 1938. The Myxophyceae of the Orissa Province, India. *Proc. Indian Acad. Sci.*, B. 8: 157–170.

Rao, C.S., 1940. The Myxophyceae of the Delhi Province, India–I. *Proc. Indian Acad. Sci.*, B. 11 (3): 125–131.

Rao, S.V.R., Rao, M.S., Singh, V.P. and Mall, L.P., 1981. Algal diversity in relation to water quality in an Indian River. *International Journal of Environmental Studies*, 16(2): 101–106.

Ratha, S.K., Naik, K. and Padhi, S.B., 2003. Epiphytic algal diversity associated with different aquatic macrophytes of freshwater ponds in and around Berhampur University Campus, Orissa. *Nature Env. Poll. Techn.*, 2(2): 205–208.

Sahai, R., Saxena, P.K. and Jabeen, S., 1985. Ecological survey of the algal flora of polluted habitats of Gorakhpur. *Phykos*, 24: 4–11.

Sahoo, D.B., 2006. Basic to applied phycology, role of algae in biotechnology, environment and rural development: Prospect and problems. In: *Abstract–International Conference on Applied Phycology,"Algae in Biotechnology and Environment."*

Sampathkumar, S., 2004. Algal diversity as indicators of pollution in a freshwater pond at Tiruvannamalai. *J. Ecobiol.*, 16(3): 207–212.

Sengar, R.M.S. and Sharma, K.D., 1982. Algal flora of Yamuna River at Agra. Chlorococcales. *Phykos*, 21: 164–165.

Shastri, Y. and Pendse, D.C., 2001. Hydrobiological study of Dahikhuta reservoir. *J. Environ. Biol.*, 22(1): 67–70.

Shukla, B.K. and Shukla, A.C., 1988. A contribution to Cyanophyceae of Kanpur. *Phykos*, 27: 206–210.

Shukla, H.M., Habib, I. and Pandey, U.C., 1988. Chaetophorales of Bareilly–I. *Phykos*, 27: 218–220.

Singh, N.B. and Pandey, D.C., 1986. A new form of *Compsopogon aeruginosa* (J.Ag.) Kuetzing. *Phykos*, 25: 84–87.

Singh, R.N., 1942. The fixation of elementary nitrogen by some of the commonest blue-green algae from paddy field soils of United Provinces and Bihar. *Indian J. Agric. Sci.*, 12: 743–756.

Singh, R.N., 1939. An investigation into the algal flora of paddy field soils of the United Provinces I. *Indian J. Agric. Sci.*, 9: 55–77.

Singh, S.P., Tiwari, G.L. and Pandey, D.C., 1981. Some interesting algal forms from Allahabad. *Phykos*, 20(1 and 2): 85–90.

Singh, V.P., 1959. Phytoplankton ecology of the inland waters of Uttar Pradesh. In: *Proc. Symp. Algol.*, (Ed.) P. Kachroo. ICAR, New Delhi, pp. 243–271.

Singh, Y., 2001. Note on Cyanobacterial flora of Rampur and its adjoining area of U.P. *Phykos*, 40(1 and 2): 75–78.

Smith, G.M., 1933. *Freshwater Algae of the United States*. New York, pp. 716.

Stoyneva, M.P. and Gaertner, G., 2004. Taxonomic and ecological notes to the list of green algal species from Bulgarian Thermomineral waters. Berichte-des-Naturwissenschaftlich Medizinischen-Vereins-in-Innsbruck, 91: 67–89.

Suseela, M.R., 2005. Freshwater algal diversity in Garuda taal, Nainital. *J. Bombay Nat. Hist. Soc.*, 102(2): 176–180.

Tiwari, A. and Chauhan, S.V.S., 2001. First report of *Sphaeroplea annulina* (Roth) Ag. from Agra. *Phykos*, 40(1 and 2): 29–31.

Tiwari, A., Upadhyay, R. and Chauhan, S.V.S., 2001. A systematic account of Chlorococcales from Kitham lake, Agra. *Phykos*, 40(1 and 2): 103–105.

Tiwari, D., Patrick, J.M. and Singh, S., 2001. Algal dynamics of the river Ganga at Kanpur. *Phykos*, 40(1 and 2): 45–51.

Tiwari, D., Singh, S. and Kaur, P., 2001. Algal Periphyton of the river Pandu. *Phykos*, 40(1 and 2): 39–43.

Tiwari, O.N., Dhar, D.W., Tiwari, G.L. and Singh, P.K., 2001. Non-heterocystous filamentous Cyanobacteria from rice fields of Uttar Pradesh, India. *Phykos*, 40(1 and 2): 61–64.

Tsarenko, P.M., 2002. The flora of Chlorococcales of Ukraine. *International J. Algae*, 4(2): 1–15.

Venkataraman, G.S., 1957. The algal flora of the ponds and puddles inside the Banaras Hindu University grounds area. *J. Bombay Nat. Hist. Soc.*, 54(4): 908–919.

Yadav, R.N. and Pandey, D.C., 1980. Observations on a new variety of *Compsopogon* (*C. aeruginosus* J. Ag. Kuetzing var. *catenatum* var. *nov.*). *Phykos*, 19(1): 15–22.

Chapter 5

Algae and Aquatic Flora in Relation to Water Pollutants

R.N. Yadav and S. Prasad

Department of Botany, R.L.S.Y. College, Bettiah – 845 438, Bihar

Introduction

The water pollution in India has reached at the climax, which is now subject of debate in almost all the viable institutions. Considerable research and development are being undertaken to assess the gravity of the situation and to develop remedial measures. In warning system to prevent adverse effects of pollutants on aquatic ecosystems are of immense importance.

Aquatic flora occupies an important position in the food chain, being the primary producers of food and energy, besides helping to maintain dissolved oxygen levels. Many aquatic plant species serve as sensitive indicators of environmental pollution and its effects. Biological monitoring is gaining ground in recent years (Yadav and Bilgrami, 1995). Apart from this significance in Biological monitoring plants can also serve as test system for ecotoxicological impact and evaluation of pollution (Patrick, 1972). Some of the species also help to maintain water quality by trapping pollutants, as in the case of heavy metals (Prosi, 1981 and Yadav, 1987). Ecosystem eutrophication due to highly nutritive effluents can also be assessed using aquatic flora (Godrej, 1982). Aquatic species like algae offer promising result in effluent treatment, biodegradation and conversion into useful substances if it is used biotechnologically.

Considerable information is available on air pollutants and terrestrial flora, but the information on water pollutants and aquatic flora is not adequate. Even in the monographs on ecotoxicology (Moriarity, 1983; Ulrick and Pankvalh 1982; Connell and Millar, 1984) which describe the overall impact of pollution, aquatic flora are not covered properly. In this review paper an attempt has been made to synthesize the work done by several workers using aquatic plants like Eichhornia, Salvinia, duck weeds and algae.

Algae

Algae constitute about 70 per cent primary productivity in aquatic ecosystem. Being primary producer they are severely affected by the pollutants which are carried into the sediments and also into the food chains and which are also responsible for their biological transformation. From an ecological point of view, the feature of the ecosystem associated with the energy utilization, nutrient cycling and size of energy reservoir within the system, if affected by the pollutants at any one point of the food chain level can affect the other organisms and overall natural balance.

The biomass decreases from the primary producer level through the primary and secondary consumer levels to the end consumer (Odum, 1971). In the aquatic ecosystem the biomass concentration at the lowest trophic level (primary producers) can bind a large part of the xenobiotics in water to the organic substances with regard to the pollutants enrichment in the food chain. The short life span of phytoplankton such as diatoms, dinoflagellates and a variety of forms from other groups of the plant kingdom constitute the base of food chain in water. The whole food chain operative in the system get affected by variety of pollutants as a result alters the number and physiology of algae.

Algae as Indicators

Algae serve an ideal test system in the study of ecological impact of pollutants and toxicity evaluation (Joubert, 1983; Van Collie *et al.*, 1981 and Yadava, 1988). Various workers have made use of algae as specific indicators (Claesson, 1984) of water pollution, so as to determine the presence of concentration of pollutants in the indigenous species (Payer *et al.*, 1975; Dunn, 1979). Metal pollution having lead and cadmium origin were determined using colonies of *Scenedesmus obliquos* as monitoring system in river basins in Thailand and Germany (Payer *et al.*, 1979). However (Bilgrami and Datta Munshi (1985) have enlisted 51 algal species from river Ganga which may be used as indicators of water quality. The pollution susceptibility of most macroalgae clearly enables them to be associated with local concentration of pollutants. Therefore, among the primary producers they are better suited as indicator organisms than the phytoplankton (Prosi, 1981). Macroalgae have larger life span thus having a larger period of exposure in the process of monitoring.

Cyanophycean algae capable of fixing N_2 is predominant in polluted water along with a few species of Chlorophyta (Walsh, 1978). A change in species diversity of phytoplankton is the first visible effect of pollution in aquatic ecosystem (Sheikh, 1981). Nitrogent fixation by cyanophytes in natural plankton associated with pollution by insecticides was siginificantly reduced (Wartsbaugh and Apperson, 1978).

Biomonitoring through Algae

Biomonitoring is now regarded as a convenient and reliable approach for evaluating the water quality (Cairns, 1970, Warren, 1971 and Palmer, 1980). Evaluation of toxicity based on algal culture (EPA, 1975, HMSO, 1982) are of paramount importance in ecotoxicological protocols. Algal species diversity indices have been employed by some workers in the recent past to assess the heal of water. Periphytonic algae such as *Stigeoclonium, Cladophora, Schizomeris* and *Chaetofora* are most suitable algae to be used for biomonitoring programme for flowing water.

Effects of Pollution on Algae

Influence of pollutants on photosynthesis and other aspects of energy utilization and incorpotation causing changes in population growth rates is well established (Walsh, 1978). Sodergren and Gelin (1983) in their study of natural oligotrophic and eutrophic lakes reported polychlorinated biphenyls inhibited 14 C uptake by phytoplankton isolated.

Mercury in the esturies, when associated with organic matter of humic acid and fulvic acid types affects the uptake and toxicity of phytoplankton (Andren, 1973).The role of these chelating agents in field conditions, in the toxicity of a large number of pollutants has been reported (Benes *et al*, 1976, Perdue *et al.*, 1976; Handrickson *et al.*, 1974). Formation of chelates implies that the amount of dissolved matter may, in some cases, regulate toxicity of metals because chelation decreases toxicity to avery great extent. As it is observed that copper dichloride pyridinate inhibited growth of *Scenedesmus quadricauda* at 0.4 mg/lt. Copper chelate were much less toxic (Lomonosov, 1969). Norin and Waern (1973) reported a very interesting case of *Cladophora glomerata*, where sewage effluents induced dwarf forms of the filamentous macro phyte by early onset of the reproduction and gamet formation resulting in a population of smaller plants with shorted generation time.

Many studies on algal species and specific pollutants have been published most have described the effects upon polpulation growth. A reduction in algal cell number was caused by low levels (0.3 to 10 ppm) of polychlorinated biphenyls (Keel *et al.*, 1972) and nickel ions (Spencer and Nichols, 1983). Statistically significant effect on cellularity of *Scenedesmus quadricauda* in the presence of *cadmium* was manifested even at as low conc. as (6.1 ppb). Mandelle (1969) reported that the growth of selected species of dinoflagellates and diatoms was inhibited by copper between 0.055 and 0.256 mg/ml. Some chemicals can affect the photosynthesis in alagae (Erickson and Hawkins, 1980). Overnel (1975) showed that light induced oxygen evolution from the fresh water species *Chlamydomonas reinhardii* was very sensitive to cadmium, methyl mercury and lead. A significantly low rate of oxygen evolution and inhibited respiration rate in *Chlorella pyrenoidosa* and reduction in batch photosynthesis when the rate of photosynthesis per cell is unchanged (Fisher, 1975), also indicate phycotoxicity. Organochlorine compounds (Moore, 1973) and alkyl benzene sulphonates reduced the utilization of bicarbonate for fixation of carbon by micro and macro algae. Holdrich (1981) has reported 50 per cent inhibition of CO_2 fixation by 1 ppb heavy metals pollutants. An increase in photosynthetic CO_2 fixation was noticed due to nutrient concentration.

In algae the altered metabolism brought about by uptake and toxicity of pollutants has also been extensively stidied. Reddy and Rao (1982) while studying Zinc smelter effluent, found that along with the dose dependenat inhibition of photosynthesis and respiration the acid phosphate and amylase activilty were completely inhibited in *Synechococcus aeruginosus* and green alga *Scenedesmus incrassatulus* while other metabolic products were reduced. The effluent concentration between 0.01 and 0.1 per cent promoted the growth of algae along with increase in pigments and carbohydrated while protein contents decreased severely in *Synechococcus spp.* A study with salts of heavy metals inhibited amylase activity even at low concentration in *Anabaena sp.* in vitro study, while the salt like sodium chloride and calcium nitrate were not harmful. However, (DeFillips *et al.*, 1981) observed effect of Zn, cd and Hg which inhibit ATP formation in turn could bring about metabolic changes. Such changes in ATP levels were also recorded in *Chlorella vulgaris* by Truhaut and his team (1980).

Lowering of protein content (Edwald *et al.*, 1976, Saroja and Bose, 1982) reduced nucleic acid synthesis (Ewald *et al.*, 1976) and carbohydrate are well established in various species of green alga. Effect of varied pollutants on photosynthesis of algae is well established. Reduction in carotene and chlorophyll pigments especially chlorophyll-a by various groups of pollutants, has been reported (Saroja and Bose, 1982).

Pollutants may affect morphology of algae. Dalmotte (1980) while studying the effect of cadmium on nitrogen metabolism on *Anabaena cylindrica* observed cellular malformation, chlorosis and significant increase in heterocyst frequency, a possible explanation could be resistance to survive through the stress by cyst formation. However, work done by Pandey and Mitra, on the role of critical nitrogen concentration for the formation of heterocyst is well quoted by Fog and Stewart. Thomas *et al.*, (1980) studied the effect of a number of heavy metals on a set of algae and found that cadmium, mercury and lead interfered with the cell division in *Thalossiosiva aestibalis* and caused disruption in cell separation and clumping. However, (Yadava and Pandey, 1980) observed the effect of colchioine on the genus *Crucigenia fenestrata* which resulted morphological changes.

Algal Bioaccumulation of Pollutants

An ecological approach to the living world is based on the synthesis and systematic analysis of the functioning of natural ecosystems. Biologically, these natural systems are successions of organization levels, each of which is more complex than the preceding one. The contamination of aquatic systems by the process involved in bio-accumulation and transfer of contaminants within the food chain starts at the primary producer level. These contaminants or pollutants may not always destroy the algae but may concentrate on or in them and thus affect other organisms higher up in the food chain (Mechean, 1957). Intake of heavy metals takes place in algae by absorbtion or adsorption to the cell wall. Fujita *et al.*,(1977) reported incorporation of mercury in close association with chloroplast and soluble fractions in fresh water diatoms *Synedra ulna*. However, numerous publication report uptake of pollutants by phytoplankton like *Microcystis*, *Scenedesmus*, *Chlorella*, and *Chlamydomonas*. *Chlorella*, an unicellular green alga is known to concentrate a large amount of metals like Cu, Mn, Cd and Zn. Their absorbtion is rapid in the beginning and not affected by pH, temperature and other metabolic inhibitors (Sakaguchi *et al.*, 1981).

Uptake and accumulation of substances is most important aspect of phytoplankton pollution dynamics. Since they are grazed upon by many animals, the phytoplanktons may be a vital link in the transfer of pollutants from water to higher trophic levels as evident by the high mortality rate of *Crassostrea virginiea* (larveae) fed upon the alga grown in heavy metal containing media (Wilkors and Ukeles, 1982). Several experimental systems with ecological food chains have been developed and it has been established that algae can accumulate larger quantities of heavy metals per unit weight than is the case of organisms in higher trophic levels (Aubert *et al.*, 1972). Field and experimental studies on the food chains are numerous (Lay *et al.*, 1984). Comprising of a wide variety of food chains components and measure not only the bioaccumulation along the trophic chains but also their productivity and energy production (Riedel and Christensen, 1979).

Many experimental studies have shown an increase in concentration of toxic substances in different trophic level of food chain (Philips *et al.*, 1980). This process of biomagnification is inevitable in case of the molecules with great ability to penetrate the biological barriers and easy to be stored in the organism. Biomagnification of pesticides and mercury to methyl mercury by algae is well documented (Moriaridy, 1983).

Biodegradation of Pollutants by Algae

Aquatic microflora has a remarkable potential to biodegrade the pollutants entering into their systems. Swisher (1970) reported that various species could degrade alkyl benzene sulphonate, linear alkyl benzene sulphonate and other detergents. This work is supported by the findings of other workers Gladhil (1974), Nava and Carmela (1979) and Davis and Gloyna (1969). Zepp and Schlotzhauer (1983) have studied the influence of algae on photolysis rates of a list of organic pollutants by green alga *Chlamydomonas*. It has been established that green and blue green algae accelerated transformation and degradation of certain polycyclic aromatic hydrocarbons, organophosphorus compounds and aniline in water by the sunlight. This algae help transformation of toxic chemicals which may have a significant effect on their concentration in water column. Thus this algae could help in controlling the pollution of the aquatic systems.

Algal Blooms

For constant chemical and biological composition of a water body, the rate of cell production must atleast approximate the rate of cell decay. An equilibrium between net photosynthesis and net respiration is prerequisite to maintain an oligotrophic lake (Varshney, 1983), with addition of pollutants, the lake water becomes rich in some chemicals, accumulate nutrients and thus gradually increases its productivity. This sudden increase in the rate of photosynthesis, causes accumulation of algae where as if respiration is higher than cell production, the biological oxygen demand (BOD) may exceed developing an anaerobic condition. In majority of cases of nutrient accumulation, nitrogen and phosphorous are the main nutrient species. Dramatic response by algae to nutrient input includes growth of both prokaryotic and eukaryotic algae. Each phosphorus atom promotes the incorporation of 7 atoms of nitrogen and 40 of carbon in aquatic algae (Watzel, 1975). The eutrophication of natural lake, in response to phosphate detergents (Godrej, 1982) gets full support from experimental model.

The species diversity index is lower in an eutrophic lake as compared to oligotrophic waters. The blue green algae like *Oscillatoria, Microcystis, Anabaena* along with certain species of diatoms attain a dominance due to over nutrient level of the aquatic system. The competitive advantage to one or two species which outnumber other species, periodically develop very high population levels resulting in algal blooms. The blooms may cause unusually severe problems of taste and odour, filter and screen clogging, and sline accumulation, some may be toxic and all may cause fish pills, when large numbers of the algae die at about the same time. In decomposition, they utilize oxygen in water, thus depriving the fish below the essential elements.

Aquatic Pteridophytes

The environmental significance of pteridophytes has been recognized only recently (Devi, 1980). Most of studies such as bracken poisoning and spore allergy are with terrestrial species. However, information on aquatic species is scanty. Aquatic pteridophytes have several salient features as compared to terrestrial species. The interest of environmentalists in aquatic ferns ecology arises mainly from the infestation of large tracts of water bodies in tropical regions all over the world by *Salvinia molesta* (Mitchell *et al.*, 1980). Another water fern that has attracted considerable attention, *i.e. Azolla* in view of its symbiotic association with *Anabaena*.

Effect of Water Pollutants on Aquatic Ferns

Concentration of pH above 10ppm in the culture medium was found to reduce the number of fronds in *Salvinia natans*. In the case of Zn, even 1 ppm caused reduction in number and size of fronds. With both metals growth effect was accompanied by metal accumulation Puckett and Burton, 1981). Cadmium was even more toxic to salvinia, even 0.1 ppm causing growth retardation and lethality (Hutchinson and Czyrska, 1975). Silver at 0.5 ppm or above, inhibited growth of *Salvinia* without accumulation of the metal (Hutchinson and Czyrska, 1975). Chandra, (1980) reported a marked increase in stomatal frequency in lower epidermis of *Pteris vittata* and *Adiantum capillus* growing in the area of polluted sewer.

Ferns as Biological Monitoring Agent

Comparatively very little use has been made of terrestrial ferns as biological monitoring of pollution (Volkman, 1978) as compared to bryophytes, Gymnosperm and Angiosperms. Infact aquatic pteridophytes have not yet been fully explored (Puckett and Burton, 1981). However, Ray and White (1979) suggested *Equisetum* as an indicator for heavy metal pollution of aquatic habitat. Experimentally it is proved that Cu, Zn, Cd, and Pb got accumulated from water in plant, especially in the rhizome.

Aquatic Angiosperms as Indicators of Pollution

Aquatic vascular plants have been used as pollution indicators, since they can accumulate metals PCB's and other pollutants from the water (Ray and White, 1976). *Lemna minor, Eichhornia spp., Utricularia vulgaris, Myriyophyllum exaebescens Nuphar* spp., *Potamogeton* and many other macrophytes have extensively been used as indicators of water quality in culture and in field conditions (Aulio and Salin, 1982).

Pollutants Effect on Aquatic Vascular Plants

Pollutants occur in surface water affect the aquatic plant communities altering physiology, metabolism, growth rate and disrupting chloroplast structure and chlorophyll content etc. indicating that vascular plants are sensitive to pollutants. Duck weeds offer a good system for such studies. Stunting of growth of *Lemna paucicostata* by heavy metals (Nasu *et al.*, 1980), a reduction in the frond count of *Spirodella polyrrhiza* (King and Coley 1984). *Elodea densa* plants exposed to low levels of methyl mercury in water accumulate substantial amount of methyl mercury in 25 days which damage the surface membrane and internal structure of the leaf chloroplast Mortimer and Czuba, (1982). In a similar study using polychlorinated biphenyls in *Spirodella* colonies, Mahanty and Fineran (1976) reported a complete internal disorganisation of chloroplast. In an aquarium system, calcium deficient condition was created incorporating macrophyte species like *Elodea canadensis, Ranunculus panicillatus* and *Myriyophyllum* spp.; were brought in contact with different boron compounds in as low as 2mg/l (which could be commonly found in common water bodies, *Myriyophyllum* species proved to be more resitent where as increased amount of boron caused significantly low photosynthesis by it. Pentachlorophenols inhibited photosynthetic O_2 production, causing no effect on dark respirational consumption of O_2 in *Lemna* minor (Huber *et al.*, 1982).

Bioaccumulation of Pollutants by Aquatic Angiosperms

Aquatic angiosperms are very good accumulator of metals and pollutants from their environment is well established. Stenner and Nickless (1975) found a high degree of metal contamination in higher water plants as compared to all other organisms. Jones and Estes (1984) working in a laboratory culture of attached plant *Potomogeton* came to the conclusion that pollutants absorbed in soil are less available to the plants as compared to that dissolved in the water in which the plant is growing. The extent and rate of accumulation are greatly affected by a number of physical and chemical factors like light, pH, of the water, solubility, movement and flow of the water and chelation of the metal compound (Engrad 1982). Several workers have shown that aquatic plants take up chemicals from the water are numerous (Chigbo *et al.*, 1982; Nakada *et al.*, 1979 and Wolvoton and Mc Donald, 1978). Differences in heavy metal accumulation in various plant organs of single plant species were found by Heydt (1977). *Ipomoaea aquatilis* growing near caustic soda factory had accumulation in the leaves and floating stems (Suckcharoen, 1978). Mange *et al.*, (1978) reported that high concentrations of chromium is toxic to duck weeds. The metal accumulation took place both in living and non-living tissue of *Spirodella polyrhiza*. This extensive power of aquatic plants to accumulate the chemicals, has been used for food chain studies and biomagnifications (Amirad and Foul, 1978). Toxicity and bioassays testing is being done by Rowe *et al.*,, 1982, EPA, 1975).

Water Plants as Pollution Depleter

The characteristics of extensive power of accumulation, these plants may serve an ideal for treating both domestic and chemical waste waters. Many countries in the world are using water hyacinth and duck weeds (*Spirodella, Lemna* and *Wolffia sp*) for many years to upgrade waste water treatment (Wolverton and Mc Donald, 1979). Water hyacinth *Eichhornia cressipes* has been most extensively used for biological treatment of polluted water (Dinges 1978, Duffer and Moyer, 1978). The controlled use of this plant not only increases the BOD removal capacity of these system but also reduces the suspended solids normally associated with sewage lagoons. By harvesting the hyacinth biomass one can minimize the potassium, calcium, sulfur and other minerals from the system. The water hyacinth can readily absorb and concentrate heavy metals such as lead, cadmium, mercury and nickle (Wolverton, 1975).

References

Amirad, T.C. and Foul, Q.L., 1978. Methods of contamination of two fresh water trophic chains by cobalt. 60: I: Direct contamination of organisms by water. *Water, Air and Soil Poll.*, pp. 475–485.

Andren, A.W., 1973. The geochemistry of mercury in three estuaries from the gulf of Mexico. *Ph.D. dissertation,* University of Florida, Gainsville, pp.140.

Aulio, K. and Salin, M., 1982. Enrichment of copper, zinc manganese and iron in five species of pond weeds (*Potamogeton* spp.). *Bull. Environ. Contam. Toxicol.*, 29: 320–325.

Benes, P., Gjessing, E.T. and Stunnes, E., 1976. Interaction between humus and trace elements in fresh waters. *Water Res.*, 10: 711–716.

Bilgrami, K.S. and Munshi, J.S. Datta, 1985. Ecology of river Ganges (Patna-Farakka): Impact of human activities. Final Technical Reported (MAB) submitted to DOEn Govt of India, Allied Press Bhagalpur, pp. 97.

Cains, J.R. and Cains, R.K., 1983. The effect of selected herbicides on Zygospore germination and growth of green alga *Chlamydomonas moewusii* (Chlorophyceae, Volvocales). *J. Phycol.*, 19: 301–305.

Chandra, P., 1980. Toxic metals in the environment and pollution effect on *Pteris vittata* and *Adiantum capillus* veneris. *Ind. J. Air Poll. Control,* 3: 27–30.

Chigbe, F.E., Smith, R.W. and Shore, F.L., 1982. Uptake of arsenic, cadmium, lead and mercury from polluted waters hycinth. *Eichhornia cvessipes, Environ. Poll.* (A). 27: 31–36.

Clark, J.R., Hasse, J.H.V., Nicholson, R.B., Cherry, D.S. and Cairns, J.R., 1981. Accumulation and depresation of metals by *Lemna perpusilla. Ecotoxicol. Environ. Safety,* 5: 87–96.

Connell, F.A., Zoltek, J.J.R., Patrinely, C.D., Furman, T. and Des Kim, J.I., 1477. Nutrient removal by water hyacinth. *J. Water Polln. Contr. Rd.*, 49: 57–65.

Daltmotte, A., 1980. Influence of cadmium on growth and nitrogen metabolism of *Anabaena cylindrica* Lemm. *J. Exp. Bot.*, 31: 1107–1118.

Davis, E.M. and Gloyna, E.F., 1969. Anionic and non-ionic surfactant sorption and degradation by aigai cultures. *J. Am. Oil Chemists. Soc.*, 46: 604–608.

De-Fillips, L.H., Hampp, R. and Ziegler, H., 1981. The effect of sublethal concentration of zinc, cadmium and mercury on *Euglena*, adenylates and energy charge 2, *Pflanzenphysiol.*, 103: 1–7.

Devi, S., 1980. Environmental pollution and cryptogamic plants, especially Pteridophytes. *Ind. J. Air. Poll. Control*, 3: 41–45.

Dinge, R., 1978. Upgrading stabilization of pond effluents by water hyacinth cultures. *J. Water Poll. Cont. Fed.*, 50: 833–845.

Duffer, W.R. and Moyer, J.E., 1978. Municipal Wastewater Aquacultures. *U.S. EPA*, 600/2: 78–110.

Edwards, M.E. and Miller, J.H., 1972. Growth regulations by ethylene in fern gametophytes. III. Inhibition of spore germination. *Am. J. Bot.*, 59: 458–465.

Engrad, P., 1982. Approach in ecotoxicologie experimentale des processus de bioaccumulation as derives dus mercure Pare les vegetaux aquatiques DEA Ecotoxicology. Univ. of Bordeaux. Talenace, France.

EPA, 1975. Test methods for assessing the effects of chemicals on plants. (Eds.) R. Rubinstein and J. Smith. EPA–560–17–75–008, Final report, U.S. Environmental Protection Agency, Washington, D.C.

Erickson, S.J. and Hawkins, C.E., 1980. Effect of hologenated organic compound on photosynthesis in esturine phytoplankton. *Bull. Environ. Contam. Toxicology*, 24: 910–915.

Ewald, W.G., French, J.E. and Champ, M.A., 1976. Toxicity of PCB to *Euglena gracilis* cell population growth, carbon fixation, chlorophyll level, oxygen consumption and protein and nucleic acid synthesis. *Bull. Environ. Contam. Toxicol.*, 16: 71–80.

Fishner, N.S., 1975. Chlorinated hydrocarbon pollutants and photosynthesis of main phytoplankton: A reassessment. *Science*, New York, 189: 434–464.

Fujta, M., Iwasaki, K. and Tapabatake, E., 1977. Intra cellular distribution of mercury in fresh water diatoms (*Synendra*). *Cells. Environ. Res.*, 14: 1–13.

Gladhill, W.E., 1974. Linear alkylbenzene sulphonate biodegradation and aquatic interaction. *Adv. Appl. Microbiol.*, 17: 265–293.

Godrej, P.J., 1982. The eutrophication of lake Tahoe emphasizing water quality. U.S. Environmental Protection Agency, Ecological Research Ser. EPA–660/3–74034, corvallis oreggon, XVII pp. 408.

Handrickson, D.W., Armstrong, D.E., Veith, G.D. and Glass, G.E., 1974. Nature of organic derivatives of selected toxic metals in natural waters. Univ. Wisconsin, Water Resources Centre. Tech. Rep. Wis. WRC, 74–07, III pp. 23.

HMSO, 1982. Methods for determination of ecotoxicity. Approved code of Practise No.8.

Holdrich, D.M., 1981. Biological effects in industrial effluent treatment. Vol. I. In: *Water and Solid Waters,* (Eds.) J.K. Waters and A. Wint. Appl. Sc. Publishers, London.

Hutchinson, T.C. and Czyrska, H., 1975. Heavy metal toxicity and synergisms to floating aquatic weeds. *Verli Intern. Verein Liminol.,* 19: 2102–2111.

Jones, T.W. and Estes, P.S., 1984. Uptake and Phytotoxicity of soil-sorbed Atrazine for submerged aquatic plant, *Potamogeton perfoliatus. Arch. Environ. Toxic.,* 13: 237–242.

Joubert, G., 1983. Detailed method for quantitative toxicity measurements using the green alga *Selenastrum capricornatum.* In: *Aquatic Toxicity,* (Ed.) J.O. Nriagu. John Wiley and Sons, New York, pp. 467–485.

Keel, J.E., Preister, L.E. and Sandiffer, S.H., 1972. DDT and Polychlorinated biphenyl arclor (1242) effects of uptake on *E. coli* growth. *Water Res.,* 6: 837–841.

King, J.M. and Coley, K.S., 1984. Toxicity of coal distillates to *Spirodella polyrrhiza. Bull. Environ. Contam. Toxicol.,* 23: 220–224.

Lay, J.P., Schaurte, W., Klien, W. and Korte, F., 1984. Influence of tetrachloroethylene on biota of aquatic system: Toxicity to phyto and zooplankton species in compartments of natural pool. *Arch. Environ. Contam Toxicol.,* 13: 135–142.

Lomonosov, M.W., 1969. The toxicity of copper complexes towards *Scenedesmus quadricauda. Mikrobiologiya,* 38: 729–731.

Mahanty, H.K. and Fineran, B.A., 1976. Effects of PCBs on ultrastructure of frond cells in aquatic plant *Spirodella oligortiza* Heglum. *N.Z.J. Bot.,* 14: 13–18.

Mandelle, E.E., 1969. The inhibitory effect of copper on marine phytoplankton. *Contrib. Mar. Sci.,* 14: 47–59.

Mange, J.H., Schmidt, J.P., Gaines, L and Turner, P., 1978. Effect of chromium on some aquatic plants. *Environ. Poll.* 16: 285–291.

Mitchell, D.S., Peter and Viner, A.B., 1980. The water fern *Salvinia molesta* in Sepik river. Papua New Guinea. *Environ. Conservation,* 7: 115–122.

Moore, S.A. Jr., 1973. Impact of pesticides on phytoplankton in Everglades estuaries South Florida. *Environmental Project, Ecological Resp.* No. 74–15, pp. 100.

Mortimer, D.C. and Czuba, M., 1982. Structural damage to leaf chlorophyll of *Elodea densa,* water plant caused by methyl mercury accumulated from water. *Ecotoxicol. Environ. Safety,* 6: 193–195.

Nakada, M., Kukaya, K., Takeshita, S., and Wada, Y., 1979. The accumulation of heavy metals in submerged plant *Elodea muttali. Bull. Environ. Contam. Toxicol.,* 22: 21–77.

Nasu, Y., Kugimoto, M., Tanaka, O. and Tarimoto, A., 1980. Comparative studies on the absorbtion of cadmium and copper in *Lemma paucicostata. Environ. Poll.,* 33: 201–210.

Norin, L.L. and Waenr, M., 1973. The zone of low algal standing crop near Stockholm. *Oikos*, 15: 179–184.

Odum, E.P., 1971. *Fundamentals of Ecology*. Philadelphia Saunders Company.

Overnell, J., 1975. The effect of heavy metals on photosynthesis and loss of cell potassium in two species of marine alage. *Mar. Biol.*, 29: 99–103.

Palmer, C.M., 1980. *Algae and Water Pollution*. Castle House Publication.

Patrick, R., 1972. Aquatic communities as indices of pollution. In: *Indicator of Environmental Quality*, (Ed.) W.A. Thomas. Plenum Press, pp. 93–100.

Payer, H.D., Sodder, C.J., Kunte, H., Karuwanna, P. Nonhof, R. and Gaf, W., 1975. Accumulation of polycyclic aromatic hydrocarbons in cultivated algae, *Naturwissensch Aften.*, 62: 536–540.

Perdue, E.M., Beck, K.C. and Reuter, J.H., 1976. Organic complexes of iron and aluminium in natural waters. *Nature*, 260: 418–420.

Phillips, G.R., Lenhart, T.E. and Gregory, R.W., 1980. Relation between trophic position and mercury accumulation among fishes from the Tongere River reservoir Montana. *Environ Res.*, 22: 73–80.

Prosi, F., 1981., Heavy metals in aquatic organisms. In: *Metal Pollution in the Aquatic Environment*, (Eds.) U. Forstner and G.T.W. Wittaman. Springer Verlag, Berlin, pp. 271–323.

Puckett, K.J. and Burten, M.A.S., 1981. The effect of trace metals on lower plants. In: *Effect of Heavy Metal Pollution on Plants*, (Ed.) N.W. Zepp. Applied Science Publisher, 2: 213–238.

Ray, S. and White, W.J., 1976., Selected aquatic plants as indicator for heavy metal pollution. *J. Environ. Sci, Hlth.*, 11: 717–725.

Ray, S. and White W.J., 1979. *Equisetum arvense* an aquatic vascular plant as a biological monitoring for heavy metal pollution. *Chemosphere*, 3: 125–128.

Reddy, T.R.K. and Rao, J.C., 1982. Toxicity of industrial effluents to physiological responses of algae. *Life Sci., Advances*, 1: 219–228.

Riedel, B. and Christensen., G., 1979. Effect of selected water toxicants and other chemicals upon AT Pase activity *in vitro*. *Bull. Environ. Contam. Toxic.*, 23: 365–368.

Rowe, E.L., Ziobre, R.J., Wang, C.J.K. and Dence, C.W., 1982. The use of an alga *Chlorella pyrenoidosa* and duck weed *Lemna perpusilla* as a test organisms for toxicity bioassay of spent bleaching liquors and their components. *Environ. Poll.*, 27: 289–296.

Sakaguchi, T., Nakajma, A. and Horikoshi, T., 1981. Studies on the accumulation of heavy metal elements in biological system XVIII. Accumulation of Molybdenum by green micro algae. *Eur. J. Appl. Microbiol. Biotech.*, 12: 84–89.

Saroja, G. and Bose, S., 1982. Efects of methyl parathion on the growth, cell size, pigment and protein content of *Chlorella protothecoides*. *Env. Poll.*, 27: 297–308.

Shiekh, K.H., 1985. The quality and some ecological effects of waste water effluents from two chemical industrial plants in Pakistan. *Environ. Conser.*, 8: 229–234.

Sodergren, A. and Gelin, C., 1983. Effect of PCBs on the rate of C-14 uptake in phytoplankton isolates from oligotrophic and eutrophic lakes. *Bull. Environ. Contam. Toxicol.*; 30: 191–198.

Spencer, D.F. and Nichols, L.H., 1983. Free nickel ions inhibit growth of two species of green algae. *Environ. Poll.*, 21: 97–104.

Suckcharoen, S., 1978. Mercury accumulation in *Ipomea aquatica* near a caustic soda factory in Thailand. *Water Air Poll.*, pp. 451–455.

Thomas, K.J., 1981. The role of aquatic weeds in changing the pattern of ecosystem in Kerala. *Environ. Conser.*, 8: 63–66.

Truhaut, R., Ferard, J.J.F. and Jouany, J.M., 1980. Cadmium IC 50 determination on *Chlorella vulgaris* involving different parameters. *Ecotoxicol. Environ. Safety*, 4: 223.

Ulrick, B. and Pankhvalh, J., 1982. *Effects of Accumulation of Air Pollution in Forest Ecosystem*. D. Reidel Publish Co.

Van Coillie, R., Visser, S.A. and Couture, P., 1981. Utilization de bioessais avec des algues pour letude des repurcussion lices la mise en cau des reservoirs. *Ann. Limnol.*, 17: 79–91.

Varshney, C.K., 1983. *Water Pollution and Management*. Wiley Eastern Limited.

Volkman, W., 1978. Man alters the cryptogamic flora. Stultgbeitr. Naturkd Ser. C., 5: 29–35.

Walsh, G.E., 1978. Toxic effects of pollutants on planktons. In: *Principles of Ecotoxicology*, (Ed.) G.C. Butler. John Wiley and Sons, pp. 257–274.

Ware, G.W. and Roan, C.C., 1970. Interaction of pesticides with aquatic microorganisms and plankton. *Residue Rev.*, 33: 15–45.

Wart Sbaugh, W.A. and Apperson, C.S., 1978. Effect of mosquito control insecticides on nitrogen fixation and growth of blue algae in natural plankton association. *Bull. Environ. Contam. Toxicol.*, 19: 640–647.

Watzel, R.G., 1975. *Limnology*. W.B. Saunders, Philadelphia.

Wilkfors, C.H. and Ukeles, R., 1982. Growth and adaptations of estuarine unicellular algae in media with excess copper, cadmium or zinc and effects of metal contaminated algal food on *Crassostrea virginia* larveae. *Marine Ecology, Progress Series*, 7: 191–206.

Wolverton, B.C., 1975. Water hyacinth for removal of cadmium and nickel from polluted water. *NASA. Tech. Mem.*, T.M.X 72721.

Wolverton, B.C. and McDonald, R.C., 1978. Water hyacinth sorption rates of lead, mercury and cadmium. *NASA, ERL*, Report No. 170.

Wolverton, B.C. and McDonald, R.C., 1980. Nutritional composition of water hyacinth grown in domestic sewage. *Economic Botany*, 34: 101–110.

Yadava, R.N., 1988. Monitoring of heavy metal pollution with the help of algae. Final Technical Report submitted to D.S.T. Govt. of India, pp. 87.

Yadava, R.N. and Bilgrami, K.S., 1995. *Monitoring of River Ganga through Biological Devices*. Kargupta and Siddiqui Pub.

Yadava, R.N. and Pandey, D.C., 1980. Studies on the morphology of the genus *Crucigenia fenestrata* Schmidle in cultural conditions and effect of colchicine on the thallus. *Bibliotheca*, Germany, 1.

Żepp, R.G. and Schlotzhauer, P.F., 1983. Influence of algae on photolysis rates of chemicals in water. *Environ. Sci. Technol.*, 17: 8–9.

Chapter 6

Chilika Lake and its Biodiversity

J. Rath[1] *and S.P. Adhikary*[2]

[1]*Department of Botany, Vishwa Bharati, Shantiniketan, W.B.*
[2]*P.G. Department of Biotechnology,*
Utkal University, Bhubaneswar – 751 004

ABSTRACT

Chilika is the largest brackish water lagoon in Asia covering an average area of 1065 km^2. The lagoon is an estuarine one and supports an unique assemblage of marine, brackish water and fresh water species. A distinct salinity gradient is observed in its different sectors which also varies with season. It is one of the hotspots of biodiversity in the country. Some rare, vulnerable and endangered species listed in the IUCN Red list of threatened species show at least a part of their life cycle in the lake. In Chilika 706 species of angiospermic plants, 102 species of algae and phytoplankton, 217 species of fishes, 61 species of Protozoa, 29 species of Platyhelminthes, 37 species of parasitic Nematodes, 31 species of Polychaetes, 3 species of Stomatopoda, 28 species of Crustacea, 30 species of Decapoda, 136 species of Mollusca, 3 species of Echicurans, one species of Protochordata, 37 species of amphibians and reptiles, 156 species of birds and 118 species of mammals were recorded. Details of this unique ecosystem and its biodiversity has been presented.

Location

Chilika is the largest brackish water lagoon situated in the east coast of India between 19°28'-19°54' N and 85° 06'–35° E covering the districts of Puri, Khurda and Ganjam of Orissa (Figure 6.1).

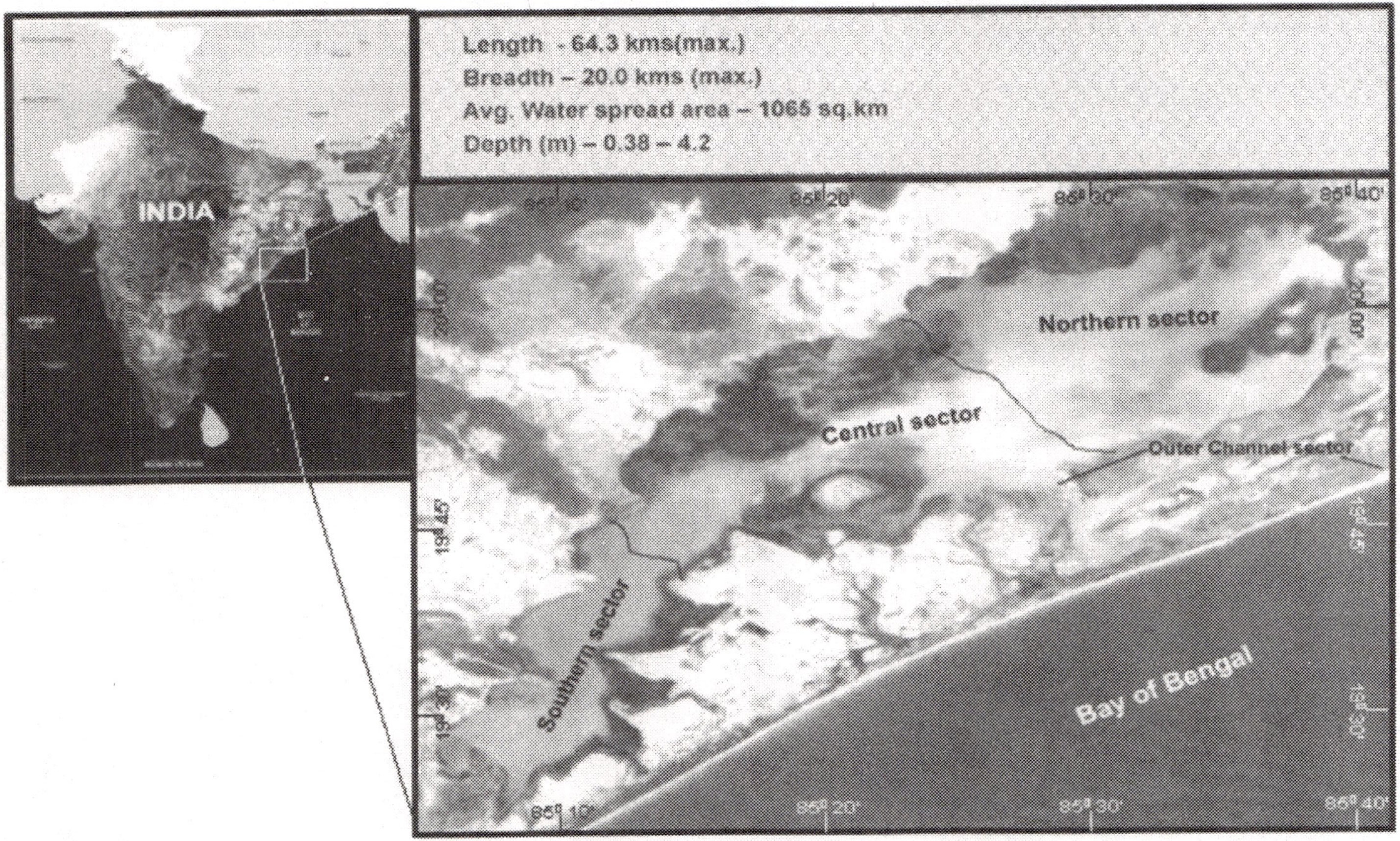

Figure 6.1: Location Map of Chilika Lagoon in the East Coast of India. The artificial separation of the lake into four different sectors has been indicated in the map.

Introductory Description of the Lake

The pear shaped lagoon has a maximum length of 64.3 km and average width of 20 km (Das and Samal, 1988). The water-spread area varies between 906 to 1065 km^2 during summer to monsoon and the water depth in general fluctuates between 0.38 to 4.2 meter. Geographically the lake is circumscribed by (i) rocky hills of eastern ghat along the western and southern margin (ii) a sixty km stretch of coastal barrier along Bay of Bengal in the eastern side, and (iii) the deltic plains of Mahanadi river in the northern part. Hammed in between the mountains and the sea, Chilika spreads itself out into a pear-shaped expanse of water. From about the middle line to the south, enormous rocks and emerald green islands jutting out of the water surface. Of these Badakuda, Sanakuda, Nalabana, Kalijai, Ghantasila, Chadheiguha, Arakhakuda and Kankadakuda etc. deserve special mention. It is connected with Bay of Bengal on its eastern side through an outlet, a 35 km long, narrow, outer channel connects the main lagoon to the Bay of Bengal near the village Motto, which cuts through liner spit that separates the lake from the sea. High tide near this inlet mouth drives in salt water through the channel during the dry months, from December to June. With onset of the rain, the rivers falls into the Northern zone are in spate, causing fresh water currents that gradually push the sea water out. As a result of these dynamics, the inlet mouth constantly changes its position. The inlet channel is connected with Chilika at Magarmukh. The other connection with the Bay of Bengal is through Palur canal on the South-eastern side. Because of the peculiar hydrology of the lake, salinity ranges and gradients are not uniform but vary with site and season. During monsoon the salinity is the lowest and highest salinity observed in the late winter and summer. The varying salinity regime and depth divides the lake into four natural sectors, *viz.* Northern sector, Central sector, Southern sector, and Outer channel sector. The rivers Daya and Bhargavi fall into Chilika in the Northern sector, which supply major fresh water input to the lake during monsoon and post monsoon period which contributes about 275,000 cusecs of fresh water to the Northern sector. This zone is relatively shallow (0.5 to 1 m) due to sediment deposits carried by these two rivers and has a salinity ranging between 0 to 10 ppt. During the monsoons (July- September), the river Daya is in spate and the water level of the lake is the highest with an average depth of 3 meters. In this sector current generated by the river out flow is directed towards the sea. The Central sector is deeper than the Northern sector with medium salinity level 5 to 20 ppt and a depth of about 1.5 to 2.5 m. The Southern sector is the deepest of about 2.5 to 3.5 m and has the highest salinity concentration with less fluctuation of salinity from 10 to 20 ppt. The river Rushikullya falls in to the Bay of Bengal close to the Southern sector while the Palur canal connects it to the sea. The fourth zone is the main mouth of Chilika lake along with the 35 km long channel called the Outer channel (salinity up to 32 ppt). It starts from the lake at Magarmukh near village Satapada, and is connected to the sea near Arakhakuda. During the monsoon season, excess fresh water input from the rivers is discharged to the sea, while during the summer (April-June) when the water level is low, seawater flows into the lake. This unique spatial and temporal salinity gradient of Chilika lake gives rise to a multitude of niches inhabited by a large diversity of plant and animal species hence is one of the hotspot of biodiversity of the country. The lagoon is the life line for more than 1.5 lakh stake holders who live in and around the lagoon and rely upon the lake's supply of fish stock and other allied activities. Based on its rich biodiversity and socio-economic importance, Chilika has been designated as a wetland of International Importance and included as one of the first two wetlands under the Ramsar convention in 1981. It is included in the list of wetlands selected for intensive conservation and management by the Ministry of Environment and Forest, Govt. of India and also identified as a priority site for conservation and management by the National wetland, mangrove and coral reef committee of the Ministry.

Origin and History of Chilika Lake

Chilika was a gulf of the Bay of Bengal during the Pleistocene period. The recession of Coastline in the Post-Pleistocene period. The recession of coastline in the Post-Pleistocene, aided by marine and fluvial activities gradually resulted in the present configuration. Where as the wave action initiated the spit-building at the southern end, the delta-building process of the Mahanadi ensured sufficient sediment load the render. The lagoon is shallow at its north-eastern extremity (Venkatratnam, 1970). Many ancient texts mention the southern sector of Chilika as being a major harbor for maritime commerce back in the days when the kings of Kalinga was known as lord of the sea. This indicates that the area was once marine and is much deeper than it is today. Geological studies indicates the coastline added along the western shores of Chilika in the Oceane era, and that the entire northeastern region of Chilika was under the sea. Since then the coastline has moved considerably east wards (Chatterjee and Goswami, 1966).

Most of the lagoons seen today were formed as a result of a world wide rise in sea levels. There was pause in the rise in the sea as about 700 years ago, when a sandy beach might have been formed near the coast at the southern sector. As the sea rose further, this sand beach grew gradually and progressed seawards to the north east, to form what now the spit of Chilika. The partial closure of an estuary outlet to the sea possibly created a lagoon system Chilika. Recently a sub-fossil shell *Ostrea virginiana* from the southern-western edge of the spit has been dated to about 4000 years ago, which indicates that how long ago the lagoon was formed (Venkatratnam, 1970). Growth of the spit at Chilika is supposed to be due to abrupt change in the direction of coastline, long shore drift, constructive wave action, flat coastal gradient and the presence or absence of strong river and tidal currents in the different areas. The islands adjacent to the spit are believed to have been built partly by waves and partly by tidal currents as tidal deltas (Venkatratnam, 1966).

Catchment Area and Drainage Basin of Chilika

The total catchment area of Chilika is 3560 km^2 of which western catchment area of Chilika is 321189 ha out of which 144439 ha lies in Ganjam and 276752 ha in Khurda district. The drainage basin of Chilika covers an area of over 4300 sq. km (ORSAC, 1988), including about 1100 sq. km of the lake it self. The watershed boundaries lie between water flowing into the Mahanadi and Chilika in the north, while in the northeast area draining into the Bhargavi river make up the watershed. In the west and southwest, the watershed boundary lies between streams flowing into the Rishikulya river and those flowing into Chilika.

The catchment area of Chilika is quite heavily populated for the most part, and the population is about 2,00,000 during 2000 AD. About 718 sq. km is covered by forests (526 sq. km) and other permanent vegetation like plantation, however, 58 per cent of the vegetation is regarded as degraded. Agricultural area (mostly dry land) accounts for 2325 sq. km, swamps 71 sq. km and wetland with grassy mud flats 91 sq km. occur mainly in the northeastern part of the catchment area. About 52 sq. km are occupied by human settlements, roads, railways etc. (Venkatratnam, 1966).

Physiographic Features

Water, its quality, quantity and distribution over time is the single most important factor affecting Chilika. A significant part of the freshwater and silt input to the lake comes from Mahanadi river system. The Daya and Bhargavi, which are distributaries of the Mahanadi, bring vast quantities of silt-laden water into the northern part of Chilika during the monsoon. In addition, the rivers Nuna, Ratnachira and Kani in the north and Malaguni, Dhanua and Salia and numerous smaller streams enter the lake all along the western margin (ORSAC, 1988). The total silt added to the lake has been estimated to be 13 million tones every year. The lake shows an overall increase in depth of about 50 cm due to the monsoon (Sahu, 1988) and freshwater flows towards the sea, overcoming the weak tidal currents at the lake mouth. The lake drainage basin is estimated to send 1760 million cubic meters into the lake every year, while the Daya and Bhargavi rivers are believed to contribute a much larger quantity of freshwater annually.

Climate

Chilika's location on the coast in the tropical zone spares it from extreme temperatures. The mean annual temperature is about 26.6-27.7°C, with a winter minimum of 21°C and a summer maximum of about 35°C (Patnaik, 1971). It rains about 1,450 mm per year at Chilika, and the rainfall over the area varies from 1,200-1,600mm. Most of the rain (80 per cent) occurs during the monsoon season between June- September. The lake area proper experiences mean wind speed ranging from 5 kmph (winter) and 10-15 kmph in the summer. However, the coastal areas experience higher average wind speed up to 25 kmph.

Salinity Fluctuation in Chilika Lagoon

Chilika undergoes a cyclical variation in salinity throughout the year, with different patterns seen in different sectors. It is this periodicity in salinity that allows freshwater as well as marine and estuarine species to thrive in the lake.

During monsoons (July to September) River Daya is in spate and water level in the lake is at its peak (about 3m depth). In the Northern sector the main currents are generated by the river in flow and directed towards the sea. They are strong enough to overcome tidal currents in the mouth. Therefore during monsoon, the Northern and Central sectors show low salinity (1.8-2.0 ppt) and in the mouth region 3.5 ppt. In the southern sector, salinity remains higher (6-10 ppt) as the freshwater input is much lower in this part of the lake. (Figure 6.2). In the Post-monsoons (October to December) the salinity remains low in the Northern sector (1.8–4.5), increases slightly

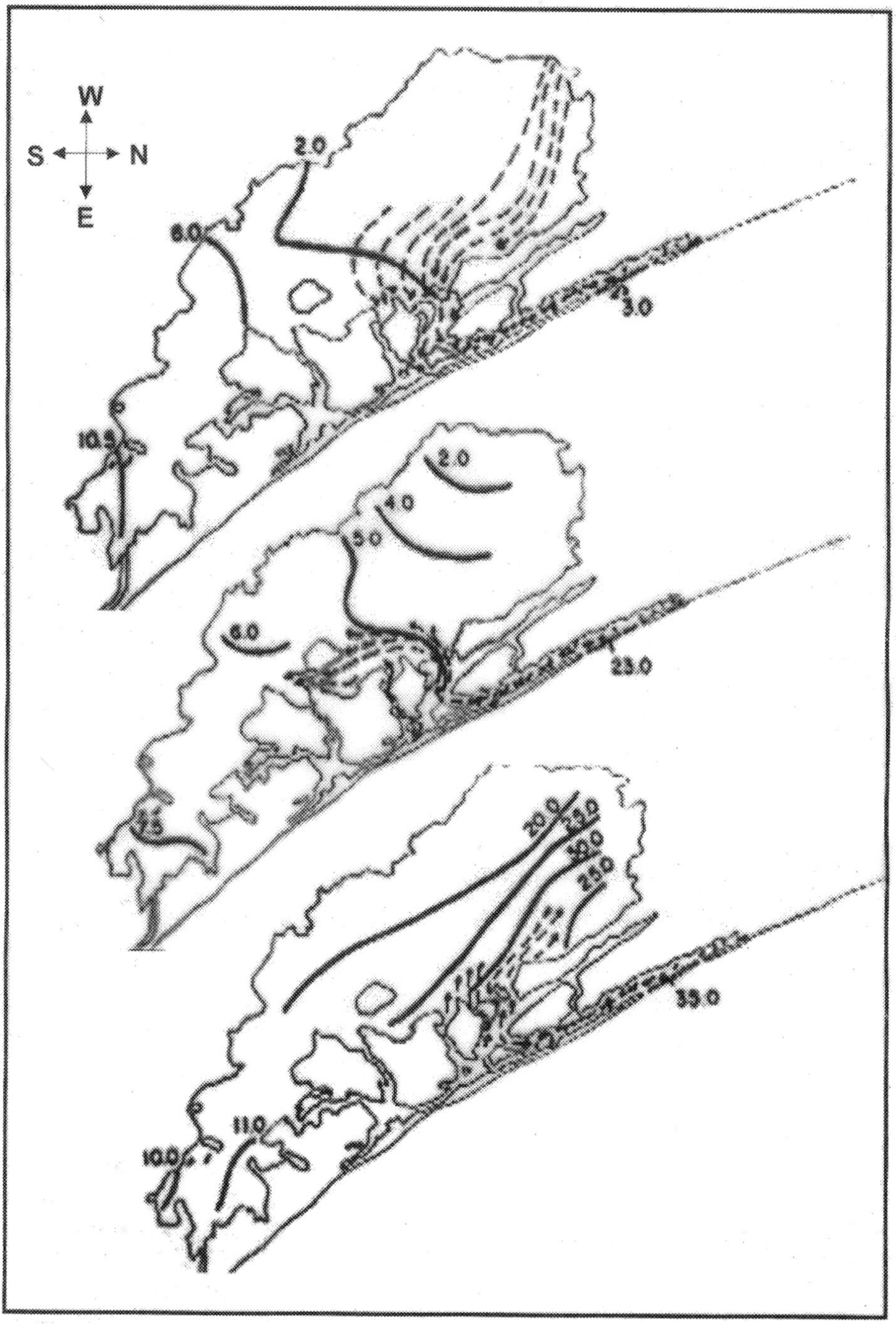

Figure 6.2: Surface Salinity Profile of Chilika Lake in Different Seasons A. Summer (January–March), B. Rainy (April–June), C. Winter (July–September). ——— = Surface salinity per cent, - - - - - - = Direction of water current.

in the central sector and in the mouth (5 ppt) but decreases in the Southern sector (7–8 ppt) as a result of slow mixing with the rest of the lake (Ramanadhan *et al.*, 1964; Banerjee and Roychaudhuri, 1966). The Outer channel shows the lowest salinity as freshwater from the body of the lake continues to flow through it (Mohanty, 1975).

During winter (January–March), almost no fresh water enters the lake. Northerly wind send tidal water in to the Outer channel which increases the salinity. In the body of the lake, a fairly small range of low salinity exists, ranging from 2-4 ppt (Northern sector), 4-6 ppt (Central sector) and 6-7.5 ppt (Southern sector).

During the summer, evaporation and prevailing southern winds cause greater salinity all over the lake. The Northern and Central sectors turns mesohaline with salinity ranging from 5.5–33 ppt and 7.5–27 ppt respectively. The Southern sector shows a slight increase (10–11 ppt) (Ramanadhan *et al.*, 1964; Banerjee and Roychaudhuri, 1966). The Outer channel is practically like seawater at this point, with salinity at 34–35 ppt (Mohanty, 1975).

Due to increased silt load, narrowing of the lake mouth and other factors, the overall average salinity for the whole lake has dropped from 22.31 ppt in 1957–1958 to 13.2 ppt in 1960–1961, and to 9.14–11.83 ppt in 1961–1964, but appears to have stabilized after the rapid drop (Sarkar, 1977). Such a decrease in salinity would obviously have grave implications for the biodiversity as well as fisheries of Chilika. Recently a new mouth was created opposite to village Sipakuda along the Outer channel and a lead channel of 200 meter wide near Magarmukh over a length of 3.2 km. After the desalination of the Outer channel as well as the opening of the new mouth increase in salinity flux by 40 per cent and tidal flux by 45 per cent into the lagoon was observed.

Biodiversity of Chilika Lake

Chilika is one of the hotspots of biodiversity in India. It's partly fresh and partly saline water character and presence of a variety of habitats in and around the lake encourage a large number of plant and animal species in the lake.

Flora

Aquatic Vegetation

Emergent Macrophytes

Emergent plants such as *Scripus littoralis, Eleocharis plantaginea* and *Cyperus* sp. grow almost throughout the year on the north and northwest margins of the lake (Patnaik, 1973a). The most conspicuous marsh plant is *Phragmites karka* which grows mainly in the north western margins.

Floating Macrophytes

The free-floating plants *Eichhornia crassipes* (water hyacinth), *Pistia stratiotes* and *Azolla pinnata* come along with flood waters in the monsoon. They remain in small patches in the eastern part of the northern sector and die out in the February-March as salinity increases (Patnaik, 1973a). Other floating plants of seasonal occurrence, seen on the north western margins during November-March are *Hydrilla verticillata,*

Vallisneria spiralis, Ceratophyllum demersum and *Polygonum* sp., *Ipomeo* sp., *Jussieua* sp. and *Marselia* sp. (Sahu, 1988).

Submerged Macrophytes

Two submerged plants dominate Chilika lake namely *Potamogeton pectinatus* and *Najas indica. Potamogeton pectinatus* locally know as Charidal. This plant tolerates a wide range of salinity. Charidal made up 78 per cent of the total weed cover of the lake in 1973. From a position of virtual absence in 1932, an area of 20 sq. km was covered by Charidal Scripus in 1973 (Patnaik, 1973a). This had increased to about 60 sq. km by 1977, 100 sq. km in 1980, up to 180 sq. km in 1985 (Figure 6.3) and still further by 2002. Charidal dominated areas made up about 125 sq.km (Chatrath, 1992). Although weeds play an important food sources for certain species and support higher concentration of benthic fauna, this unchecked spread of weeds can be disastrous for Chilika.

The other dominant submerged plant is *Najas indica,* which was estimated to form 14 per cent of the weed cover in 1973 (Patnaik, 1973a). Locally known as Chinguria the plant grows in patches mixed with Charidal *(P. pectinatus).* Other submerged plants seen in Chilika include *Halophila ovata* (7 per cent of weed cover, locally known as Kamari, *H. beccarii* and *Ruppia maritime* (Patnaik, 1973a).

While macrophytes dominate large parts of the Northern and Southern sector, the Outer Channel and deeper parts of the Central sector are relatively free of them. According to satellite image weed coverage areas in 1973, 1977, 1985 and 1993 are 20, 60, 200 and 368 sq.km. respectively.

Non-aquatic Vegetation

Depending on the habitat conditions the non-aquatic vegetation of Chilika can be described under two categories. Littoral Scrub Jungles and the Vegetations on the Sand dunes.

Littoral Scrub Jungles

The mangroves which were once plentiful towards the margin of the lake and its uninhabited islands like Badakuda, Sanakuda, Cherriakuda and Ghantasila etc are no longer found due to ruthless cutting. The only vegetation now exist are the littoral scrub jungals along the rocky faces, representing the remnants of one time dominant tidal swamp forests. The vegetation is completely devoid of arboreous species of mangroves through some of their associates like *Aegiceras corniculatum, Salvadora persica, Pongamia pinnata, Cassipourea ceylanica* etc could be observed in denuded condition giving the appearance of low shrubs. Other important species constituting this type of vegetation are *Azima tetracantha, Colubrina asiatica, Carmona retusa, Carissa spinarum, Opuntia dillenii* etc. climbers/twiners are fairly common like *Cissus quadrangularis, Pentatropis capensis, Capparis roxburghii, Capparis zeylanica, Derris trifoliate* are quite abundant. *Crateva adansoni, Lepisanthes tetraphylla,* species of *Ficus* and *Streblus asper* are the other tree species found.

The ground flora composed of a large number of herbaceous members like *Cressa cretica, Indigofera asplathoides, Heliotropium curassavicum* forms the vegetation in such habitats. *Solarium trilobatum* is fairly common among bushes and scrub jungles (Panda and Patnaik, 1988).

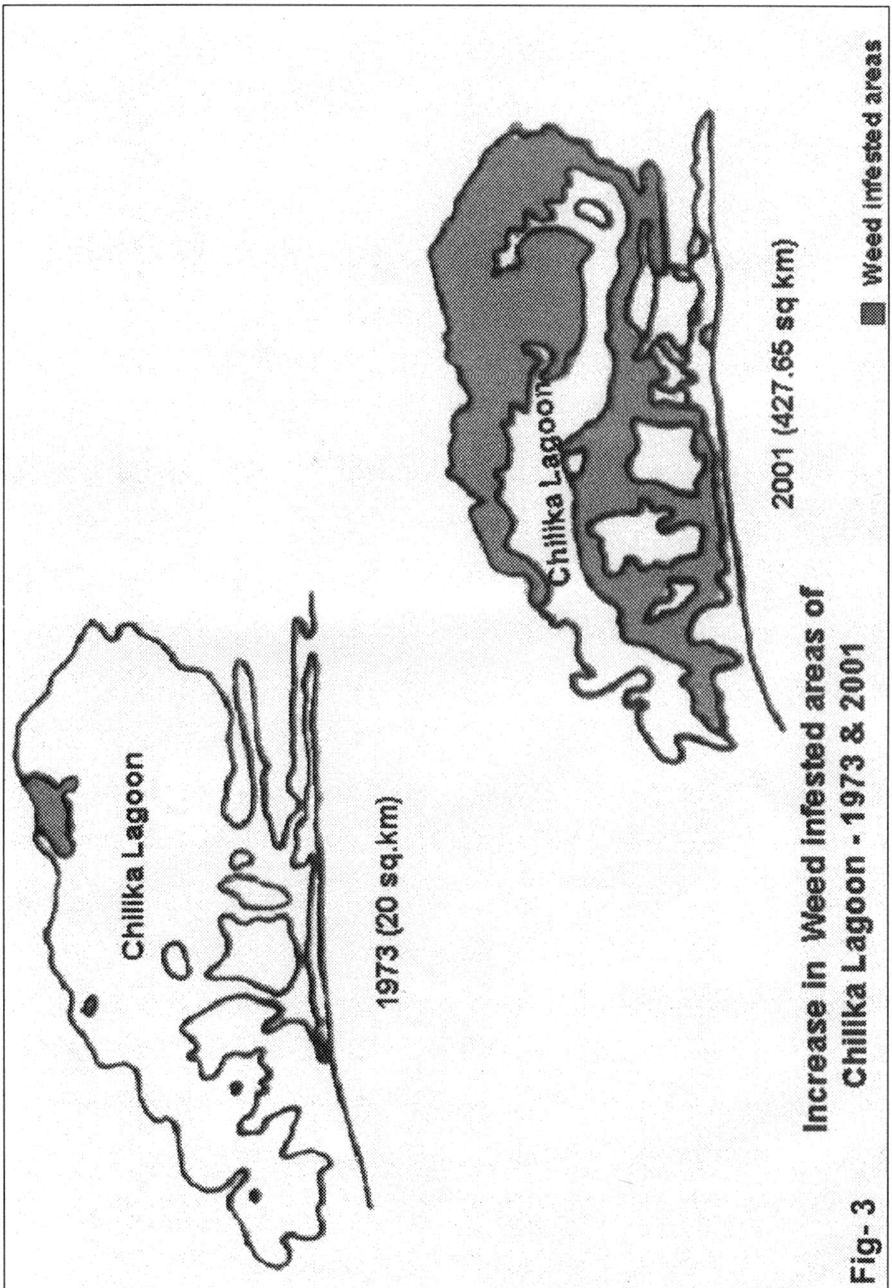

Figure 6.3: Map Showing Weed Infested Areas of Chilika Lake during 1973 and 2001

Vegetation on Sand Dunes

The scorching sun, shifting sands and salt-laden winds on the beaches, sandy Islands and the spit support a variety of plant including broad-leaved runners, herbs, shrubs, trees and reeds and sedges along the water's edge. *Hydrophylax maritime, Ipomoea pes-caprae* are some of the broad leaved runners on the dunes. Other dominant herbaceous elements constitute *Cyperus arenarium, Bulbostylis barbata, Hedyotis graminifolia, Becopa monnieri, Crotalaria nana, Gisekia pharnaceoides* etc. *Pandanus odoratissimum* attains very large dimension due to congenial edaphic factor and is found in a continuous belt along the fringes of the lake. A stretch of land lying in between Bay of Bengal and Chilika lake has been planted with *Casuarina equisetifolia* which exhibit luxuriant growth and adequate natural regeneration. Most of these plants found are salt-tolerant. The edges of the island have several marshy plants. Extensive flowering occurs at different times of the year, and the island appears to change color with the flower (Panigrahi, 1988).

Algal Flora

The algal flora of Chilika lake and its distribution play a very important role in the brackish water environment. Through the number of forms recorded were few, their distribution in space and time was greatly influenced by the peculiar hydrographic conditions of the lake. The algae found in Chilika include benthic algae and those associated with aquatic weeds. Twelve species of Cyanophyta, twenty three species of Chlorophyta and four species of Rhodophyta have been reported to constitute the benthic algal population (Rath and Adhikary, 2002). *Chaetomorpha linum* is the most dominant and widely distributed algae in the lake followed by *Enteromorpha intestinalis* and *Gracilaria verrucosa.*

Northern sector is the largest and shallowest and showed great fluctuation of salinity (0.31 to 12.24 per cent) because of river discharge, and that's why some red algae like *Polysiphonia, Ceramium* and *Gracilaria* were not growing in this sector but found in rocky islands of southern sector. However forms like *Chara, Nitella, Anabaena, Spirulina* etc have been recorded from comparatively freshwater regions of Northern sectors. In the Central sector the rocks and hillocks which remain submerged had good growth of algae like *Gracilaria verrucosa, Ceramium* sp., *Enteromorpha intestinalis, E. compressa* and *Chaetomorpha linum.* The Nalabana Island of this sector which gets submerged during monsoon develop good quantity of algae like *Spirogyra, Enteromorpha* and *Chaetomorpha.*

The Southern sector was comparatively small and deep and showed the least salinity variation. The transparency was comparatively high and highest algal density was obtained during January to March. In this sector *Lyngbya aestuarii* was dominant throughout except during October to December when *Chaetomorpha linum* was the dominant. The rocky islands like Badakuda, Sanakuda, Bird's island, Somola etc of this sector had luxuriant growth of *Enteromorpha, Ceramium, Polysiphonia* and *Gracilaria* on the submerged rocks.

The Outer channel sector being the link between the sea and the main lake, tidal effect was felt and during monsoon period flood water flushed out to the sea through this Channel. The algal flora in general is comparatively poor because of constant to and fro flow of water and fluctuation in salinity. However in the sheltered areas of islands, there was good growth of *Enteromorpha, Chaetomorpha, Gracilaria* and *Lyngbya* species.

Phytoptankton of Chilika lake consisted of freshwater, brackish water and marine forms which appear and disappear on the salinity conditions. The phytoplankton population in Chilika lake was constituted mainly by diatoms, blue-green algae, green algae and dinoflagellates. The phytoplankton in the lake as a whole showed two peak seasons. The principal one being from March to June (summer). The second peak season of short duration in November (winter) was possibly due to rapid vegetative multiplication consequent on rise of salinity and richness of nutrient in the water after monsoon (Jhingran, 1963; Patnaik, 1973b). Out of the total 89 species of phytoplankton found in the lake, 47 species frequently occurred. These belong to the division Cyanophyta, Chlorophyta, Bacillariophyta and Dinophyta, and the numbers of species under these groups were 5, 5, 34 and 3 species respectively. *Cocconis pediculus, Gyrosigma acuminata, Mastigloia minuta, Pinnularia alpina, Pinnularia nobilis* and *Pleurosigma normanii* all belongs to Bacillariophyta and *Lyngbya aesturii* under Cyanophyta group were found occurring in all the sectors of the lake throughout the year. Sector specific distribution of the plankton species was also observed only in the Outer Channel sector of the lake. These species are *Bacterastrum furcatum, Chaetoceros ebenii and Chaetoceros paradoxus* belongs to Bacillariophytes occurred in the Outer channel during summer and winter seasons, however, were absent in the rainy season showing their preference to higher salinity levels. *Bacterastrum furcatum* was especially occurring in higher number in the Outer channel during summer season when the salinity exceeds 30 ppt, however, three other diatoms *Chaetoceros affinis, Coscinodiscus centralis* and *Odentila mobilensis* occurred in maximum number in Outer channel of the lake during summer season. Their biomass in terms of number of cells/liter was 1421, 1372 and 1256, respectively, however, in the other two seasons another Bacillariophyte member *Coscinodiscus centralis* was the most dominant plankton in the sector with its number being 1170 and 520 no/liter during winter and rainy season respectively. Distribution of phytoplankton and their relative occurrence in different sectors showed a wide range of variations in physiochemical features and the habitat preferences of these organisms. In the Southern sector *Coscinodiscus centralis* was also dominant in the summer season, but in the rainy season *Anabaena flos-aquae* and in the winter *Pleurosigma normaniii* were the most dominant species showing the number 314, 221 225 no/liter respectively. In the Central sector also higher number of *Coscinodiscus centralis* was found in the water column in the summer season (285 no/liter) and in the rainy season number of *Anabaena flos-aquae* rose to 1081 no/liter. However, in the winter another diatom species *Thalasionema nitzchiodes* (176 no/liter) was the dominant. In the Northern sector *Anabaena flos-aquae* was the dominant plankton in the rainy season (1684 no/liter). In the summer and winter season the diatom *Cylindrothica clostratum* was the most abundant organism.

Regarding the total biomass of phytoplankton in the Chilika lake it was highest during summer (17,748 no/liter) followed by in winter season (15,005 no/liter) and least in rainy season (14,051 no/liter). With respect to quantity of biomass of phytoplankton in different sectors, their number was highest in Outer channel sector (20,556 no/liter) followed by in the Northern sector (10,194 no/liter), Central sector (8300 no/liter) and lowest in Southern sector of the lake 7754 no/liter. In Southern and Outer channel sector with increase in salinity biomass of phytoplankton increases but in Northern and Central sector with increase in salinity phytoplankton biomass decreases which indicates that in Outer channel and Southern sector marine phytoplankton were dominant and in contrast in Central and Northern sector fresh water phytoplankton were dominant. Though there are few exception of dominance of a few particular species in the respective seasons, only in the Outer channel sector the number of phytoplankton was highest during the summer season. The phytoplanktons which occur in Northern sector were mostly dominated by a blue green alga *Anabaena flos-aquae* and the number of diatom in this sector was quite low. The heterogeneity in the composition of Phytoplankton in different sectors during the same period is explained by the fact that most of the phytoplankton populations in Chilika lake are allochthonous in origin.

Fauna

The pioneering work on the fauna of the lagoon was carried out by Annandale and Kemp (Annandale and Kemp, 1915a; 1915b; 1916; Annandale, 1922; 1924). They have recorded their findings in 13 volumes published in form of memories of Indian museum. This is the first comprehensive report on the faunal resources of Chilika lagoon, which has not been further followed up in systematic manner. Several organizations and individuals have subsequently worked on different groups of plants and animal species, prominent among those include Central Inland Fisheries Research Institute, Zoological Survey of India and academic and research institutes in Orissa and Andhra Pradesh. These investigations have brought out information on various taxonomic groups of plants and animals in and around Chilika lagoon. Choudhury (1916a, 1916b, 1916c, 1917, 1923) and Hora (1923) reported 118 species of fishes. The survey of Zoological Survey of India (1995) recorded 217 species of Fishes, and in addition, 61 species of Protozoa, 29 species of Platyhelminthes, 37 species of parasitic Nematodes, 31 species of Polychaetes, 3 species of Stomatopoda, 28 species of Brachyara Crustacea, 30 species of Decapoda, 136 species of Mollusca, 3 species of Echicurans, one species of Protochordata, 37 species of Amphibians and Reptiles, 156 species of birds and 118 species of Mammals. The diverse fauna of this lagoon can be broadly described as pelagic, benthic and phytal based on their taxonomy, place of occurrence and ecological status. They can be grouped as zooplankton, zoo-benthos, phytal fauna, icthyofauna, avifauna and mammals. The most recent account of fauna of Chilika has been published by Zoological Survey of India (1995) documenting 800 species. Starting from Sarkar (1977), several investigations have been made dealing with the species composition and seasonal distribution of zooplankton in Chilika (Devasundaram and Roy, 1954; Rajan, 1964; Jhingran and Natarajan, 1966; Patnaik, 1986; Sarma *et al.*, 1988). In total 26 faunal groups comprising 170 species have been identified from this lagoon. Copepods constitute the most dominant group having 43-70 per cent of the total zooplankton communities.

Benthic Fauna

Much work has been done on the systematics and ecology of zoo benthos of this lagoon. In total, 117 species of benthos have been identified (Subha Rao *et al.*, 1988). The prominent groups include foraminifers, nematodes, polychaetes, copepods, isopods, amphipods and gastropods.

Phytal Fauna

The macrophytes and seaweeds of the lagoon provide conducive habitats for the growth of many phytal fauna. A total of 77 invertebrates occurring on macro algae and other macrophytes have been identified from the lagoon.

Avifauna

Chilika is known world wide for its resident as well as migratory birds. It falls on the major fly route of the migratory birds. 205 species of birds have been recorded from this lagoon by Bombay Natural History Society (2002) of which 97 are intercontinental migrants coming from Caspian sea, Baikal, Aral Sea, remote parts of Russia, Kirghiz Steppes of Magnolia, Central and South east Asia, Ladakh and the great Himalayas. Some rare birds from Pakistan, Iran and Rann of Kutch also constitute a part of its avian fauna. The more important bird species of the lagoon are *Anser clypeata, Aytha ferina, Anas querquedula, Anas Penelope, Anas fuligula, Anser strepera, Limnodromus semipalmatus* and *Eurynorhynchus pygmeus.* The other interesting and threatened species are Spoonbilled Sandpiper *Eurynorhynchus pygmeus* and the Asian dowitcher *Limnodromus semipalmatus,* Spotbilled Pelican *Pelecanus philippensis,* Dalmatian Pelican *Pelecanus crispus* and Pallas Fishing Eagle *Haliaeetus leucoryphus.* Each year more than one million migratory shore birds, ducks, terns congregate in large numbers at Chilika for feeding and resting.

Mammalian Fauna

The mammalian fauna of the lagoon include aquatic species like Dungong *(Dungong dungong),* Irrawady dolphin *(Orcaella brevirostris)* and many terrestrial ones such as *Antelope cervicapra, Axis axis, Hyena hyena* and *Canius aureus.* Considerable decline has occurred in the population density *of Antelope cervicapra, Axis axis,* and *Dungong dungong* and not spotted for quite some time. The Irrawaddy dolphin is a somewhat elusive species, found in various large rivers, bays and estuaries in South and Southeast Asia. Its IUCN Red data book status is 'undetermined', but suspected to be rare, vulnerable or endangered and it is listed in CITES Appendix II. The status of Irrawady dolphin in Chilika is not yet studied.

Fish Fauna

Fisheries of Chilika have got an immense significance from the socio economic point of view. Over 1.5 lakh fisher folk living in 122 villages in and around the lagoon earn their livelihood primarily from the fish and shellfish harvest. As per the survey by the ZSI-1995, a total of 217 species of fish from 147 genera, 71 families under 15 orders have so far been reported from the lagoon. The faunal composition of finfish has undergone many changes. Jhingran and Natrajan, had identified 158 species, but Subha Rao *et al.* (1988) enlisted 159 species belonging to 83 genera and 48 families.

Biswas (1995) on the other hand, reported 166 species of which 115 were brackish water, 24 limnetic and the rest 27 marine. Out of 166 species, 80 species have been listed as commercially important. The more important species are *Mugil cephalus, Mugil macrolepis, Liza parsia, Lates calcaarifer, Hilsa ilisha, Nematolosa nasusa, Pseudosiena coiber, Eleuthorenima tetradactylus, Mystus gulio, Silago sihama, Anabus testudineus, Chanos chanos, Tachysurus caelatus, Therapon jarbua, Liza troschelli* and *Gerres setifer.* A total of 30 species of prawns have been recorded from the lagoon of which, 6 species *viz. Penaeus monodon, P. semisulcatus, P. indicus, Metapenaeus monoceros, M. affinis* and *M. dobsoni* are considered very important. Of late, occurrence of *Macrobrachium malcolmsonii* has also been reported. The crab population is represented by four species *viz. Scylla serrata, Neptunus pelagicus, Varuna littoralis, Paratelphus* sp. Existence of edible oysters *viz. Ostrea talpus, Meretrix* sp. and *Arca* sp. were also reported but have been wiped out due to the drastic fall in salinity.

Primary Production Rate

Primary production is measured by finding out the total amount of plant matter produced every year. However phytoplankton primary production has not been measured at Chilika. Preliminary studies on the primary production of the lake indicate that the northern sector is much more productive than the rest of the lake (Asthana, 1979; Mishra *et al.*, 1988). Based on the observation at Rambha and Balugaon, Misra *et al.* (1988) reported a bimodal pattern of production with a peak in may and the other in September. The Central sector has comparatively higher benthic algal biomass than the Northern sector probably due to lesser variation in salinity, higher transparency and abundance of sheltered areas.

Threat to the Lagoon Ecosystem

The lagoon is at present facing the problems like–siltation, shrinkage of water spread area, chocking of the inlet as well as the outer channel connecting the sea, decrease in salinity, weed infestation, decrease in fishery resources and an overall loss of biodiversity.

Siltation

Siltation is attributed as most detrimental to the lagoon ecosystem. As per the monitoring done by the Chilika Development Authority recently 3.65 million cubic meters of silt is pumped in the lagoon through the 52 number of rivers and the rivulets. The catchment of the lagoon (excluding the Mahanadi drainage) is estimated to be 3, 21,189 ha. out of which 37 per cent is arable land, 32 per cent is fallow land, 26 per cent is forest land and 5 per cent is waste land.

Shifting of the Mouth and Choking of the Inlet and Outer Channel

Various studies indicate mat due to the littoral drift the mouth of the lagoon is shifting in North Eastern direction. The cross section of the outer channel is also significantly reducing due to shoal formation along the outer channel which leads to considerable hydraulic head loss and poor flushing action. The recent bathymetry of the Magarmukh (the gate-way between Lake and outer channel) indicates that the summer depth is alarmingly low *i.e.*, only 03 meter. This is preventing the discharge of fresh water and sediment during monsoon to sea through outer channel, thereby causing water logging in the adjoining villages in the Northern Sector. This is also reducing the prospect of auto-recruitment of the fish and other animals into the lagoon.

Shrinkage of Water-Spread Area

The water-spread area as calculated from Survey of India Topo-sheet (1972-73 was 824 sq.kms.–(summer). The Land Sat-5 data of 25th May 1986 indicates the water-spread area to be 790 sq.kms. The shrinkage of water-spread from 1973 to 1986 is from 824 to 790 sq.kms. From the latest satellite imagery the water spread area taken during May 2000 was computed to be 745 km^2.

Weed Infestation

Analysis of Satellite data reveal that during the years 1973, 1977, 1985 and 1993 the weed covered area in the lagoon were 20, 60, 200 and 398 Sq.kms respectively. The weeds are emergent, floating and submerged type. The predominant weeds are *Potomogeton pectinatus* mostly found in Central and fringes of Southern Sector. *Shoenoplectus littoralis* is predominant in Northern Sector.

Decrease in Fish-Landing

The species composition of fish in Chilika Lagoon is complex in the sense that it includes fresh water, brackish, catadramous and anadromous fish species. A large percentage of the commercial fish species are migratory (either catadramous or anadromous). Jhingran (1963) estimated that 63-75 per cent of the annual fish production was contributed by migratory species. This percentage is reduced to 30 per cent due to relative increase in fresh water species. A large percentage of the fish species feed off plant material and detritus. Because of this trophic charcteristic the Chilika Lagoon is highly productive as estimated in terms of grams of Carbon fixed on an aerial basis. As per calculation of Biswas, the annual fish production from Chilika average is about 90 Kg/ha during the 1980's (Range–65-122 Kg/ha).

The total fish landing on an average was 6,000 MTS but it is gradually declining. The fish landing of 1999-2000 was 1737 MTS. (Source: Fishery and Animal Resource Development Dept.) This decline can be attributed to composite factors like over fishing beyond carrying capacity, and obstruction of migratory route of economic species (due to chocking of mouth and outer channel as well as at Magarmukh). The Orissa Marine Fisheries Regulation Act prevents operation of off-shore vessels within 5 km of coast as the fish recruitment into Chilika Lake is mainly by spawners which concentrate within 5 kms of restricted fishing zone. There is also provision for restriction of fishing in Palur canal and Outer channel. The decline is also attributed to decrease in salinity level of lagoon.

Notes on the Remarkable Changes of Biota in the Lake

As an estuarine lagoon, Chilika is destined to be converted into a marsh and slowly to dry land over geological time. However, a vastly accelerated rate of eutrophication is being seen in Chilika today caused by human activity. The major environmental changes in Chilika are:

- ☆ Shifting of the Mouth and Sand Casting
- ☆ Silting of the lake
- ☆ Changes in overall salinity concentration of the lake
- ☆ Aquatic weed infestation
- ☆ Downward trend of Fish population
- ☆ Changes in species composition of avifauna
- ☆ Pollution of the lake

Several reptiles whose presence had been reported earlier (Annandale and Kemp, 1915a; Annandale, 1922) were not seen in the ZSI survey from 1985-1987. These include Crocodiles, Sea turtles, Gharials etc which causes concern. Most of these reptiles threatened or endangered in India. Mangroves which were once plentiful found along the margin of the lake and on uninhabited islands, are not found now due to ruthless cutting. The other associated mangroves which are now threaten are-*Cassipourea ceylanica, Derris indica, Aegiceras curniculatum* etc. The agar producing red algae *Gracilaria verrucosa* which occurred in the Southern and Central sectors of the lake until two decades ago is now restricted to the Pathara region of the Central sector and *Gracilaria lichenoides* was not observed in any part of the lake during 1980 (Mohanty, 1980). The total absence of *Gracilaria* sp. in the Northern sector over years was probably because of siltation, comparatively high turbidity and decrease in salinity of the lake.

Institutional Measures for Upgrading Chilika Lake Environment

The Ministry of Environment and Forest is actively involved in conserving and improving Chilika and its environment. They have provided funds for soil conservation efforts in the dranage basin, habitat improvement of Nalaban Island, preparation of environmental education and publicity material on Chilika, and setting up an information database on the lagoon. In addition, the ministry is spending research on various aspects of Chilika under the Man and Biosphere and other program. The ministry suspended the Chilika Aquatic Farms Ltd located at Balugaon. Due to concerns over environmental impacts it is trying to get the lagoon notified under the environmental protection act so that all major developmental activity would require environmental clearance. A Chilika Development Authority was created in 1992 by the Ministry of Environment and Forest to look over all related issues on the lagoon. Recently an Interpretation center at Satapada and a wetland Research center was also established at Balugaon for long term research on different aspects of Chilika lake.

References

Annandale, N. and Kemp, S., 1915a. Fauna of Chilika lake. *Mem Indian Mus.*, 5: 1–118.

Annandale, N. and Kemp, S., 1915b. Fauna of Chilika lake: Aquatic insects other than Coleoptrea with notes on some marginal species. *Mem Indian Mus.*, 5: 175–189.

Annandale, N. and Kemp, S., 1916. Fauna of Chilika lake: Mollusca Gastropoda and Lamellibran Chiata. *Mem Indian Mus.*, 5: 327–366.

Annandale, N., 1924. Fauna of the Chilika lake: Mollusca Gastropoda (revision). *Mem. Indian Mus.*, 5: 853–873.

Annandale, N., 1922. Introduction to the study of the fauna of an island in Chilika lake. *Mem. Asiat. Soc. Bengal*, 7: 257–287.

Asthana, V., 1979. Limnological studies of lake Chilika, Orissa. Final Report. Indian Programm on Man and Biosphere (MAB). Project No. 112. Department of Science and Technology, New Delhi.

Banerjee, A.C. and Roychaudhuri, N.C., 1966. Observations on some physico-chemical features of the Chilika lake. *Indian Journal of Fisheries*, 13(1–2): 359–429.

Biswas, K.P., 1995. *Ecological and Fisheries Development in Wetlands: A Study of Chilika Lagoon*. Daya Publishing House, New Delhi, 205 pp.

Chatrath, K.J.S., 1992. The cxhallenge of Chilika. In: *Wetlands of India* (Ed.) K.J.S. Chatrath. Ashish Publishing House, New Delhi, pp. 135–169.

Chatterjee, A. and Goswami, S., 1966. Puri-Chilika coastal area: A geographic study. *Geographical Review of India*, pp. 18–33.

Choudhury, B.L., 1916a. Description of two new fishes from Chilika lake. *Rec. Indian Mus.*, 12: 105–109.

Choudhury, B.L., 1916b. Fauna of Chilika lake: Fish Part I. *Mem. Indian Mus.*, 5: 403–439.

Choudhury, B.L., 1916c. Fauna of Chilika lake: Fish Part II. *Mem. Indian Mus.*, 5: 441–458.

Choudhury, B.L., 1917. Fauna of Chilika lake: Fish Part III. *Mem. Indian Mus.*, 5: 491–508.

Choudhury, B.L., 1923. Fauna of Chilika lake: Fish Part IV. *Mem. Indian Mus.*, 5: 711–736.

Das, N.K. and Samal, R.C., 1988. Environmental survey of Chilika. In: *Chilika: The Pride of our Wetland Heritage*, (Ed.) S.N. Patro. Orissa Environmental Society, Bhubaneswar, pp. 96–103.

Devasundaram, M.P. and Roy, J.C., 1954. A preliminary study of the plankton of Chilika lake for the year 1950 and 1951. In: *IPFC Proceeding Cymp. Brackishwater Plankton in Indo-Pacific*, Bangkok, 25–26 January, pp. 48–54.

Hora, S.L., 1923. Fauna of Chilika lake: Fish Part V. *Mem. Indian Mus.*, 5: 737–769.

Jhingran, V.G. and Natarajan, A.V., 1966. Final report on the fisheries of the Chilika lake (1957–65). *Bulletin, Central Inland Fisheries Research Institute*, pp. 8.

Jhingran, V.G., 1963. Report on fisheries of Chilika lake 1957–1960. *Bulletin, Central Inland Fisheries Research Institute*, pp. 1.

Mishra, P.M., Sahu, B.K. and Panigrahi, R.C., 1988. Studies in the hydrobiological features of Chilika lake, Orissa. Final Report. Department of Environment, Government of Orissa. Bhubaneswar.

Mohanty, P.R., 1980. Seaweed culture with particular reference to *Gracilaria verrucosa* (Huds) Papenfu in Chilika lake 1–11. In: *Summer Institute on Brackish Water Capture and Fisheries*, Central Inland Fisheries Research Institute (ICAR), Barrackpore, West Bengal, p. 1–11.

Mohanty, S.K., 1975. Some observations on the physicochemical features of the outer channel of the Chilika lake during 1971–73. *Bulletin of the Department of Marine Sciences*, University of Cochin, 7(3): 543–559.

ORSAC, 1988. *Interm Report on Study of Chilika Lake Resources and Environment*. Orissa Remote Sensing Application Center, Bhubaneswar.

Panda, P.C. and Patnaik S.N., 1988. A contribution to the flora and vegetation of Chilika Wildlife Sanctuary, Orissa. In: *Chilika: The Pride of our Wetland Heritage*, (Ed.) S.N. Patro. Orissa Environmental Society, Bhubaneswar, pp. 63–80.

Panigrahi, G., 1988. Vegetation and flora of Chilika lagoon. In: *Chilika: The Pride of our Wetland Heritage*, (Ed.) S.N. Patro. Orissa Environmental Society, Bhubaneswar, pp. 63–80.

Patnaik, P.K., 1986. On the spatial distribution of zooplankton in Chilika lake (Brackish water lagoon). *Geobios News Reports*, 5: 31–34.

Patnaik, S., 1973a. A study of the aquatic plants of Chilika lake. In: *Proceedings of the National Academy of Sciences of India*, 43B(1–2): 53–65.

Patnaik, S., 1973b. Observation on the seasonal fluctuation of plankton in the Chilika lake. *Indian Journal of Fisheries*, 20(1): 43–56.

Patnaik, S., 1971. Seasonal abundance and distribution of bottom fauna of the Chilika lake. *Journal of the Inland Fisheries Society of India*, 8: 38–48.

Rajan, S., 1964. Environmental studies of the Chilika lake: Feeding spectrum of fishes. *Indian Journal of Fisheries*, 2(A): 521–532.

Ramanadhan, R., Reddy, M.P.M and Murty, A.V.S., 1964. Limnology of the Chilika lake. *Journal of the Marine Biological Association of India*, 6(2): 183–201.

Rath, J. and Adhikary, S.P., 2002. Growth and spectral characteristics of four algae and agar yield of *Gracilaria verrucosa* at different salinity gradients of Chilika lake. *Seaweeds Res Utiln.*, 24: 41–46.

Sahu, B.N., 1988. Geomorphology, evolution and recession of Chilika lake. In: *Abstract, Seminar on Conservation and Management of Chilika*. Department of Science, Technology and Environment, Government of Orissa, Bhubaneswar.

Sarkar, S.K., 1977. Further studies on seasonal and spatial variations of salinity in Chilika lake. *Journal of the Inland Fisheries Society of India*, 2(2): 1–9

Sarma, A.L.N., Rao, D.G., Satapathy, S. and Pattanaik, P.K., 1988. Invertebrate community ecology in the context of conservation and development of Chilika lagoon (Bay of Bengal). In: *Paper Presented at the National Seminar on Conservation and Management of Chilika*, Department of Science, Technology and Environment, Government of Orissa, Bhubaneswar.

Subha Rao, M.V., Sekhar, P.S.R., Prasada Rao, P.V.S.V. and Rao, K.K., 1988. Ichthyofauna of Chilika lake. In: *Abstract, National Seminar on Conservation and Management of Chilika*. Department of Science, Technology and Environment, Government of Orissa, Bhubaneswar.

Venkatratnam, K., 1970. Formation of the barrier spit and other sand ridges near Chilika lake on the east coast of India. *Marine Geology*, 9: 101–116.

Venkatratnam, K., 1966. Formation of the spit and the development of the Chilika lake: A coastal lagoon (abstract). *Int. Indian Ocean Exp.*, WEWSI, 4(2).

Chapter 7

Algal Phenolics: Aspects and Prospects with Special Reference to Freshwater Green and Blue-green Algae

A.P. Sahay[1], P.K. Das[2] and Prem Kumar Prasad[3]

[1]*Department of Botany, Rameshwar College, Muzaffarpur – 842 001, Bihar*
[2]*Depatment of Botany, P.R. College, Sonepur – 841 101, Saran, Bihar*
[3]*Principal, B.M. College, Rahika, Madhubani, Bihar*

ABSTRACT

Algal phenolics are one of the secondary metabolites which have not received due attention so far. Existing information about it mainly centre around the phenolics of brown algae. Attempt has been made in this communication to discuss its maximum possible aspects giving emphasis over the phenolics of fresh water green and blue-green algae. Biosynthesis of phenolic acid and its bioconversions, methodologies underlying its extraction, isolation, purifications and identifications have been discussed in detail. Level of phenols reported in green and blue-green algae has been summarized, and phenol producing potency of these groups has been discussed in relation to brown as well as red algae. Functions and uses have been described and its possible role in the blooming of cyanophytes in a eutrophic water bodies have been discussed. Future course of work has been also discussed especially to explore its physiological and biotechnological importance.

Introduction

Phenolics are one of the secondary metabolites synthesized by plants ranging from algae to higher plants. It has been extensively studied in the past few decades largely due to its established roles in metabolic processes, industries, pharmaceuticals, defence mechanism and aging. However, these studies are restricted mostly to the higher plants (Ibrahim and Edgar, 1976; Saleh and El-Lakany, 1979; Rao *et al.*, 1978; Sinha, 1980, Dogra and Sinha, 1983; Cambie *et al.*, 1983; Meravy, 1987; Ambasta and Hamidi, 1988; Scalbert and Haslam, 1987; Scalbert *et al.*, 1988; Ernst *et al.*, 1991) and to some extent to the Phaeophyceae (Haug and Larsen, 1958 a,b,c; Munda, 1960, 1962a,b; Glombitza, 1977; Ragan, 1976, 1984, 1985a,b; Ragan and Craigie, 1976; Glombitza and Sattler, 1973; Joshi and Gowda, 1975: Ragan and Jensen, 1977, 78; Ronnberg and Rockolahti, 1986). Fresh water green and blue-green algae, on the other hand, have received least attention with respect to their phenolics and hence, available informations are sketchy and fragmentary. A good beginning has been initiated in the Phycological Laboratory, B.R.A Bihar University, Muzaffarpur, under the guidance of Dr. B. N. Verma mainly due to Shukla (1989), Sahay (1990), Verma (1991), Kund (1992), Kund and Verma (1994), Sahay *et al.* (1995) Das (1997). The objective of the present communication is to summarize these findings and to review the works so far done in this specific area of study.

Biosyntehsis of Plant Phenolics and its Bioconversions

Tracer techniques have been very useful in tracking the path of phenol biosynthesis in plants, and although phenol biogenesis seems to proceed along three different metabolic pathways (1) the shikimic acid pathway (2) the acetate melonate pathway and (3) the acetate mevalonate pathway. The most common and significant pathway in both higher plants and microorganisms seems to be that involving shikimic acid. This pathway is the source of phenolic substances and also of aromatic amino acids such as phenyl alanine, tryptophan and tyrosine. Dehydroquinic acid formed from carbohydrate is dehydrated to form dehydroshikimic acid. It is converted to shikimic acid which then combines with a C_3 compound (possibly pyruvic acid) to form prephenic acid. Phenyl alanine is formed from prephenic acid via phenylpyruvic acid. Formation of tyrosine from prephenic acid by way of p-hydroxyphenyl pyruvic acid has also been reported.

Phenolic acid so formed occurs rarely in free state in the living cell. They are present either in conjugated form or in polymerized state. In the simplest instance, they are bound to sugar, a β-D-glucopyranosides but a wide array of other bound forms are also known. The other forms of binding include sulphation or acylation of free hydroxyl groups.

Conjugation has generally been found in an organism to be associated with the following three alternatives

(*a*) To make the compound more soluble,

(*b*) To detoxify exogenous materials and

(*c*) To convert physiologically important compounds into reserve substances

These alternatives may exhibit partial overlapping. Despite their metabolic importance and extensive research carried out for many years, several problems are still open in the area of conjunction reactions in plant cell system. Unialgal cell culture can be proved an ideal experimental material for carrying out such study due to its easier experimental handling and short incubation.

The other process which reduces the chance of phenolic compounds to interfere with the cellular enzyme reactions is polymerization. As the molecular weight of resulted polyphenols increases, their transport within the cell is considerably diminished. Such polymerizations also ensure that inhibitory concentrations of phenolic aglycones are not attained in cells. Polymer formation greatly depends on the substitution pattern of the substrates (Barz, 1975) and both phenolases and peroxidases (Stafford. 1974) may be regarded as the enzymes responsible. The two main types of Phenolic polymers in plants are the lignin and tannins. None of these polymers have been reported either in any green or blue-green algae but the presence of lignin has been reported in brown algae (Golman *et al.*, 1973; Reznikov and Mikhaseva, 1976; Reznikov *et al.*,1978; Dọvgan *et al.*,1982 a,b). Demonstration of lignin in brown seaweeds is consequently of considerable phylogenetic importance.

Localization of Phenols

Most of the polyphenols have been reported to be localized in brown algae within a variety of subcellular vacuoles/vesicles called "physodes" (Chadefaud, 1932). These physodes, measuring 0.1–10 µm in diameter or occasionally greater, are abundantly found in epidermal, outer cortical, apical and/or meristematic cells. Distribution of these physodes in various tissues of brown algae has been described by Ragan and Glombitza (1986). Instead of vacuoles and vesicles, localization of phenolic compounds in plastids, endoplasmic reticulum and golgi apparatus has also been reported in higher plants (Schnepf, 1976).

Light microscopic examinations of plants for phenolic localization have utilized visualizing agents as ferric chloride or ammonia vapors (Beckman *et al.* 1972). aluminum chloride induced shifts in ultraviolet microscopy (Charriere–Ladreix, 1976; Guerin, 1971) and so many nitroso dies (Beckman and Muller, 1970). This type of examination shows highly distinctive distributional patterns of colorless phenols.

Electron microscopy can also be much helpful in examining the localization of phenolics in cells tissues in plants on account of its osmophilic nature but problem is largely due to (a) distinguishing phenols from other osmophilic material and (b) retaining them within the appropriate subcellular compartment during killing and embedding the tissues in preparation for sectioning. For this reasons, ultra structural localization of phenolics is restricted mainly due to cells producing such copious amounts that some are retained during processing the tissue (Mueller and Rodehorst, 1977).

The above mentioned localization of phenolics in specialized vesicles/cellular organelles has not been reported so far in any green and blue-green algae. It may be presumed to be present either in central vacuoles or structurally bound to cell wall in periplasmic space, however this assumption needs confirmation.

Extracellular Phenolics

There is no definite cytological evidence for extra cellular release of polyphenols from brown algae. But macroscopic studies confirm that the polyphenols may be released either directly (from the intact thallus to the surrounding medium) or indirectly through erosion or shedding of macroscopic parts of thallus (Newell and Lucas, 1981) and may occur from healthy algae or after injury. Direct release of phenols by exponentially growing green and blue-green algae has been reported by Shukla (1989) and Sahay (1990). Release of phenols by these actively growing algae appears to be an essential phenomenon related to maintain equilibrium between intra and extracellular environment (Mohan and Mukerjee, 1979). Its level inside the cell and in the surrounding medium might be acting as a limiting factor in the regulation of growth of the alga either alone or in association with the Phytohormones.

The chemical nature of the excreted polyphenols has received little attention. Ragan and Craigie (1980) suggested that the polyphenols released from *Fucus vesiculosus* were closely related to their high molecular weight polyphloroglucinol fraction, but detailed structural studies have not been conducted. Green and blue-green microalgae have been reported to excrete phenols both in free and bound forms (Sahay, 1990). Such leaching in the natural habitats might be affecting the aquatic flora in a variety of ways on account of their phytoalexinic, astringent and allelopathic properties. Therefore, a systematic work on the chemical and physiological characterization of these extracellular phenolic substances may yield many useful informations of applied nature.

Extraction, Purification and Elucidation of Structure

Methods available for extraction, purification and characterization of phenols/ polyphenols in general have been reviewed by Verma and Sahay (1994) and have been described in details with reference to brown and red algae by Ragan and Glombitza (1986).

Polyphenols can be extracted from fresh or frozen materials using aqueous alcohols (Glombitza *et al.*, 1976; Ragan and Craigie, 1976) or aqueous acetone (Ragan and Jensen. 1977; Zavodnik and Jensen, 1981). It is preferable, however, to carry out the extraction under an inert atmosphere in reduced light.

Alcohol or acetone extract is concentrated to small volume under vacuo, precipitated lipophilic materials can be removed by filtration or extracted into chloroform or hexane (Glombitza *et al.*, 1976). Polyphenols can then be extracted from the aqueous phase into ethylacetate. Aqueous phase can also be directly taken to dryness and the dry material is redissolved in slightly aqueous methanol and loaded onto a dry-packed column of micro crystalline cellulose (Avicel). Elution with toluene removes residual pigments, and low molecular weight polyphenols can be eluted with acetone. Subsequent elution with methanol removes more oligomers and higher polymers (Ragan and Craigie, 1976).

The free phenols can be separated by TLC on microcrystalline cellulose with butanol : benzen : acetic acid : water or formic acid: water systems, and on silica gel with chloroform: dioxane (Glombitza *et al.*, 1973) or chloroform : methanol : water systems (Ragan and Craigie, 1978; Gregson and Daly, 1982). As the free phenols are subject to air-oxidation, it is often preferable to acetylate the ethyl acetate-extract immediately, using acetic anhydride pyridine (Glombitza and Rosener, 1974).

Initial fractionation of acetylated polyphenols can be done by preparative HPLC (Glombitza *et al.*, 1978) or thick-layer chromatography on silica gel using chloroform: acetone (Glombitza and Rosener. 1974). For final separations or for confirmation of purity TLC is preferable on silica gel using chloroform: acetone or chloroform: carbon tetrachloride (Glombitza *et al.*, 1975, 1977) or HPLC on silica gel with chloroform: methanol gradients (Koch *et al.*, 1980). The peracetates can be located on TLC by their UV-quenching against a fluorescent background (F254) or after spraying with Fast Blue Salt B or Vanillin-sulphuric acid (Glombitza and Rosener, 1974).

The techniques most useful for elucidating structure of polyphenols include electron impact mass spectrometry (ElMS), field desorption mass spectrometry (FDMS) and ^{1}H and C^{13} nuclear magnetic resonance (NMR) spectroscopy (Ragan and Glombitza, 1986). Ultraviolet (UV) and infrared (IR) spectra provide, at best, only limited structural information in most cases and phlorotannin has so far been only examined by X-ray crystallography. (Gregson and Daly, 1982).

Existing knowledge of fresh water green and blue-green algal phenolics is limited only to its quantitative estimations (Table 7.1) after extracting through ethanol/ methanol/acetone etc. and detection by using ferric chloride potassium ferricyanide reagent (Singh *et al.*, 1978), Folin Ciocaltean reagent (Bray and Thorpe, 1954).

Structural elucidation of the phenolics synthesized as well as excreted by the freshwater green and blue-green algae is restricted to its major fractions studied as per the methods described essentially by Zadernowski and Kozlowska (1983) with some modifications after Meravy (1987). The different cellular and extra-cellular fractions studied are as follows:

(A) Cellular Phenolic Fractions

F_1–Free phenolic acids, coumarines and flavanoid aglycones.

F_2–Esterically bound phenolic acids

F_3–All glycosides

F_4- Phenolic acids bound to cell structure.

(B) Extracellular Phenolics (phenolics in the culture filtrate)

FM_1–Free phenolic acids

FM_2–Bound phenolic compounds

Levels of Phenolics in Green and Blue-Green Algae

Determination of the absolute content of phenols in flesh water green and blue-green algae is of direct importance in assessing its role and significance in microalgal physiology and in elucidating the importance of algal phenolics contributed to the freshwater ecosystem.

Table 7.1: Intracellular Phenolic Content Reported in Different Taxa of Blue-green Algae as Ascertained by Barclay *et al.* (1987)

Name of Taxa	*Amount of Phenolics (Per cent dry wt.)*
Prochloron 16 species	1.8 - 7.1
Anabaena flosaquae	0.92
A. cylindrica	0.91
A. inequalis	0.90
A. apirodes	0.58
Nostoc muscorum	1.05
N.commune	0.34
Microcoleus vaginatus	0.62
Gleocapsa sp.	0.03
Synechococcus leopoliensis	0.83

Barclay *et al.*, (1987) investigated as many as 16 species of *Prochloron* and species belonging to 5 other genera of blue-green algae for their intracellular phenolic contents (Table 7.1). They, however, did not quantify the amount of phenolics excreted by the algae investigated upon. Table 7.2 indicates the quantity of both intra and extracellular phenolics determined in 5 species of green and 9 species of blue-green algae (Shukla, 1989; Sahay, 1990; Sahay *et al.*, 1995 and Das, 1997). Level of phenolics in the table under reference is optimum value determined at the growth phase indicated against the respective taxa.

As indicated in the Tables 7.1 and 7.2 level of phenolics falls within the range in green algae from 0.825 per cent (*Stichococcus* sp) to 4.21 per cent dry wt. (*Oedogonium* sp.) and in blue-green algae from 0.03 per cent (*Gleocapsa* sp.) to 15.314 per cent dry wt. (*Oscillatoria curviceps* var. *anguesta*). Reported levels of phenols in about 100 species of brown algae vary from 0 per cent (below the level of detection, usually around 0.02 per cent) to about 20 per cent dry wt.

Data available on the level of phenolics in green and blue-green algae (Tables 7.1 and 7.2) remain far from sufficient to draw any definite conclusion on their phenol producing potency. However, the range of phenolics determined in the limited number of blue-green algae is quite matching with those in brown algae. Therefore, green and blue-green algae appear in no way less potent phenol producers than the brown algae.

There appears to be no literature depicting quantity of phenols excreted by brown algae. But it has been quantified in 6 taxa of green and 8 taxa of blue-green algae (Table 7.2). As indicated in the table under reference, the quantity of extracellular phenolics in culture filtrate of the algae indicated against the respective taxa ranged from 0.23 to 11.75 mg/100mg dry wt. amongst green algae and from 0.29 to 48.515 mg/100 mg dry wt. amongst blue-green algae. Thus the blue-green algae are not only ahead to the green algae in synthesizing the phenolic compounds but also in contributing phenolics to the growing medium. It is further apparent from the Table

7.2 that the amount of both cellular and extracellular phenolics determined by Sahay (1990) in *Merismopedia glauca* is exceedingly low as compared to the amount determined by Das (1997). This quantitative variation in cellular and extracellular phenolic content possibly relates to the environmental verities as well as different techniques adopted by them. Sahay (1990) followed the method of Singh *et al.* (1978) and Das followed the method of Meravy (1987).

Table 7.2: Quantity of Intracellular and Extracellular Phenolics in Different Taxa (mg/100mg dry wt. of algae)

Name of Taxa	*Intra Cellular Phenolis*	*Extra Cellular Phenolis*	*Total*	*Stage of Growth Phase*	*Author Year*
Scenedesmus bijugatus	0.890	4.763	5.653	Early exponential	Das, P.K. (Pers. comm.)
Stichococcus sp.	0.826	11.75	12.576	Early exponential	"
Oedogonium sp.	4.210	8.48	12.69	Late exponential	"
Nostoc linckia	2.704	48.51	51.21	early exponential	"
Merismopedia glauca	0.120	7.04	7.17	late static	"
Oscillatoria curviceps var. *angusta*	15.310	24.99	40.31	late exponential	Sahay *et al.*, 1995
Merismopedia glauca	2.800	0.88	3.68	static	Sahay, 1990
Protococcus viridis	2.240	0.31	2.55	static	"
Oocystis lacustris	3.300	0.23	3.53	late exponential	"
Ankistrodesmus falcatus	3.260	0.89	4.15	late exponential	"
Microcystis elabens	1.590	0.29	1.88	middle exponential	Shukla, 1989
Synechocystis pevalekii	6.700	0.48	7.18	middle exponential	"
Chroococcus minor	7.150	1.47	8.62	late exponential	"
Oscillatoria subbrevis	1.030	11.20	12.23	Late exponential	"

Relative percentage of phenolic fractions of cellular and extracellular phenolics has been ascertained only in 3 taxa of green as well as, blue-green algae by Das (1997), the details of which have been summarized in Table 7.3. Evidently, bulk of the cellular phenolics in green algae is invariably predominated by the phenolic acids either in free state or bound to cell structures–coumarines and flavanoid, aglycones (F_1 and F_4 fractions) and esterically bound phenolic acids and glycosides (F_2 and F_3 fractions) in blue-green algae. Predominance of bound phenolic compounds (FM_2) in the culture filtrate was noted as the common feature of both green and blue-green algae. Such bound phenohic compounds were identified as β-cyclocitral in *Microcystis* (Juttner, 1981) and as mesityloxin and 6- methylpent- 5 en-2-one (terpenoids) in *Anabaena cylindrica* (Juttner, 1983).

Table 7.3: Relative Percentage of Intra and Extracellular Phenolics

Name of the Taxa	Intracellular				Extracellular	
	F1	F2	F3	F4	FM1	FM2
Scenedesmus bijugatus	32.25	20.34	21.46	25.95	32.02	67.98
Stichococcus sp.	28.087	23.487	16.465	31.961	29.787	70.213
Oedogonium sp.	42.75	21.37	7.36	28.50	10.38	89.62
Merismopedia glauca	22.80	41.04	15.31	20.85	29.36	70.64
Nostoc linckia	3.07	36.69	29.32	30.92	31.08	68.92
Oscillatoria curvideps var. *angusta*	7.38	42.47	28.85	21.30	11.27	88.73

Biology of Phenolics

Phenolics have been reported to accumulate in actively growing cells invariably upto early/late exponential growth phase after an initial decline at the onset of exponential (Shukla, 1989; Sahay, 1990; Sahay *et al.*, 1995). Similar accumulation was also reported earlier in higher plants (Ibrahim and Edgar, 1976; Shah *et al.*, 1976 and Rao *et al.*, 1983). The rise continues even upto static growth phase as in *Merismopedia glauca* and *Protocuccus virdis* (Sahay, 1990). Beyond this, the level of phenols gradually decreases with age. The initial fall in level of phenols may be presumed due to its excretion for maintaining equilibrium between cellular and extracellular environment and rise in its level during exponential possibly relates to the excessive production of aromatic amino acids essential for the biosynthesis of phenols (Kund, 1992).

As against the age bound decline in the level of cellular phenolics extracellular phenolics increased concomitantly with age irrespective of growth phase showing thereby that once the phenolics excreted outside in the growing medium were not reabsorbed for recycling. On the other hand, Sahay *et al.* (1995) have recently reported that the relative percentage of extracellular phenolics in *Oscillatoria curviceps* var. *angusta* gradually fall and that of cellular phenolics rose during exponential. It suggests reabsorption and recycling of the extracellular phenolics to meet the augmented demand of physiologically active exponential population. Post exponential increase in the level of extracellular phenolics may be assumed due to decreased cell wall permeability and degradation by enzymes present in the growing medium and/or at the cell surface (Berlin and Barz, 1975; Hrubcova *et al.*, 1981; Pospisil *et al.*, 1983). Sahay *et al.* (1995) have further reported an apparent change in the relative percentage of the four phenolic fractions with age. The extract obtained from the population at early exponential was predominated by free phenolic acids, coumarines and flavanoid aglycones (F_1) followed by esterically bound phenolic acids (F_2), phenolic acids bound to cell structures (F_4) and glycosides (F_3). This sequence of predominance is changed to $F_2 > F_3 > F_4 > F_1$ at mid-exponential, $F_1 > F_4 > F_3 > F_2$ at static growth and $F_4 > F_1 > F_3 > F_2$ during decline. This growth phase bound qualitative variations in phenolic compounds appear due to specific requirements associated with growth phase and conjugation of phenolic substances.

Sahay *et al.* (1995) reported further that the bulk of extracellular phenolics are consistently predominated by the bound phenolic compounds (FM_2) which increase with age even beyond static growth phase. FM_1 fraction (free phenolic acids) do increase concomitantly but decrease sharply only after 5th day of incubation (early exponential) coinciding with decrease in relative percentage of extracellular phenolics followed by apparent increase in the level of cellular phenolics. Therefore, it can be concluded that it is the FM_1 fraction of extracellular phenolics which are reabsorbed for recycling. Functional role of the phenolic substances in algal cell physiology is not yet studied. But, growth phase specific qualitative variation in the phenolic compounds and reabsorption of physiologically active phenolic acids during exponential apparently suggest its active involvement in growth physiology as reported in higher plants (Thimann *et al.*, 1962; Lee and Skoog, 1965; Ray and Laloraya, 1984; Ray, 1986). This assumption can be further strengthened from the fact that certain phases of intensive growth are accompanied by substantial qualitative and/ or quantitative changes in the spectrum of phenolics and these can only usually be interpreted by assuming catabolic processes to be involved. Thus, various phenols and flavanoids present in seeds or young seedlings disappear completely upon germination. (Barz, 1975; Barz and Hosel, 1975; Harborne, 1962 and Zenk, 1967).

Three types of reactions have been proposed to be involved in the aforesaid catabolic processes:

1. Interconversions which are involved in biosynthetic sequences.
2. Catabolism where the products are converted to primary metabolic constituents.
3. Oxidative polymerization reactions leading to insoluble structures of high molecular weights.

All the three metabolic routes may occur simultaneously in plant cell system and their ratio will thus depend on the various parameters regulating cellular metabolism (Zenk, 1967).

Phenol metabolism, therefore, constitutes a complex metabolic grid which requires a detailed biochemical analysis. Microalgal cultures undoubtedly can provide comparatively more suitable systems than higher plant cell suspension culture for such studies because of its easy handling and low culture cost.

Factors Influencing Biosynthesis of Phenolics

A number of physical as well as chemical factors have been studied in relation to their effect on phenol production mainly in brown algae (Craigie and Mclachlan, 1964: Sieburth, 1969; Hornsey and Hide, 1976; Ragan and Jensen, 1978; Pedersen, 1984 and Ragan and Glombitza, 1986) and to some extent in green as well as blue-green algae. These findings may be described under respective heads as follows:

pH

It has been reported to influence indirectly through regulating the activities of the enzyme polyphenol oxidase in *Nitella mirabilis* (Holst and Yopp, 1976). This effect is however, substrate dependent.

Seasonal Variations

Its highly variable effects have been recorded by different workers (Larsen and Haug, 1960; Munda, 1962a,b; Geiselman, 1980) in brown algae. These conflicting observations have been attributed mainly to the interference of local environmental factors which are more instrumental in regulating phenol production in brown algae (Ragan and Glombitza, 1986). Salinity, temperature and light are the three factors which are supposed to vary from habitat to habitat and they have been reported to play a significant role in regulating the phenol production in Phaeophyceae (Munda, 1960; Joshi and Gowda, 1975; Pedersen, 1984 and Ronnberg and Ruokolahti, 1986).

Organic Nitrogen

The micro-green as well as blue-green algae are short lived completing their growth cycle invariably within single rotation of seasons. Hence, there is no question of any seasonal impact on the accumulation of phenols in these microalgae. Bloom forming algae in eutrophic water bodies is, however, exception to this. Their presence can be marked year after year and are exposed to the effect of seasonal variations, while working on the limnology of an eutrophic pond enjoying Cyanophycean bloom, Kund (1992) recorded distinct annual variations in the level of phenolic content (MV 0.12±0.03 ppm). It was recorded lowest in the month of October (0.05 ppm). Gradually it increased reaching its optimum level in the month of June (0.26 ppm) and then decreased gradually till October. Its level was recorded significantly low during October–February (0.5 to 0.07 ppm) and average (0. 13 to 0. 14 ppm) in rest of the months except June when it was very high. There was no external source of phenolics in the pond.

Accumulation of phenolics in the pond did not show significant correlation with any of the seventeen physico-chemical parameters studied by Kund (loc.cit) except organic nitrogen (r= 0.74**) which in turn also exhibited significant relationship with cyanophycean population predominated by *Microcystis aeruginosa*.

Preferential growth of *M.aeruginosa* in the pond water was further confirmed by comparing its growth with other indigenous green alga (*Scenedesmus bijugatus*). Simultaneously in the Hughe's inorganic medium and in the sintered glass filtered pond water of June month containing 0.2 ppm phenols (Kund and Verma, 1994) was inhibited by 87 per cent and that of the second was accelerated by 30 per cent in the pond water against inorganic medium. This was also followed by the decrease in the level of phenols by 64 per cent in *S.bijugatus* and increase by 57 per cent in *M.aeruginosa*. Therefore it can be concluded that eutrophication (organic nitrogen) induces phenol production on the one hand and growth of the pollution loving blue-green algae on the other. Positive effect of nitrogen on phenol production by plants has also been reported by Fitzgeral (1969) and this has been further confirmed by Meravy (1987).

** Significant.

Chloride

It has been reported to bear direct relationship with accumulation of phenolics in the stagnant water bodies. Munda (1960) also holds the view that individual plant is capable of altering their polyphenol levels in response to changes in salinity.

Light

Formation of terpenoids (Mesityloxide) in *Anabaena cylindrica* has been reported to be light dependent (Juttner *et al.*, 1983). Level of polyphenols decreases when daily hours of sunshine increase, the levels thereafter increase steadily despite further increase in day length (Ragan and Jensen, 1978). They have further reported a good temporal correlation between polyphenol level and developmental (reproductive) state of the algae. However, photosynthesis is not directly involved in its formation as indicated by the small effect of DCMU. Indeed, cessation of its production may be due to depletion of a precursor of terpenoid which can not be synthesized in the dark.

Nutrient Status and the Relative Levels

Different primary metabolites have been reported to generate apparent effect on accumulation of phenolic substances in higher plants (Shah and Mehta, 1978 a,b) and this has been further confirmed with reference to *Protococcus viridis* by Sahay (1990). Fall in the level of phenolic substances during static growth phase in later case has been attributed either to the diversion of free carbohydrates towards the synthesis of primary metabolites which is initiated by the increase in the level of nitrogen in growing medium and/or age bound conversion of free carbohydrates into bound ones. Level of carbohydrate in the cell has been also reported to act as limiting factor in premature decline in the phenol production (Shah and Mehta, 1978a).

Phenols as Allelopathin

The tremendous fluctuations in abundance of phytoplankton in all kinds of water bodies have intrigued phycologists for many years. Many different suggestions have been recorded to explain this preferential abundance of algae (Fogg and Westlake, 1955; Rice, 1974; Wolf and Rice, 1979) and it has been finally concluded by Fogg *et al.*, (1965) "the extensive liberation of extracellular products by phytoplankton suggests a strong potential for growth effects on other microorganisms and co-existing algae". Various compounds including toxins have been claimed to be present as growth inhibitory factor in the conditioned growing medium of different algae (Harris, 1971; Khailov, 1974; Murphy *et al.*, 1976; Nozawa, 1968; Proctor, 1957; Fogg and Westlake, 1955) and phenolic compounds are one of them (Fenical, 1975; Craigie and McLachlum, 1964; Ragan and Craigie, 1978 and Katayama, 1962).

Keating(1977) reported convincing evidence of algal allelopathy in algal succession in a eutrophic lake and it was further confirmed by Wolfe and Rice (1979). They apprehended production of some allelochemical by the predominant algal forms to suppress the growth of other coexisting algae but they could not characterize this compound. Kund and Verma (1994) identified three broad categories of algae in a eutrophic pond–Harahi at Darbhanga (Bihar, India) sharing the relative percentage indicated against each

1. Cyanophyceae:
   (*a*) Non N_2 fixer - 79 per cent
   (*b*) N_2 fixer - 14 per cent
2. Chloropbyceae - 07 per cent

Bulk of non-N_2 fixer cyanophytes was dominated alone by *Microcystis aeruginosa* sharing its 82 per cent which also formed 76 per cent of the bulk of total phytoplankton. Three representative members from each of the three categories–*M. aeruginosa, Anabaena circularis* and *Chlorococcum* sp. were isolated from the eutrophic pond, purified and cultured under laboratory conditions. All the three test algae were screened for their phenolic contents (Table 7.4). Growth of all the three test algae was ascertained in the conditioned pond water (sterilized by vacuum filtration through ultra fine sintered glass filters) of exponentially growing population of each other (Table 7.5). Sterilized pond water was used as tile growth medium through out the experiment in view of providing natural conditions to the experiment. Growth of *A. circularis* and *Chlorococcum* sp. was apparently inhibited the conditioned culture filtrate of *M. aeruginosa* carrying 1.98 ppm of phenols. But the growth of *M. aeruginosa* was neither inhibited in the conditioned culture filtrate of *Anabaena circularis* nor of *Chlorococcum* sp. in which the level of phenols was relatively very low *i.e.* 0.44 and 0.22 ppm respectively.

Table 7.4: Extracellular Phenolics of the Three Test Organisms

Name of the Taxa	*Extracellular Phenolics (mg/100 ml)*
Chlorococcum sp.	0.22
Anabaena circularis	0.44
Microcystis robusta	1.98

Table 7.5: Growth Contact of the Three Test Organisms in the Culture Filtrates of each Other

Culture Filtrate	*S. bijugatus*	*A. circularis*	*N. robusta*
Chlorococcum sp. (1.1 ppm of phenolics)	–	0.157	0.22
Anabaena circularis (0.63 ppm of phenolics)	0.055	–	0.24
Microcystis robusta (1.25 ppm of phenolics)	0.0	0.099	–
Control (growth in Hugh's)	0.088	0.217	0.135

The growth inhibitory effect of conditioned pond water was further confirmed through the plate diffusion method. No growth inhibition zone was observed in the lawn of *M. aeruginosa* either through the conditioned pond water of *A. circularis* or of *Chlorococcum* sp. But a distinct growth inhibitory zone measuring 18-20mm in diameter appeared in the lawn of *Chlorococcum* sp. and 12-19 mm in the lawn of *Anabaena* sp. after treatment with the pond water conditioned by *M. aeruginosa.*

The scraps of growth inhibitory zone were inoculated in 5ml of nutrient broth and incubated at 28+2°C in dark. The turbidity appeared within 14 hrs. Its microscopic examination revealed presence of cocci and small rod shaped bacteria of gram positive nature. On plating at suitable dilutions, isolated colonies with entire margins appeared on the agar surface. Individual colonies were suitably grown in nutrient broth and were added to the well on the lawn of both *Anabaena* sp. and *Chlorococcum* sp. No growth inhibitory zone appeared in either case confirming that the antimicrobial compound was not of any bacterial origin but of algal origin.

FM_1 and FM_2 fractions of phenolic compounds contributed by *M.aeruginosa* in the growing medium were extracted and were added independently to the wells on the lawn of *Chlorococcum* sp. and *Anabaena* sp. Growth inhibition zones formed in both cases confirming that the allelopathic property of *M.aeruginosa* is due to the free as well as bound phenolic compounds excreted in the surrounding medium. Structural elucidation of these compounds was not made by Kund and Verma (1994) but characterization of the phenolics excreted by different groups of algae has been done. The presence of terpenoid compounds–Mesityloxide and 6-methylhept-5-en-2-one have been reported in the medium of *Anabaena cylindrica* by Juttner *et al.* (1983) but their physiological effects was not studied. Ragan and Craigie (1978) identified some 50 Phenolic compounds from Phaeophyceae, Rhodophyceae, Cyanophyceae. Chlorophyceae, Bacillariophyceae, Xanthophyceae, Chrysophyceae which are important allelopathic agent. Katayama (1962) has identified the terpenes and phenols produced by several green, brown and red algae and also determined the antibacterial activities against *Staphylococcus aureus* and *Escherichia coli.* A number of brominated phenols produced by red algae are allelopathic in nature.

Halogenated phenols, monoterpenes, diterpenes, sesquiterpenes have been reported to be produced by different red algae as allelochermical and their growth inhibitory effects have been tested against various kinds of organisms (Fenical, 1975). *Fucus vesiculosus* is also known to release flavanoids or catechin type tannins that inhibit the growth of several marine algae (Craigie and Mc.Lachlan, 1964). Chemistry and biological activities of the phenolic compounds released by marine algae and their biological activities have been thoroughly reviewed by Bhakuni and Silva (1974).

Apparently the phenolic substances released by algae in freshwater bodies have not received due attention in regard to their identifications and biological effects. Since freshwater bodies are intricately associated with the day to day activities of human being in various ways and allelopathic effect of the phenols contributed by the algae growing therein is not only confined to algae versus algae or algae versus bacteria in the ecosystem, their far reaching effects along the long grazing chain can not be ignored. Therefore, extensive studies are required to ascertain (i) the level of phenols contributed by algae especially in eutrophic water bodies. (ii) identification of the phenols present and (iii) biological effects of phenols identified.

Functions and Uses

The functions and uses of algal phenolics known to this date are limited. Preliminary informations gathered so far on this aspect can be read categorically as follows:

1. The high concentration of phenolic compounds in the *Prochloron* spp. (Cyanophyceae, a symbiont of *Didemnum candidum*) have been reported to serve as protective agent (Barclay *et al.*, 1987) in the same way as they act in many higher plants.
2. Methanol extracts of *Didemnum molle* containing *Prochloron* exhibit cytostatic activity when tested in cultures of L5 178, mouse lymphoma cells. This activity is attributed to the phenolic compounds contributed by *Prochloron* (Barclay *et al.*, 1987).
3. Electron micrographic studies have revealed phagocytosis of *Prochloron* cells by amoebocytes present in the ascidian *Lissoclimum voeltzkowi*. *Prochloron* has been reported to protect itself against phagocytosis by excreting phenols. This hypothesis, however, requires thorough investigations of *Prochloron* present on the surface of ascidians for their phenolic content.
4. Polyphenols have been reported to be involved in protection of *Fucus vesiculosus* against grazing by isopod *Idotea baltica* (Ilvessalo, 1986; Ronnberg and Ruokolahti, 1986) and in aging of *Ascophyllum nodosum* (Pedersen, 1984). Age specific increase/decrease in phenolic content (Shukla, 1989; Sahay, 1990) as well as in relative percentage of different phenolic fractions (Sahay *et al.*, 1995) have been reported in a number of green and blue-green algae. This further confirms the relationship of phenolics with aging. However, these preliminary informations require confirmation with wider coverage of test organisms and also the chemical as well as biological characterization of the phenols in question for better understanding of its broad based role and for exploring avenues of their proper utilization in the interest of society.
5. Phlorotannins in physodes (Phaeophyceae) intercept ions of toxic heavy metals entering the brown algal cells, thereby allowing the alga to grow in otherwise toxic environments (Skipnes,1974). Because polyphenols are released from the alga, it is possible that the chelated metal ions could be released simultaneously (Myklestad *et al.*, 1978; Eide *et al.*, 1980). Consequently polyphenols could contribute to algal fitness by cleansing the cytoplasm of toxic ions. This property of physode containing polyphenols can be exploited for concentrating out the heavy metals present in the sea water.
6. Polyphenols in brown algae provide a low-level defense against bacteria (Conover and Sieburth, 1964; Sieburth and Conover, 1965) and may be responsible for the lower numbers of yeast epiphytic on brown algae compared with red and green algae (Seshadri and Sieburth, 1971). Because the high-molecular weight phlorotaninins are weak antibacterial and

antifungal (Ragan, 1985b), the intermediate and low- molecular weight reactions are responsible for more of these activities which is significant in the ecological context, as bacterial slime could promote larval settlement and attract grazers. Polyphenol also limit secondary bacterial infection of injured algae (Fagerberg and Dawes, 1976). However, sufficiently comprehensive studies have not been carried out in this direction.

7. Brown algae contain compounds which discourage grazing by herbivores (Hunger, 1902). This compound was identified as tannins by Stahl (1888). Laboratory experiments examining antifeedant activity of isolated phlorotannnis and field oriented observations correlating polyphenol and tannins levels with herbivory have further confirmed Hunger's conclusion.
8. The release of polyphenols upon injury to the alga is probably an early stage of the "wound response" whereby polyphenols released from physodes would precipitate protein of the alga it self and/or of the invader (fungus, herbivore, etc.), thereby sealing off the injured regions and limiting bacterial infection (Fagerberg and Dawes, 1976). Similar mechanisms are thought to contribute to the fitness of terrestrial plants (McKey, 1979).
9. Allelopathic action of the phenol excreted by a number of algae has been demonstrated against large number of test organisms including bacteria, fungi and algae. This has been also involved in blooming of cyanophycean population under eutrophic condition by triggering the growth preferably of pollution loving taxa at one hand and inducing the synthesis/excretion of phenols for maneuvering dominance over coexisting algae at the other (Kund and Verma, 1994).
10. *Nostoc muscorum* has been reported to contain phenolic compounds with growth inhibitory effect on many microorganisms of economic importance. Its methanolic extract containing phenolics, identification of which is still awaited, has been reported to evoke significant growth inhibition to human pathogens *Candida albicans* and *Staphylococcus aureus* (Cairn *et al.*, 1990)

Future Scope of Work

Desirable works required to be undertaken to decipher various aspects of algal phenolics have already been discussed at appropriate places in the text. This section has been earmarked especially to highlights some relevant problems prevailing in the field of algal phenolics.

Phenolics are one of the most important compounds produced by algae. Their functions and uses known today are limited and are mainly restricted to the marine brown or red algae. Explorations made so far have given sufficient indications that the green and blue-green algae are no less potent phenol producers and hence offer a good scope of exploiting these micro algae for phenol production in mass. For this, their thorough screening is a prerequisite for selection of suitable phenol producers followed by development of techniques for their mass culture.

The different abiotic factors *e.g.* pH, temperature, light, salinity and nutritional status are known largely to affect the phenol producing capacities of the concerned algae. Therefore, this needs to be suitably manipulated in a specific manner for duly enhancing as well as accumulating capacities of algae in question. Sophisticated and less time consuming laboratory technique has to be developed for easy and better qualitative and quantitative estimations of algal phenolic products.

Phenolic compounds synthesized by the algae are excreted to a considerable extent in the surrounding environment. Its accumulation in the natural habitat often exceeds maximum limit prescribed by WHO (Kund and Verma, 1994) and generate far reaching effect along the long grazing chain (biological magnification) in a variety of ways on account of their phytoalexinic, astringent and allelopathic properties. Unfortunately no attention has been paid to this alarming situation. Eutrophication is a global problem and blooming of algae is associated with it that contributes quite good amount of phenohic compounds in the water bodies. Therefore, systematic studies are required.

1. To screen out the algae for their phenol producing potency
2. Chemical characterization of the phenolic compound synthesized/excreted by algae
3. Biological evaluation of algal phenolics.

Quantity and quality of phenolic compounds synthesized/excreted by the algae depends upon the physiological requirements which vary with age. Therefore, from an economic point of view emphasis should also be given to determine the threshold age factor responsible for the optimum phenol production.

References

Ambasta, K K. and Hamidi, M.K., 1988. Variability of total phenolics in different parts of *Parthenium hysterophorous. Environment and Ecology*, 6(2): 484–485.

Barclay, W.R., Kennish, J.M., Goodrich, V.M. and Fall, R., 1987. High levels of phenolic compounds in *Prochloron* species. *Photochemistry*, 26: 739–743.

Barz, W., 1975. Abbau von Flavonoiden und Isoflavonoiden-ein vederblick. *Ber. Deutch. Bot. Ges.*, 88: 71–81.

Barz, W. and Hosel, W., 1975. Flavnoid metabolism. In: *The Flavnoids*, (Eds.) J.B. Harborne, T.J. Mabry and H. Mabry. Chapman and Hall, London, Chapter 17.

Beckman, C.H. and Mueller, W.C., 1970. Distribution of phenols in specialized cells of banana roots. *Phytopathology*. 60: 79–82.

Beckman, C.H., Mueller, W.C. and McHardy, W.E., 1972. Localization of stored phenols in plant hairs. *Physiol. Plant Path.*, 2: 69–74.

Berlin, J. and Barz, W., 1975. Oxidative decarboxylation of parahydroxy benzoic acids by peroxidases under *in vivo* and *in vitro* conditions. *Z. Naturforschg*, 30C: 650–658.

Bhakuni, D.S. and Silva, M., 1974. Biodynamic substances from marine flora. *Bot. Mar.*, 17: 40–51.

Bray, H.G. and Thorpe, W.V., 1954. Determination of Phenols. In: *Methods of Biochemical Analysis*, (Ed.) D. Glick, Academic Press, New York, 21: 27–52.

Cambie, R.C., Cox, R.E., Croft, K.D. and Sidwel, D., 1983. Phenolic diterpenoids of some podocarps. *Phytochemistry*, 22(5): 1163–1166.

Cano, M.S.de., Mule, M.C.Z de, Caire, G.Z. de and Halperin, D.R. de, 1990. Inhibition of *Candida albicans* and *Staphylocccus aureus* by phenolic compounds from the terrestrial cyanobacterium *Nostoc muscorum*. *J. Applied Phycol.*, 2: 79–81.

Chadefaud, M., 1932. Sur les physodes des Pheophycees. *CR. Acad. Sci.*, 194: 1675–1677.

Charriere-Ladreix, Y., 1976. Repartition intracellulaire du secretat flavonique de *Populus nigra* L. *Planta*, 129: 167–174.

Conover, J.T. and Sieburth, J.Mc N., 1964. Effect of *Sargassum* distribution on its epibiota and antibacterial activity. *Bot. Marina*, 6: 147–157.

Craigie, J.S. and McLachlan, J., 1964. Excretion of coloured ultraviolet absorbing substances by marine algae. *Can. J. Bot.*, 42: 23–33.

Das, P.K., 1997. Screening of algae for their phenolics. *Ph.D. Thesis*, Bihar University, Muzaffarpur, Bihar.

Dogra, J.V.V. and Sinha, S.K.P., 1983. Variability of total phenolics in maturing leaves and fruits of three medicinal plants used by Santhal tribes. *Herba Hungarica*, 22(3): 25–31.

Dovgan, I.V. Kreitsberg, Z.N. and Medvedeva, E., 1982a. Isolation and investigation of the lignin of the brown algae *Cystoseira barbata* Khim. *Drev.*, 5: 67–7 1.

Dovgan, I.V., Kreitsberg, Z.N. and Medvedeva, E., 1982b. Products of the nitrobenzene oxidation of the alga *Cystoseira barbata*. Khim. *Drev.*, 5: 72–73.

Eide, I., Myklestad, S. and Melsom, S., 1980. Long-term uptake and release of heavy metals by *Ascophyllum nodosum* (L.) Le Jol. (Phaeophyceae) *in situ*. *Environ. Poll. Ser*: A23: 19–28.

Ernst, W.H.O., Kuiters, A.T., Nelissen, J.M. and Tolsma, D.J., 1991. Seasonal variation in phenolics in several savanna tree species in Botswana. *Acta Bol. Neerl.*, 40(1): 63–74.

Fagerberg. W.R. and Dawes, C.J., 1976. Studies on Sargassum I. A light microscopic examination of the wound regeneration process in mature stipes of *S. filipendula*. *Am. J. Bot.*, 63: 110–119.

Fenical, W., 1975. Halogenation in the Rhodophyta: A review. *J. Phycol.*, 11: 245–259.

Fitzgeral, G.P., 1969. Some factors in competition and antagonism among bacteria, algae and aquatic weeds. *J. Phycol.*, 5: 351–359.

Fogg, G.E. and Westlake, D.F., 1955. The importance of extracellular products of algae in fresh water. *Proc. Int. ver Theor. Angew Limnol. Mitt.*, 12: 219–232.

Fogg, G.E., Nalewajko, C. and Watt, W.D., 1965. Extracellular products of phytoplankton photosynthesis. In: *Proc. Roy Soc.*, London Ser. B, 162: 517–534.

Geiselman, J.A., 1980. Ecology of chemical defenses of algae against the herbivorous snail, *Littorina littorea* in the New England rocky intertidal community. *Thesis*, MIT/woods Hole xiv+209pp.

Glombitza, K.W., 1977. Highly hydroxylated phenols of the Phaeophyceae. In: *Marine Natural Products Chemistry* (Eds.) D.J. Faulkner and W.H. Finical. Plenum Press, New York, pp. 191–204.

Glombitza, K.W., Forster, M. and Eckhardt, G., 1978. Polyhydroxyphenyletheraus der Phaeophyceae *Sargassum muticum*. *Phytochemistry*, 17: 579–580.

Glombitza, K.W., Koach, M. and Eckhardt, G., 1976. Polyhydroxyphenyletheraus *Laminaria ochroleuca*. *Phytochemistry*, 15: 1082–1983.

Glombitza, K.W., Rauwald, H.W. and Eckhardt, G., 1977. Phlorotannine aus *Himanthalia elongata*. *Phytochemistiy*, 16: 1614.

Glombitza, K.W., Rauwald, H.W. and Eckhardt, G., 1975. Fucole, Polyhydroxyoligophenyle aus *Fucus vesiculosus*. *Phytochemistry*, 14: 1403–1405.

Glombitza, K.W. and Rosener, H.V., 1974. Bifuhalol: Ein Diphenylather aus *Bifurcaria bifurcata*. *Phytochemistry*, 13: 1245– 1247.

Glombitza, K.W. and Sattler, E., 1973. Trifuhalol. ein neuer Triphenyldiather aus *Halidrys siliquosa*. *Tetrahedron Lett.*, 14: 4277–4280.

Golman, L.P., Mikhaseva, M.F. and Reznikov, V.M., 1973. Infrared spectra of lignin preparations of pteridophytes and seaweed. *Dokl. Akad. Nauk. BSSR*, 17: 1031–1033.

Gregson, R.P. and Daly, J.J., 1982. Polyhydroxy biphenyl ethero from the brown alga *Cystophora congesta*. *Aust. J. Chem.*, 35: 649–657.

Guerin, H.P., Delaveau, P.G. and Paris, R.R., 1971. Localisations histochemiques. II. Precedes simples de localisation de pigments flavonique. Application a quelqes Phanerogames. *Bull. Soc. Bot. Fr.*, 118: 29–36.

Harborne, J.B., 1962. The flavonol glycosides of wild and cultivated potatoes. Biochem. J., 84: 100–104.

Harris, D.O., 1971. A model system for the study of growth inhibitors. *Arah. Protistenkd.*, 113: 230–234.

Haug, A. and Larsen, B., 1958a. Brunfarging av tangmel og tangextrakter. Norsk lnst, for Tang-og Tareforsk. *Rep.*, 22: 1–18.

Haug, A. and Larsen, B., 1958b. Phenolic compounds in brown algae. I. The presence of reducing compounds in *Ascophyllum nodosum* (L.) Le Jol. *Acta Chem. Scand.*, 12: 650–657.

Haug, A. and Larsen, B., 1958c. Influence of habitat on the chemical composition of *Ascophyllum nodosum*. *Nature, (London)*, 181: 1224–1225.

Holst, R.W. and Yopp, J.H., 1976. An algal polyphenol oxidase: Characterization of the O-diphenol-oxidase from the charophyte, *Nitella mirabilis*. *Phycologia*, 15(2): 119–124.

Hornsey, I.S. and Hide, D., 1976. The production of antimicrobial compounds by British marine algae. II. Seasonal variation in production of antibiotics. *Brit. Phycol. J.*, 11: 63–67.

Hrubcova, M., Cvikrova, M. and Pospisil, F., 1981. Changes of 3,4-diliydroxyhenzoic and 4-hydroxybenzoic acids in *Nicotiana tobacum* cell suspension culture. *Biol. Plant.*, 23: 318–320.

Hunger, F.W.T., 1902. Uber das Assimilation sprodukt der Dictyotaceen. Jahrb. *Wiss Bot.*, 38: 70–82.

Ibrahim, R.K. and Edgar, D., 1976. Phenolic synthesis in *Perilla* cell suspension cultures, 15: 129– 131

Ilvessalo, H., 1986. Levakilkki (*Idotea baltica*) rakkolevan (*Fucus vesiculosus*) hebivorina. *M.Sc. Thesis*, Department of Biology, University of Turku.

Joshi, G.V. and Gowda, C.A., 1975. Seasonal variations in chemical composition of *Sargassum ilicifolium* Gurn. and Sea water. *Indian. J. Mar. Sci.*, 4: 165–168.

Juttner, F., 1981. Biologically active compounds released during algal blooms. Verhandlungen der Internationalen Vereinigung fur theorentische und angewandte. *Limnologie*, 21: 227–230.

Juttner, F., Leonhardt, J. and Mohren, S., 1983. Environmental factors affecting the formation of mesityloxide, dimethylallylic alcohol and other volatile compounds excreted by *Anabaena cylindrica*. *J. General Microbiology*, 129: 407–412.

Katayama, T., 1962. Volatile constituents. In: *Physiology and Biochemistry of Algae*, (Ed.) R.A. Lewin. Academic Press, New York, pp. 467–473.

Keating, K.I., 1977. Allelopathic influence on blue-green bloom sequence in an eutrophic lake. *Science*, 196: 885–887.

Khailov, K.M., 1974. Objectives problems and certain methodological conditions in the study of coastal communities and ecosystems. In: *Biochemical Trophodynamics in Marine Coastal Ecosystems*, (Ed.) K.M. Khailov. Naukova Dumka, Kiev, pp. 14– 27.

Koch, M., Glombitza, K.W. and Eckhardt, G., 1980. Phlorotannins of phaeophyceae *Laminaria ochroleuca*. *Phytochemistiy*, 19: 1821–1823.

Kund, K., 1992. Biochemical analysis of certain alga growing in polluted pond. *Ph.D. Thesis*, Bihar University, Muzaffarpur, Bihar.

Kund, K. and Verma, B.N., 1994. Biological assessment of algal bloom with special reference to phenolics. In: *Advances in Plant Science Research*. Int. Book Distributors, Dehradun, pp. 165–178.

Larsen, B. and Haug, A., 1960. Salinitet og voksesteds variasjon. Norsk Inst. Tang. og Tareforsk Forelob. Rep., 112: 1–14.

Lee, T.T. and Skoog, F., 1965. Effect of substituted Phenols on bud formation and growth of tobacco tissue cultures. *Physiol. Plant.*, 18: 386–402.

McKey, D., 1979. The distribution of secondary compounds within plants. In: *Herbivores: Their Interaction with Secondary Plant Metabolites*, (Eds.) G.A. Rosenthal and D.H. Janzen. Academic Press, New York, pp. 55–133.

Meravy, L., 1987. Phenolic substances in tissue cultures of *Centaurium erythraea. Biol. Plant*, 29: 81–87.

Mohan, M. and Mukerjee, K.G., 1979. Some biologically active extracellular products of blue-green algae. *Phykos*, 18(1–2): 73–82.

Mueller, W.C. and Rodehorst, F., 1977. The effect of some alkaloids on the ultrastructure of phenolic containing cells in the endodermis of cotton roots. In: *35th Ann. Proc. Electron Microscopy Soc. Amer.*, Boston Mass, pp. 544–545.

Munda, I., 1960. Der Einfluss von Chlorinitat auf die chiemische Zusammensetzung, Wachstum und Frukitifizierungstatigkeit von einign Fucaceen. Norsk Inst. Tang– og Tareforsk. Forelab. Rep., 3: 1–87.

Munda, I., 1962a. Odnos sadrzaja fizoda ireduktivnih supstanca u nekim jadranskim smedjim algama. *Acta Adriat.*, 10: 3–11.

Munda, I., 1962b. Geographical and seasonal variations in the chemical composition of some Adriatic brown algae. *Nova Hedw*, 4: 263–274.

Murphy, T.P., Lean, D.R.S. and Nalewajko, C., 1976. Blue-green algae: The excretion of Fe-selective chelators enables them to dominate other algae. *Science*, 192: 900.

Myklestad, S., Eide, I. and Melsom, S., 1978. Exchange of heavy metals in *Ascophyllum nodosum* (L.) Le Jol. *in situ* by means of transplanting experiments. *Env. Poll.*, 16: 277–284.

Newell, R.C. and Lucas, M.I., 1981. The quantitative significance of dissolved and particulate organic matter released during fragmentation of kelp in coastal water. Kieler Meerreforsch. *Sonderh*, 5: 356–369.

Nozawa, K., 1968. The effect of Peridinium toxins on other algae. *Bull Misaki, Mar. Biol. Inst. Kyoto Univ.*, 12: 21–24.

Pedersen, A., 1984. Studies on phenol content and heavy metal uptake in Fucoids. *Hydrobiologia*, 116/117, 11: 498–504.

Pedersen, M., Saenger, P. and Fries, L., 1974. Simple brominated phenols in red algae. *Phytochemistry* 13: 2273–2279.

Pospisil, F., Cvikrona, M. and Hrubcova, M., 1983. Oxidation of gallic acid by an enzyme preparation isolated from culture medium of *Nicotiana tabacum* cell suspension. *Biol. Plant*, 25: 373–377.

Proctor, V.W., 1957. Studies of algal antibiosis using *Haematococcus* and *Chlamydomonas*. *Limnol. Oceanogr.*, 2: 125–139.

Ragan, M.A., 1976. Physodes and the phenolic compounds of brown algae. Composition and significance of Physodes *in vivo*. *Bot. Mar.*, 19: 145–154.

Ragan, M.A., 1984. *Fucus* "lignin": A reassessment. *Phytochemistry*, 23: 2029–2032.

Ragan, M.A., 1985a. Brown algal polyphenols: synthesis of "fucophlorethol A" octamethyl ether (2.2′, 4.6,6′-pentamethoxy-4′-(2.4.6-trimethoxyphenoxy biphenyl). *Can.J. Chem.*, 63: 291–293:

Ragan, M.A., 1985b. The high-molecular weight polyphloroglucinols of the marine brown alga *Fucus vesiculosus* L. degradative analysis. *Can.J. Chem.*, 63: 294–303.

Ragan, M.A. and Craigie, J.S., 1976. Physodes and the phenolic compounds of brown algae: Isolation and characterization of *phloroglucinol polymers* from *Fucus vesiculosus* (L.). *Can. J. Biochem.*, 54: 66–73.

Ragan, M.A. and Craigie, J.S., 1978. Phenolic compounds in brown and red algae. In: *Handbook of Phycological Methods*, (Eds.) J.A. Hellebust and J.S. Craigie. Cambridge Univ. Press, Cambridge, 2: 157–179.

Ragan, M.A. and Craigie, J.S., 1980. Quantitative studies on brown algal phenols IV. Ultraviolet spectrophotometery of extracted polyphenols and implications for measuring dissolved organic matter in sea water. *J. Exp. Mar. Biol. Ecol.*, 46: 231–239.

Ragan, M.A. and Glombitza, K.W., 1986. Phlorotannine, brown algal polyphenols. In: *Progress in Phycological Research*, (Eds.) Round and Chapman. Biopress Ltd., Bristol, 4: 129–241.

Ragan, M.A. and Jensen, A., 1977. Quantitative studies on brown algal phenols. I. Estimation of absolute polyphenol content of *Ascophyllum nodosum* (L.) Le Jol. and *Fucus vesiculosus* L. *J. Exp. Mar. Biol. Ecol.*, 30: 209–221.

Ragan, M.A. and Jensen, A., 1978. Quantitative studies on brown algal phenols. II. Seasonal variation in polyphenol content of *Ascophyllum nodosum* (L.) Le Jol. and *Fucus vesiculosus* L. *J. Exp. Mar. Biol. Ecol.*, 34: 245–258.

Rao, K.N., Reddy, A.N., Rao, A.V.N. and Krishna, H.R.G., 1978. Phenolic compounds in Chick pea (*Cicer arietinum* L). *Indian J. Exp. Biol.*, 16: 1213– 214.

Rao, S., Chennaveeraiah, M.S. and Nataraja, K., 1983. Effect of indole-3-acetic acid on penolic accumulation and peroxidase activity in cell cultures of *Musseanda Rosea* L. *Indian J. Exp. Biol.*, 21: 653–655.

Ray, S.D., 1986. GA-ABA-phenol interaction in the control of growth: Phenolic compounds as effective modulators of GA–ABA interaction in radish seedlings. *Biol. Plant*, 28: 361–369.

Ray, S.D. and Laloraya, M.M., 1984. Interaction of gibberelic acid, abscisic acid and phenolic compounds in the control of hypocotyl growth of *Amaranthus caudatus* seedlings. *Can. J. Bot.*, 62: 2047–2052.

Reznikov, V.M. and Mikhaseva, M.F., 1976. Lignin of lower plants. IV. Isolation and investigation of the lignin of algae. *Khim. Drev*, 4: 76–79.

Reznikov, V.M., Mikhaseva, M.F. and Zilbergleit, M.A., 1978. Lignin of the alga *Fucus vesiculosus*. *Khim Prir: Soedin.*, 5: 645–647.

Rice, E.L., 1974. *Allelopathy*. Academic Press, New York, pp. 353.

Ronnberg, O. and Ruokolahti, C., 1986. Seasonal variation of algal epiphytes and phenolic content of *Fucus vesiculosus* in a northern Baltic archipelago. *Ann. Bot. Fennici*, 23: 317–323.

Sahay, A.P., 1990. Studies on the metabolites at different growth phases of certain algae. *Ph.D. Thesis*. Bihar University, Muzaffarpur, Bihar.

Sahay, A.P., Das, P.K. and Verma, B.N., 1995. Changes in the content of phenolic substances during growth of a cyanobacterium *Oscillatoria curviceps* var. *angusta*. *Phykos*, 34(1 and 2): 9–16.

Saleh, N.A.M. and El-Lakany, H., 1979. A quantitative variation in the flavonoids and phenolics of some *Casuarina* species. *Biochem. Syst. Ecol.*, 7: 13–15.

Scalbert, A., Monties, B. and Favre, J.M., 1988. Polyphenols of *Quercus robur:* Adult tree and in vitro grown cali and shoots. *Phytochemistiy*, 27(11): 3483–3488.

Scalbert, A. and Haslam, E., 1987. Polyphenols and chemical defence of the leaves of *Quercus robur*. *Phytochemistry*, 26(12): 3191–3195

Schnepf, E., 1976. Morphology and Cytology of storage spaces. In: *Secondary Metabolism and Coevolution*, (Eds.) M. Luckner, K. Mothes and L. Nover. Deutsch Akad. Naturforsch. Leopoldina, Halle, pp. 23–44.

Seshadri, R. and Sieburth, J.McN., 1971. Cultural estimation of yeasts on seaweeds. *Appl. Microbiol.*, 22: 507–515.

Shah, R.R., Shylaja, R. and Mehta, A.R., 1976. Distribution of polyphenol and development of peroxidase activity during germination in *Crotalaria*. *Indian J Exp. Biol.*, 14: 299–301.

Shah, R.R. and Mehta, A.R., 1978a. Effect of carbohydrates on growth, formation of phenolic com pounds and related enzymes in callus cultures of Crotalaria. *Indian J. Exp. Biol.*, 16: 768–770.

Shah, R.R. and Mehta, A.R., 1978b. Effect of nitrogen nutrition on growth and phenolic accumulation in *Crotalaria* tissues cultured *in vitro*. *Indian J. Exp. Biol.*, 16: 1144–1147.

Shukla, C.P., 1989. Biochemical studies on certain members of blue-green algae. *Ph.D. Thesis*. Bihar University, Muzaffarpur, Bihar.

Sieburth, J.McN., 1969. Studies on algal substances in the Sea. III. The production of extracellular organic matter by littoral marine algae. *J.Exp. Mar Biol. Ecol.*, 3: 290–309.

Sieburth, J.McN. and Conover, J.T., 1965. *Sargassum* tannin: An antibiotic which retards fouling. *Nature (London)*, 208: 52–53.

Singh, M., Singh, S.S. and Sanwal, G.G., 1978. A new colorimetric method for the determination of phenolics. *Indian. J. Exp. Biol.*, 16: 712–714.

Sinha, S.K.P., 1980. Some biochemical and physiological studies on certain medicinal plants of Euphorbiaceae. *Ph.D. Theis*, Bhagalpur University, Bhagalpur, Bihar.

Skipnes, O.M., 1974. Opptak og lokalisering. av sink *Ascophyllum nodosun*, (L.) Le Jolis. *Thesis*, Univ. Trondheim. N.T.H.pp. vi+313.

Stafford, H.A., 1974. The metabolism of aromatic compounds. *Ann. Rev. Plant Physiol.*, 25: 459–486.

Stahl, E., 1988. Pflanzen und Schnecken. Biologische studie uber die Schutzmittel der pflanzen gegen Schneckenfra B. Jena. *Z.Med. Naturwiss*, 22: 557–684.

Thiman, K.V., Tomaszewski, M. and Porter, W.L., 1962. Growth promoting activity of caffeic acid. *Nature*, 193: 1203.

Verma, B.N., 1991. Biochemical and Biological evaluation of algae inhabiting wild water in North Bihar. Project report (Govt. of India, Ministry of Environ, and Forest, Ref. 19/4/87, RE).

Verma, B.N. and Sahay, A.P., 1994. Algal phenolics: A protocol. In: *Microbes and Environment*. Narendra Publishing House, Delhi, pp. 51–74.

Wolf, J.M. and Rice, E.L., 1979. Allelopathic interactions between algae. *J.Chem. Ecol.*, 5: 533–542.

Zadernowski, R. and Kozlowska, H., 1983. Phenolic acids in soybean and rapeseed flours. *Lebensum Wiss. Technol.*, 16: 110–114.

Zavodnik, N. and Jensen, A., 1981. Studies on phenolic compounds of some brown algae. In: *Proc. Int. Seaweed Symp.*, 8: 655–660.

Zenk, M.H., 1967. Biochemie und Physiologie Sekundarer pflanzenstoffe. *Ber. Deutch. Bot.Ges.*, 80: 573–591.

Chapter 8

Ultraviolet-B (UV-B) Radiation: Effects and Protective Mechanisms in Cyanobacteria

Madhu B. Tyagi[1], Prabhat N. Jha[2] and Ashok Kumar[3]

[1]*Department of Botany, MMV*
[2]*Department of Biosciences, BITS, Pilani, Rajasthan*
[3]*School of Biotechnology,*
Banaras Hindu University, Varanasi – 221 005

ABSTRACT

The continuing degradation of stratospheric ozone layer mainly due to release of man- made atmospheric pollutants such as chlorofluorocarbons (CFCs) has resulted in the increase of solar ultraviolet-B radiation (UV-B; 280-320 nm). UV-B is a small (less than 1 per cent of total energy) but highly active component of the solar spectrum that can penetrate to biologically significant depths in lakes, ponds, rivers and oceans. It has been demonstrated that UV-B can cause wide ranging effects including mutagenesis, destruction of proteins/enzymes, inhibition of a number of vital metabolic processes and formation of thymine dimers in DNA leading to death of a number of organisms. Cyanobacteria being strict photoautotrophic organisms always grow in lighted habitats and thereby always sense and face the stress of solar UV-B radiation. Damaging effects of increased UV-B radiation has been reported in a number of cyanobacterial species. These effects include inhibition of growth and survival, decrease in photosynthetic pigments, loss of PS II and RuBisco activity, inactivation of nitrogenase enzyme and inhibitory effects on several other metabolic processes. Among photosynthetic pigments, phycobiliproteins seem to be the primary target of action since phycobilisomes are completely disintegrated after UV-B exposure.

The degree of damaging effects brought about by UV-B varies in different species suggesting the existence of protective mechanisms against the above harmful stress in many species. A number of species counteract the damaging effects of UV-B by synthesizing UV protective compounds such as mycosporine-like amino acids (MAAs) and the sheath pigment, scytonemin. These compounds are known to act as sunscreen pigment and their synthesis is induced by UV-B treatment to the cells. In this contribution an attempt has been made to highlight the main effects of UV-B radiation on cyanobacteria and the role of protection strategies in mitigating UV-B damage.

Keywords: *UV-B, Cyanobacteria, Survival, Enzymes, UV-absorbing compounds, MAAs, Scytonemin, Photoreactivation, DNA repair.*

Introduction

Ultraviolet radiation is an important component of solar radiation having many deleterious effects on biological systems (Hader *et al.*, 1998; Vincent and Roy, 1993). Based on wavelength, it has been arbitrarily divided into UV-A (320-400 nm), UV-B (280-320 nm) and UV-C (<280 nm). Prior to their attenuation by the earth's atmosphere these wavelengths represent 6.3, 1.5 and 0.5 per cent of the total solar irradiance corresponding to an intesity of 85.7, 21.7 and 6.4 W/m^2 respectively (Frederick *et al.*, 1989). Among different kinds of UV radiations UV-C is strongly mutagenic and lethal whereas UV-A is biologically least reactive component. Ozone present in the stratosphere acts as an efficient screen for short wavelength UV radiation absorbing all UV-C radiation, but allowing all UV-A to reach the earth's surface. On the other hand it allows some UV-B radiation to reach the ground surface. Ozone column affects both the intensity and quality of UV-B radiation; with decrease in ozone column, the intensity increases and the wavelength composition is shifted to proportionately shorter wavelength (Frederick *et al.*, 1998; Madronich *et al.*, 1998).

Continuing depletion of stratospheric ozone caused by the anthropogenic emission of chlorofluorocarbons and other gaseous pollutants into the atmosphere has resulted in a specific increase in UV-B radiation (Crutzen,1992; Molina and Molina, 1992; Madronich *et al.*, 1998). Such increases in UV-B radiation are most pronouched in Antarctica, but similar trends have been reported at other latitudes also (Kerr and McElroy, 1993). It is felt that if ozone concentration goes on decreasing, it may encompass a broader range of latitudes in next few decades (Madronich *et al.*, 1998). Besides ozone, to a certain extent, variables such as cloud cover, sun-angle, surface retlectanls, aerosols and various tropospheric pollutants also affect the intensity of UV-B radiation reaching the ground thereby leading to a marked latitudinal and seasonal variations in intensity varying from 0.4–2.5 W/m^2 (Madronich *et al.*, 1998).

The continued increase in UV-B radiation has been demonstrated to affect both primary production and higher trophic levels in biological food webs causing instabilities in ecosystems (Smith *et al.*, 1992; Hader *et al*, 1998). Besides exerting a variety of adverse effects on terrestrial vegetation, UV-B radiation is also known to penetrate fairly deep in freshwater and marine environments (Smith *et al.*, 1992) where it affects the plankton in the eupholic zone as well as benthic organisms (Hader *et al.*, 1998). Cyanobacteria are a large and heterogeneous group of prokaryotic microorganisms which form an important component of the nitrogen-fixing microbial population both in aquatic and terrestrial ecosystems (Stewart, 1980). They are capable of growing in diverse habitats ranging from hot springs to Arctic and Antarctic climates (Stanier and Cohen-Bazire, 1977) and are the only organisms capable of simultaneously fixing nitrogen and carrying out oxygenic photosynthesis.

The damaging effects of UV-B radiation on cyanobacteria are of special concern because they form a dominant flora of nitrogen-fixing microorganisms in tropical rice-fields and enrich the fertility of soil (Singh, 1961). As such any adverse effect on their existence may lead to an imbalance in nitrogen status of entire ecosystem (Sinha and Hader, 1998). Besides rice-fields, since most of the habitats occupied by cyanobacteria are directly exposed to sunlight, they are highly vulnerable to the effects of increased solar UV-B radiation. Phylogenetically the group originated during pre-cambrian era before the existence of ozone shield, it is presumed that they faced more intense solar UV- radiation as compared to other organisms. In general they are more tolerant to environmental stresses than arc eukaryotic algae, and are believed to have faced a wide spectrum of environmental stresses.

Like other photosynthetic organisms cyanobacteria sense and respond to UV-B radiation although the degree of response significantly vary from species to species. In general the damaging effects of UV-B radiation on cyanobacteria can be classified into two main categories, damage to DNA leading to mutations, and damage to physiological and biochemical processes. Following absorption of UV radiation by DNA formation of a variety of photoproducts may occur. These photoproducts include: cyclobutane- type dimers of thymine, cytosine and uracil, pyrimidine adducts, photohydrates, and DNA- protein cross links. On the other hand physiological changes seen after increased UV-B radiation exposure may include, growth inhibition followed by loss of survival, inhibition of photosynthetic processes, loss of enzymatic activity, increased synthesis of UV-B absorbing pigments, induction of stress protein, and others (Hader and Worrest, 1991; Quesada *et al.*, 1999; Sinha *at al.*, 2002).These changes could result from a number of primary UV-B events such as direct photosynthetic damage, loss of permeability/membrane changes, pigment destruction, protein/enzyme inactivation, reduced DNA and protein synthesis, reduced uptake of nutrients, hormone inactivation and signal transduction through phycochrome or signal transduction via a specific UV-B photoreceptor (Vincent and Roy, 1993; Sinha and Hader, 2002; Kumar *et al.*, 2003). Above mentioned events are partly speculative and to determine precisely which factors are involved, one has to critically determine the action spectrum and the kinetics of response for each event. In this review, an attempt has been made to give a brief account of UV-B–mediated effects on various physiological and biochemical processes of cyanobacteria. Together with this, the protective mechanisms operative in cyanobacteria so as to counteract the damaging effects of UV-B radiation are also presented under separate head.

UV-B Mediated Effects on Cyanobacteria

Studies conducted so far reveal that UV-B radiation affects the growth, survival, pigmentation, orientation, general metabolism, photosynthesis, nitrogen fixation and assimilation of nitrogen in diverse cyanobacterial species (Table 8.1). There is still considerable discussion about the primary targets of UV-B radiation damage but it is generally felt that the deleterious effects are in part due to the direct effects of UV-B radiation on membrane component (Sinha and Hader, 2002), DNA (Quesada *et al.*, 1999), enzymes (Dohler *et al.*, 1986), growth regulators and the photosynthetic apparatus, particularly the reaction center of photosystem II and RuBisco (Wilson *et al.*, 1995; Sinha and Hader, 2002) or to its indirect effect through the formation of reactive species such as super oxide (O_2), hydroxyl radicals (OH^-) or hydrogen peroxide (H_2O_2) (He and Hader, 2002). Although DNA has an absorption spectrum that peaks at 260 nm, a significant absorption also occurs in the UV-B range (280-320 nm). This results in production of cytotoxic lesions such as pyrimidine base pairs. Such lesions, if left uncorrected, interfere with DNA replication and transcription ultimately leading to mutation or death. Proteins also show considerable absorption in the UV-B region, which is primarily due to the presence of aromatic amino acids such as tryptophan and tyrosine. Such UV-B-induced damage/denaturation of various proteins/enzymes may also cause changes in the structure and function ultimately leading to changes in membrane permeability and functioning of various enzymes. The known effects of UV-B radiation on several metabolic events of cyanobacteria are briefly discussed below.

Table 8.1: UV-B-Induced Effects in Cyanobacteria

Process	*Effects*	*References*
Movement and orientation	Inhibition; impairment of phototaxis and photophobic responses	Hader *et al.* (1998); Donkor and Hader (1991): Donkor *et al.* (1993); Hader and Worrest (1991)
Growth and survival	Inhibition and killing	Newton *et al.* (1979); Tyagi *et al.* (1992); Sinha *et al.* (1995a,b); Quesada and Vincent (1997); Sinha and Hader (2002); Kumar *et al.*(2003)
Cell differentiation	Delayed heterocyst differentiation and akinete formation; increased akinete production; changes in spacing pattern of heterocysts	Blakefield and Harris (1994) Sinha *et al.* (1996).
Cell membrane	Membrane disruption, altered membrane permeability, peroxidation of membrane lipids; release of ^{14}C-labelled compounds; opening of membrane bound calcium channels	Dohler *et al.* (1986); Sinha *et al.* (1997); Rai *et al.* (1998); Richter *et al.* (1999)
Photosynthetic pigments		
Chlorophyll-a	Decrease, photobleaching	Tyagi *et al.* (1992); Quesada and Vincent (1997); Sinha and Hader (2002)
Phycobiliproteins	Decrease in phycobilin content and disintegration of phycobilisomes	Sinha *et al.* (1995 a,b; 1997)

Contd...

Table 8.1–Contd...

Process	*Effects*	*References*
Carotenoids	Increase in carotenoids especially echinenone and myxoxanthophyll	Vincent and Roy (1993) Ehling-Schulz *et al.*(1997)
Photosynthesis Reaction centres	Impairment: degradation of D_1 and D_2 polypeptides	Sass *et al.* (1997); Campbell *et al.* (1998)
Photosystem II	Instant decrease of O_2 evolution	Hader and Worrest(1991)
Rubisco	Inhibition	Wilson *et al.* (1994); Sinha *et al.* (1997)
$^{14}CO_2$ uptake	Inhibition	Newton *et al.* (1979); Tyagi *et al.* (1992); Sinha *et al.* (1995); Cockell and Rothschild (1999)
Nitrogen metabolism NO_3^- uptake	Inhibition	Dohler *et al.* (1986); Rai *et al.* (1998)
Nitrate reductase	Stimulation of *in-vivo* activity	Tyagi *et al.* (1992)
Nitrogenase	Inhibition of *in-vivo* activity leading to inactivation	Newton *et al.* (1979); Tyagi *et al.* (1992); Sinha *et al.* (1996); Kumar *et al.* (2003)
Glutamine synthetase	Inhibition of *in vivo* activity	Sinha *et al.* (1995a)
Cellular ammo acids pool	Increase	*Dohler et al.* (1986)
General protein profile	Elimination of low molecular weight protein bands	Sinha *et al.* (1995b); Sinha and Hader (2002)
Protein synthesis	Inhibition	Kumar *et al.* (2003)
DNA	Thymine-dimer formation.	Sinha *et al.* (2001)

Movement/Orientation

As such cyanobacteria employ motility and orientation mechanisms to escape harmful UV conditions and high light intensities but it has been reported that these avoidance mechanisms themselves are severely affected by UV-B radiation, resulting in a decrease or loss of their ability to escape from harmful UV conditions (Donkor and Hader,1991; Donkor *et al.*, 1993). Filamentous cyanobacteria such as *Anabaena variabilis, Oscillatoria tenuis* and *Phormidium uncinatum* use gliding motility to optimize their position in their microhabitat. Donkor *et al.*(1993) found a drastic reduction in the percentage of motile filaments of the above cyanobacteria within 10-30 min of exposure to unfiltered solar radiation at Ghana (4.30° N). Removal of UV-B radiation by an appropriate long-pass or band-pass filters resulted in higher percentage of motility in most cases suggesting the inhibitory effect of UV-B radiation on gliding motility of these cyanobacteria. Likewise, inhibition of phototactic orientation and photophobic responses reduce the ability of the organisms to orient themselves in their photo-environment. In the event of failure to quickly respond to changes in the photo-environment their death may occur (Hader and Worrest, 1991).

Cellular Differentiation

Differentiation of vegetative cells into heterocysts or akinetes is also affected by UV-B radiation. Blakefield and Harris (1994) observed delayed heterocyst differentiation and akinete formation in *Anabaena aqualis* following exposure to UV-B radiation although increase in akinete production was noted. It has been suggested that UV-B exposure alters the C: N ratio which in turn affects the spacing pattern of heterocysts in a filament (Sinha *et al.*, 1996). A decrease in the level of major heterocyst polypeptides (26, 54 and 55 kDa) following UV-B treatment has been noted which is presumed to be responsible for the disruption of multi-layered heterocyst wall, the essential component for maintaining the active form of nitrogenase. Role of UV-B radiation in the opening of specific calcium channels in the heterocysts of *Anabaena* sp. has also been demonstrated (Richter *et al.*, 1999). Whether this opening is mediated through signal transduction pathway and/or has any significance in heterocyst formation and function is yet to be proved.

Growth and Survival

UV-B radiation has been shown to severely affect the growth and survival of several rice-field and Antarctica cyanobacteria (Tyagi *et al.*, 1992; Sinha *et al.*, 1995a, 1996; Quesada and Vincent, 1997). Most of these studies have been performed under laboratory conditions using artificial lamps specifically emitting UV-B wavelengths. Long-term studies in natural habitats under ambient levels of UV-B radiation are lacking. Various species differ with respect to their tolerance to UV-B radiation and even closely related strains show differential sensitivity. Complete killing and loss of survival within 120-180 min of UV-B exposure has been reported for several cyanobacteria (Tyagi *et al.*,1992; Sinha *et al.*, 1995a). Strains such as *Nostoc commune* and *Scytonema* sp. whose filaments are embedded in a mucilage sheath are more tolerant to UV-B radiation than the species which do not have such mucilage covering (Sinha *et al.*, 1996). In a *Nostoc* strain (isolated from the rice-fields of Varanasi) the phycoerythrin rich (brown) form was found to be more tolerant to UV-B radiation than the phycocyanin rich (blue-green) form of this strain (Tyagi *et al.*, 1992). Two Antarctica cyanobacteria, *Phormidium murrayi* and *Oscillatoria priestleyi,* isolated from a common habitat showed differential sensitivity to UV-B radiation. In contrast to complete suppression of growth in *Oscillatoria priestleyi,* only a 62 per cent inhibition of growth of *Phormidium murrayi* was observed following exposure to the same quantum of UV-B intensity (Quesada and Vincent, 1997)

Pigment Content/Spectroscopic Analysis

Several workers have demonstrated that UV-B radiation causes photobleaching of photosynthetic pigments, a decrease in phycocyanin/chlorophyll *a* ratio and disintegration of phycobilisome complex leading to the release of phycobiliprotein pigment complex (Hader and Worrest 1991; Tyagi *et al.*, 1992; Sinha *et al.*, 1995b, 1997; Quesada and Vincent 1997). Phycobiliprotein seems to be the most sensitive pigment most probably due to its water soluble property. *In vivo* absorption spectra of *O. priesteyi* and *P.murrayi* pigments showed a strong response to UV-B exposure. During a 4 day growth assay, phycocyanin and to a lesser extent chl *a* dropped with increasing UV-B flux (Quesada and Vincent 1997). Sinha *et al.* (1995b, 1997) have reported a decrease in the phycobiliprotein content and disassembly of phycobilisomal complex following UV-B irradiation in a number of cyanobacteria. The pattern of fluorescence emission spectra of phycobiliproteins observed after UV-B irradiation suggests an impairment of energy transfer from the accessory pigments to the photosynthetic reaction centers (Sinha *et al.*, 1995b, 1997). Exposure of cyanobacterial cultures to intense UV-B radiation causes bleaching of phycocyanin and phycoerythrin which is in accordance with the disintegration of phycobilisomes.

Photosynthetic Events/Processes

Photosynthetic parameters such as $14CO_2$ uptake, O_2 evolution and ribulose-1, 5 bisphosphate carboxylose (RuBisco) activity are greatly inhibited following exposure to UV-B radiation (Wilson *et al.*, 1995; Quesada and Vincent, 1997; Sinha *et al.*, 1997). Photosynthetic O_2 evolution of a *Phormidium* strain from Lake Baikal declined within minutes of exposure to solar irradiation (Hader and Worrest, 1991). Several workers have reported an inhibition of $^{14}CO_2$ uptake by a number of cyanobacteria following exposure to UV-B irradiation (Newton *et al.*, 1979; Tyagi *et al.*, 1992; Sinha *et al.*, 1996). In a study of diurnal variability of the effect of solar UV-radiation on photosynthesis in an algal mat comprising of *Lyngbya aestuarii*, Cockell and Rothschild (1999) found a mean increase in the total primary productivity (measured in terms of $^{14}CO_2$ uptake) over the day when solar UV-B was screened out. It is felt that RuBisco may undergo a number of modifications such as photo-degradation, fragmentation and denaturation of polypeptide chain, change in active site and increased solubility of membrane proteins (Sinha and Hader, 1998). In addition supply of ATP and $NADPH_2$ may also be inhibited which may cause inhibition of CO_2-fixing ability. The D_1 and D_2 polypeptides which are the major constituents of PS-II reaction centers are also degraded, even after exposure to intermediate levels of UV-B radiation (Sass *et al.*, 1997; Campbell *et al.*, 1998). UV-B radiation may also induce modifications and photo-degradation of membrane proteins, which may subsequently affect the integrity of thylakoid membrane thereby causing the damage of photosynthetic components which are other wise essential for active CO_2 fixation.

Nitrogen Metabolism

Among all the key enzymes involved in nitrogen metabolism, the nitrogen-fixing enzyme nitrogenase has been found to be extremely sensitive to UV-B radiation (Newton *et al.*, 1979; Tyagi *et al.*, 1992; Kumar *et al.*, 1996a; Kumar *et al.*, 2003)). In a *Nostoc* sp, nitrogenase activity was lost within 45 min of UV-B exposure whereas nitrate reductase and glutamine synthetase activity remained more or less unaffected (Tyagi *et al.*, 1992). Kumar *et al.* (2003) reported complete inactivation of nitrogenase activity within 25-40 min of UV-B exposure in a number of cyanobacteria including *Anabaena* sp., *Nostoc* spp., *Calothrix* sp. and *Scytonema* sp.(Figure 8.1, Table 8.2). Employing inhibitors of protein synthesis and PS-II activity they have demonstrated that restoration of nitrogenase activity in UV-B-treated cultures occurs only after fresh synthesis of nitrogenase polypeptide. The inactivation of activity most probably occurs due to complete damage to the nitrogenase polypeptide. It has been suggested that the aromatic amino acids present or the native structure of nitrogenase is responsible for the extremely susceptible nature of nitrogenase to UV-B radiation.

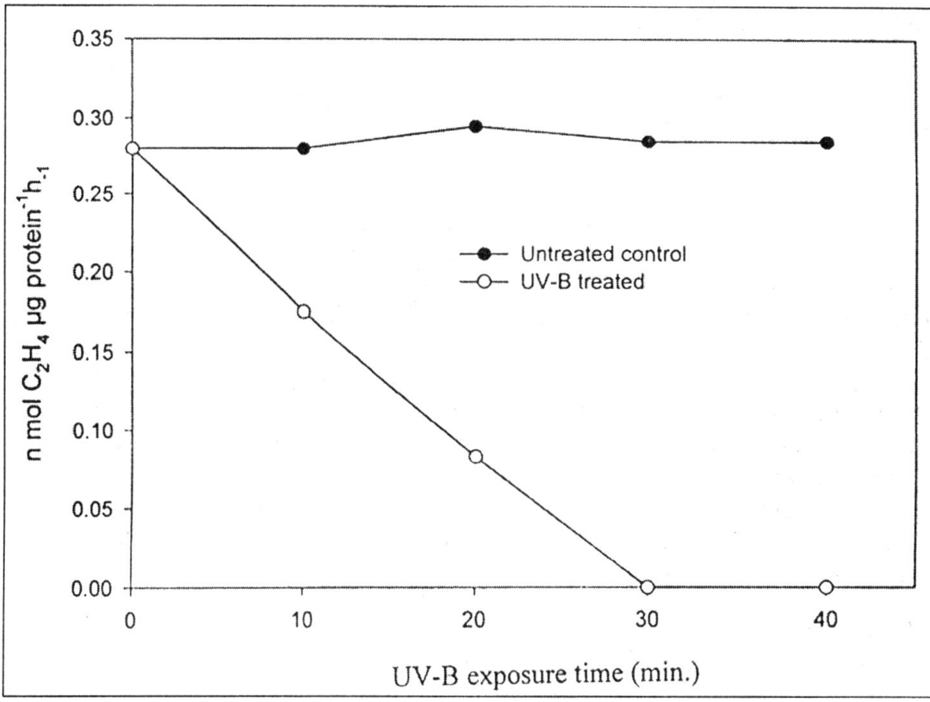

Figure 8.1: Effect of UV-B Treatment on Nitrogenase Activity. Actively N_2-fixing culture of *Anabaena* BT_2 was exposed to UV-B (2.5 Wm^{-2}) for different time periods and thereafter C_2H_2 reduction assay was made under fluorescent light. Results are based on average of three experiments performed independently under identical conditions.

They have further suggested that the estimation of nitrogenase activity may be used as a marker enzyme for assessing the impact of UV-B radiation in diazotrophic species.

Table 8.2: Test for the Inactivation of Nitrogenase Activity in Different Cyanobacteria following Exposure of UV-B radiation

Organisms[a]	*Time (min) Required for the Inactivation of Nitrogenase Activity*[b]
Anabaena BT 2	30
A. doliolum	30
A. cycadeae	25
Calothrix sp.	40
N. spongiaeforme	30
N. muscorum	25
N. commune	30
Scytonema sp.	40

a Equal volume of cultures having an optical density of 0.25 (at 663 nm) from all the species were exposed to UV-B (2.5 Wm^{-2}) for desired time period. Acetylene reduction assay was made at an interval of 10 min following UV-B exposure.

b Results represent the average of three independent experiments.

General Protein Profile

Exposure to UV-B radiation is known to causes a subtle change in the protein profile of several cyanobacteria (Sinha *et al.*, 1995b; Sinha and Hader, 2002). Increased duration of UV-B radiation exposure brings about a linear decrease in the number and quantity of protein bands of a number of cyanobacteria (Sinha *et al.*, 1997). In case of a *Nostoc* sp, the low molecular weight protein of around 20 kDa (α- β-monomers of phycocyanin) were the most affected one. It is interesting to find that UV-B radiation has a drastic effect specifically on low molecular weight proteins. In a comparative study on the effects of UV-B treatment on total protein profile of several cyanobacteria, Sinha *et al.* (1995a,b), reported complete loss of protein bands between 14.5 and 45 kDa after 90 and 120 min of UV-B exposure in *Nostoc carmium* and *Anabaena* sp. However protein bands of relatively higher molecular weight *viz.*, 55 and 66 kDa were unaffected even after 120 min of UV-B exposure. In species such as *N. commune* and *Scytonema* sp, which contain a mucilage covering/sheath, complete elimination of bands occurred only after 150 min of UV-B exposure (Sinha *et al.*, 1995a,b). The increased tolerance to UV-B radiation in certain species may be ascribed to the presence of mucilage covering or photoprotectans such as scytonemin and mycosporine-like amino acids (MAAs) in the cell.

Effect on DNA

Although DNA has an absorption spectrum that peaks at 260 nm, but a significant absorption also occurs between 280-320 nm. A variety of lesions in DNA after UV-B radiation treatment have been identified. These lesions result either due to direct absorption of UV-B radiation by DNA or from free radicals and reactive oxygen species formed by various photochemical processes. The main class of UV-induced lesions consist of dimeric photoproducts such as cis-syn cyclobutane pyrimidine dimers (CPDs) and pyrimidine (6-4) pyrimidone photoproducts (6-4 PPs) (Mitchell and Karentz,1993; Friedberg *et al.*, 1995; Lindahl and Wood, 2000). Dimers have been reported to inhibit the progress of DNA polymerase, and mammalian RNA polymerase is known to stall at both CPDs and 6-4 PPs (Brill, 1995). Photoproducts formation interferes with DNA synthesis and RNA transcription, ultimately leading to mutation or death of the organism.

The damaging effects of UV-B radiation on DNA and formation of various photoproducts are well known in a number of organisms including the Antarctic marine planktonic diatom communities (Karentz *et al.*, 1991a; Buma *et al.*, 1995). However little information is available regarding the specific effects of UV-B radiation on DNA of cyanobacteria, although few studies have been conducted with UV-C radiation (Levine and Thiel, 1987; Eker *et al.*, 1990). Sinha *et al.* (2001a) have demonstrated that exposure to UV-B radiation (10 W/m^2) could induce the formation of thymine dimers in three rice-field cyanobacteria. *Nostoc, Anabaena* and *Scytonema* sp. The frequency of dimer formation increased with increase in exposure time and a 120 min exposure of these cyanobacteria to UV-B radiation led to dimer frequency of 35-40 per cent. By using blotting and chemi-luminescence method these workers successfully developed a relatively quick, simple and efficient method for the quantitative analysis of thymine dimers in aquatic primary producers including cyanobacteria (Sinha *et al.*, 2001a).

Protection/Recovery Mechanisms

Since cyanobacteria are believed to have originated in the precambrian era at a time when ozone shield was absent (Schopf, 1992), they presumably faced high fluxes of UV radiation which must have acted as an evolutionary pressure leading to selection for efficient UV radiation protecting mechanisms. Tolerance of cyanobacteria to intense sunlight as well as UV radiation might have contributed to their success during early stages of colonization. As a consequence they have developed effective mechanisms to counteract the damaging effects of UV-B radiation. In case of microbes behavioral responses as well as external and internal mechanisms of protection against UV-induced damages have been reported and it has been proposed that several of these mechanisms for UV-protection and/or repair of UV-induced damage might have worked singly or in various combinations (Pierson *et al.*, 1993). These defense strategies (Table 8.3) include the avoidance of brightly lit habitats, UV-screening by cellular and extracellular photoprotectants, quenching reaction for phototoxic products and the efficient repair mechanisms which include DNA photorepair, light independent nucleotide excision repair and the UV-A/blue light-mediated repair of photosynthetic apparatus (Vincent and Quesada 1994; Han *et al.*, 2001). Protective strategies operating especially in cyanobacteria are summarized below.

Table 8.3: UV-B Protective Mechanisms in Cyanobacteria

Type of Protective Strategy	*Mode of Protection*	*References*
Avoidance	Thick mucilagenous matting and vertical stratification, presence of mucilage covering	Vincent and Quesada (1994); Sinha *et al.* (1995a); Quesada and Vincent (1997)
	Escape from ultraviolet radiation by rapid migration and light dependent orientation strategies	Hader and Worrest (1991); Bebout and Garcia-Pichel (1995); Donkor *et al.* (1993); Quesada and Vincent(1997)
	Increased production of akinetes	Blakefield and Harris (1994)
Quenching	Increased production of carotenoids such as echinenone and myxoxanthophylls	Vincent and Quesada (1994); Ehling Schulz *et al.* (1997)
	Enzymatic defence against super-oxides and other reactive products	*Shibata et al.* (1991)
Synthesis of ultraviolet absorbing compounds/ pigments	Screening by MAAs and scytonemin	Garcia Pichel and Castenholz (1991); Garcia-Pichel *et al.* (1993); Ehling Schulz *et al.* (1997); Groniger *et al.* (2000); Sinha *et al.* (2001b);
	Screening by other UV-absorbing compounds	de Chazal and Smith (1994); Kumar *et al.* (1996b)
Repair mechanisms	UV-A/blue light- guided repair of photosynthetic apparatus	Han *et al.* (2001)
	UV-A/blue light- mediated photo-reactivation of DNA by photolyase	Vincent and Roy (1993); Sinha and Hader (2002)

Avoidance

Motile organisms use very precise orientation mechanisms to adjust their position within the water column in response to signals like light, gravity, chemical and temperature gradients and magnetic field of the earth. Motility and orientation strategies such as negative phototaxis, photokinetic and photophobic reactions and movements up and down in a water column or formation of microbial mat are useful mechanisms to counteract the adverse impact of UV-B radiation (Hader and Worrest, 1991; Donkor *et al.*, 1993). Gliding ability of Oscillatorean trichomes in microbial mat and optimization of vertical movement of bloom-forming cyanobacteria in a water column by gas vasicles (Reynolds *et al.*, 1987) are key strategies which protect cyanobacteria against UV-B radiation. As such the sinking and floating behaviour by a combination of gas vacuole and ballast act as an important means of protection against UV-B damage in a number of cyanobacteria. In Antarctica many cyanobacterial species form thick mucilage mats where the deep strata containing much of the biomass are completely screened from UV as well as short photosynthetically active radiation (PAR) (Vincent and Quesada, 1994). Cyanobacterial populations form crusts, films and massive mats attaining up to several centimeters in thickness and more frequently associated with deep pigmentation. Such morphological adaptations allow cyanobacteria to thrive and tolerate high PAR as well as elevated level of UV-B intensity. The motile trichomes of *O. priestleyi* remain well below the microbial mat surface during the continuous daylight of Antarctica summer and rise to the surface only after prolonged period of shading (Quesada and Vincent, 1997). Photosynthetic microbial blankets are commonly formed in environments where solar UV is a significant stress (Bebout and Garcia-Pichel, 1995). These benthic communities are rich in cyanobacteria and develop commonly in tropical and subtropical shallow water bodies or wetlands. Within the microbial blankets, individuals of cyanobacteria optimize their position by some sorts of vertical movement which seems to be induced and regulated by both visible and UV-B radiation (Bebout and Garcia-Pichel, 1995). Exposure to moderate doses of UV-B radiation to surface layers of microbial mats, dominated by *Microcoleus chthonoplastes*, from a salt lake in Egypt caused migration in response to added UV-B radiation (Bebout and Garcia-Pichel 1995). One of the survival strategics, reported by Blakefield and Harris (1994) is the increase in total number of akinetes produced, besides some delay in heterocyst production and akinete formation in certain species of cyanobacteria. Under such conditions the increased number of akinetes enhances the chance of survival.

Quenching

The interaction of UV-radiation with oxygen and organic compounds can produce highly reactive toxic intermediates such as super oxide (O_2), hydroxyl radicals (OH") or hydrogen peroxide. In cyanobacteria there are two quenching mechanisms namely, (*i*) possession of high concentration of carotenoids and, (*ii*) existence of enzymatic defenses against super oxides and related oxidants, that allow them to detoxify the highly reactive photochemical reactive products (Vincent and Quesada, 1994; Karsten and Garcia-Pichel, 1996).

Carotenoids are photosynthetic accessory pigments having a number of physiological roles. Besides acting as a light harvesting pigment they can also neutralize singlet state oxygen and also participate in xanthophyll cycle which consists of a set of reactions that dissipate excess solar energy. In some Antarctican benthic mats of cyanobacteria consisting of *Nostoc commune* and Oscillatorean sp, the surface layers contain high amounts of carotenoids such as myxoxanthophyll and canaxanthin, which give bright orange or pink colouration and act as photoprotectants (Vincent and Quesada, 1994). Liquid cultures of the terrestrial cyanobacterium, *Nostoc commune* derived from field material showed an increase in carotenoids especially echinenone and myxoxanthophylls following a 5 hr exposure to UV-B radiation (1 W/m^2) (Ehling-Schulz *et al.*, 1997). The total carotenoid to chlorophyll ratio after one day of UV-B irradiation was found to be 34-40 per cent higher than that in the untreated control. The presence of echinenone and myxoxanthophylls has also been demonstrated in the outer membrane of *Synechocystis* sp (Jurgens and Weckesser, 1995). It has been proposed that myxoxanthophyll and echinenone are outer membrane bound UV-B photoprotectants of *Nostoc commune* whose induction by UV-B radiation is a fast active SOS response whereas other UV- screening pigments are passive UV-screens produced against long-time exposure to UV-B radiation (Ehling-Schulz *et al.*, 1997).

Many species of cyanobacteria are known to possess superoxide dismutase which scavanges superoxide radicals and convert it to hydrogen peroxide. The hydrogen peroxide so produced is then converted to water and O_2 via a combined catalase-peroxidase system (Tel-or *et al.*, 1986). Since many of the effects of UV-B radiation are mediated by reactive oxygen species, these enzymatic defense strategies may have a role as protectant against UV-B radiation. Irradiation of *Anacystis nidulans* culture with near UV radiation (295-390 nm) has been shown to induce shock proteins which may include superoxide dismutase and related proteins (Shibata *et al.*, 1991). However the distribution of such enzymatic defense system in various cyanobacteria and their relationship with UV-B conditions has not yet been explored (Canini *et al.*, 1992; Vincent and Roy 1993).

Screening

An important mechanism to prevent UV-induced photodamage, is the synthesis of UV-absorbing compounds. Phenylpropanoids, mainly flavonoid derivatives and melanins are known to protect higher plants and animals, respectively, from harmful UV radiation. In cyanobacteria photoprotectants such as scytonemin and mycosporine-like amino acids (MAAs), having strong absorption in the UV-A and/ or UV-B region of the spectrum play an important role in allowing these organisms to grow and survive in habitats exposed to strong irradiation. In a terrestrial *Nostoc commune* strain, subjected to regular cycles of desiccation and rewetting, UV-absorbing compounds play a key role as UV-protectants (Ehling-Schulz *et al.* 1997). Role of different screening compounds in UV-B protection is discussed below.

Scytonemin

Scytonemin is a yellow-brown coloured pigment occurring in the extracellular sheath of a number of terrestrial cyanobacteria especially in those species growing in habitats exposed to intense solar radiation (Garcia-Piehel and Castenholz, 1991; 1993). It is a lipid soluble compound having *in vivo* absorption maximum at 370 run in the UV-A region but shows considerable absorption in the UV-B region also. Purified scytonemin shows absorption maximum at 386 nm although there being appreciable absorbance at 252, 278 and 300 nm. It is a novel dimeric pigment composed of indole and phenolic subunits (Table 8.4) with a molecular mass of 544 Da (Proteau *et al.*, 1993). Microspectrophotometric measurements of the transmittance of pigmented sheath and the quenching of ultraviolet excitation of phycocyanin fluorescence suggest that the pigment is effective in shielding the cells from incoming UV radiation and is able to reduce the incident UV-A radiation entering the cells to around 90 per cent (Garcia-Pichel and Castenholz, 1991; Garcia-Pichel *et al.*, 1992; Brenowitz and Castenholz, 1997).

Table 8.4: Photoprotective Compounds Reported from Cyanobacteria

Compound	*λ max (nm)*	*Molecular Structure*
MAAs		
Asterina–330	330	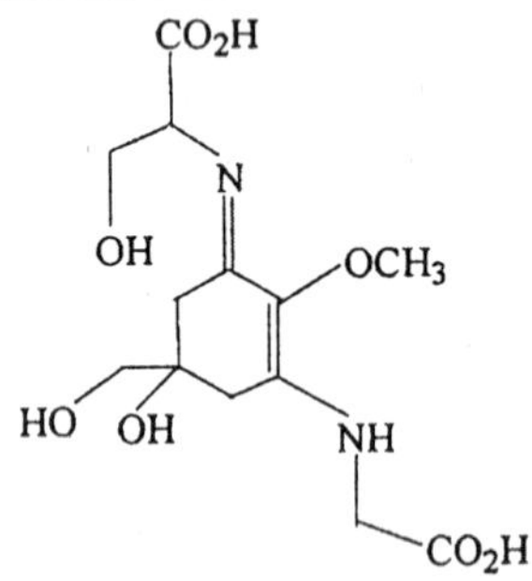
Mycosporine-glycine	310	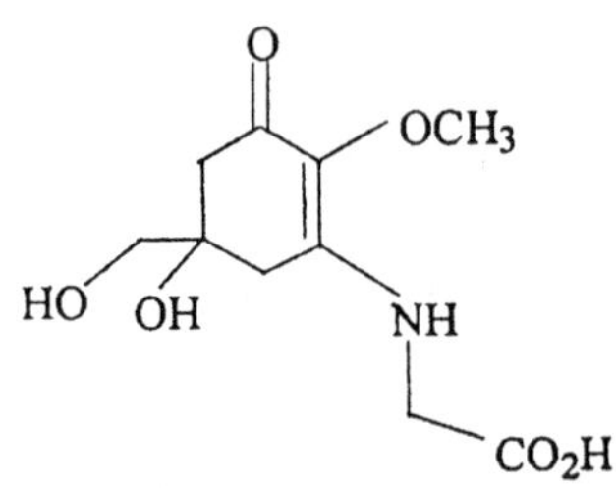
Palythinol	332	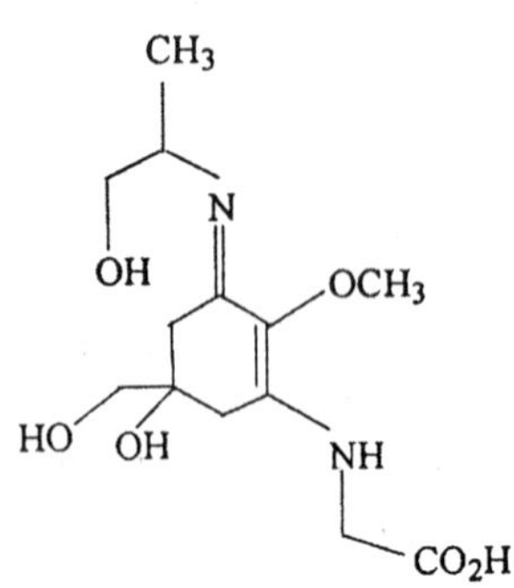

Contd...

Table 8.4–Contd...

Compound	*λ max (nm)*	*Molecular Structure*
Shinorine	334	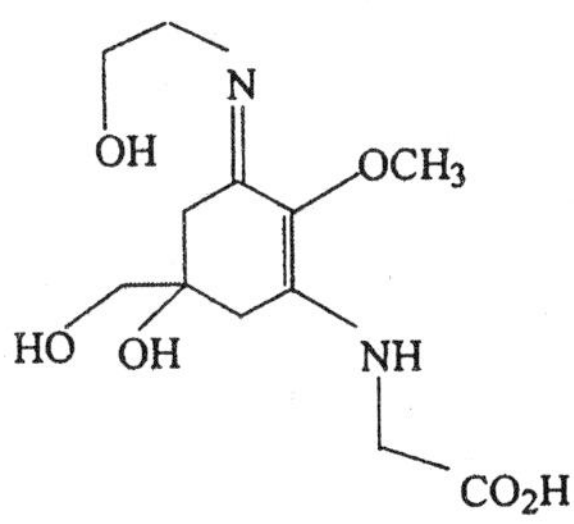
Scytonemin	386	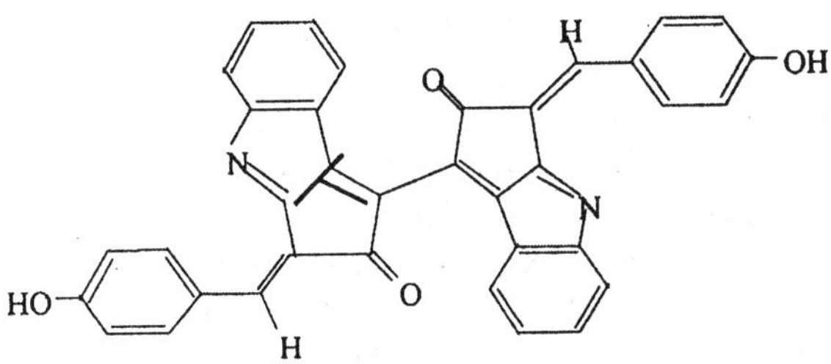

Modified from Groniger *et al.* (2000)

In *Chlorogloeopsis and Diplococcon* the rate of synthesis of scytonemin could be induced by exposure to UV-A or blue light and physiological removal of the scytonemin containing envelope resulted in a loss of resistance to photobleaching by UV-A radiation (Garcia-Pichel *et al.*, 1992). In a natural population of *Calothrix* sp. receiving solar irradiance, a high scytonemin content is required for uninhibited photosynthesis under high UV-B flux (Brenowitz and Castenholz, 1997). Scytonemin may constitute as much as 5 per cent of total cellular dry weight or even higher in some naturally occurring cyanobacteria (Castenholz, 1997). Since rapid photodegradation of scytonemin does not occur, it is highly stable and able to persist very long in terrestrial cyanobacterial crusts or dried mats (Garcia-Pichel *et al.*, 1992; Brenowitz and Castenholz, 1997; Quesada *et al.*, 1999). This property may be helpful for cyanobacteria which face intermittent periods of physiological activity such as desiccation and freezing. It has been proposed that appearance of scytonemin during the precambrian era must have helped the cyanobacteria in the colonization of exposed, shallow water and terrestrial habitats (Quesda *et* al.,1999). Besides having a role as UV protectant, scytonemin may have additional roles such as protection against pathogenic attack, bacterial decomposition or herbivore grazing (Sinha and Hader, 2002).

Mycosporine-Like Amino Acids (MAAs)

The other class of photoprotectants are colourless, water soluble MAAs which are located within the cell and have absorption maxima in the UV-range (between 310-360 nm). First identified in fungi, these compounds are now reported to be present in a number of taxonomically diverse groups of organisms from eukaryotic and prokaryotie classes. They are present in cyanobacteria (Garcia-Pichel *et al.*, 1993; Sinha *et al.*, 1998,1999, 2001b), eukaryotic algae (Carreto *et al.*, 1990; Karentz *et al.*, 1991b), marine heterotrophic bacteria and marine invertebrates {Karentz *et al.*, 1991b), fish (Dunlap *et al.*, 1989} and in a wide variety of other marine organisms (Karentz *et al.*, 1991b; Dunlap and Yamamoto, 1995; McClintock and Karentz, 1997; Dunlap and Shick, 1998). Among cyanobacteria, MAAs have been reported from species growing, in freshwater, marine or terrestrial habitats (Garcia-Pichel and Castenholz, 1993; Garcia-Pichel *et al.*, 1993; Sinha *et al.*, 1998). Their presence has also been reported from strains growing in Antarctica (Quesada and Vincent, 1997) as well as from a community of halophilic species (Oren, 1997).

Structurally these compounds contain cyclohexenone or cyclohexenime chromophore conjugated with the nitrogen substituent of an amino acid or its imino alcohol (Table 8.4) and have an average molecular weight of 300 (Sinha *et al.*, 1998; Cockell and Knowland, 1999). In a survey of 20 strains of cyanobacteria (belonging to 13 genera) isolated from exposed habitats, Garcia-Pichel and Castenholz (1993) demonstrated the presence of MAAs in 13 strains. In all 13 distinct compounds were found in these strains. The specific MAA content in these isolates was found to vary between 0.16 to 0.84 per cent of the total dry weight. In a terrestrial *Gloeocapsa* strain, the intracellular concentration of MAA was found directly related to the intensity of the UV-B radiation received by the cells (Garcia-Pichel *et al.*, 1993). Moreover high content of MAA inside the cells conferred increased resistance to UV-B radiation. Sinha *et al.* (2001b) have demonstrated the induction of MAAs by solar ultraviolet-B radiation in three cyanobacteria namely, *Anabaena* sp., *Scytonema* sp. and *Nostoc commune.* They reported that PAR and UV-A were ineffective in causing induction of MAAs. All these cyanobacteria possessed only a single type of MAA which was identified as shinorine. Shinorine is a bisubstituted MAA containing both glycine and serine groups with absorption maximum at 334 nm (Table 8.4).

The presence of MAAs in cyanobacteria is restricted not only to non-planktonic species rather typical MAA-like absorbance bands in field samples of planktonic cyanobacteria such as *Gloeotrichea, Aphanizomenon* and *Microcystis* sp. have also been demonstrated (Garcia-Pichel *et al*, 1993). Field populations of halotolerant cyanobacteria are also known to possess high concentrations of MAAs (Oren, 1997). It has been suggested that the content of MAAs in the cell is regulated by osmotic mechanisms (Oren, 1997). The conclusive evidence for the exclusive role of MAAs as a sunscreen pigment is lacking and hence forth it has been speculated that besides acting as sunscreens; they may transfer energy to the photo synthetic reaction centers (Sivalingam *et al.*, 1976; Sinha *et al.*, 1999), may have a role in osmotic regulation (Oren, 1997) and, they may act as antioxidants and prevent cellular damage resulting from UV-induced production of highly reactive oxygen species (Dunlap and Yamamoto, 1995). Interestingly, in laboratory-grown cultures of *Chlorogloeopsis* sp., both UV and osmotic stress are able to induce and regulate the synthesis of MAAs (Portwich and Garcia-Pichel, 1999, 2000). In this species, a specific UV-B photoreceptor, probably pterin, may mediate MAA induction (Portwich and Garcia-Pichel 2000). However, in *Anabaena* sp the induction of MAAs (identified as shinorine) is solely under photocontrol and occurs only by UV radiation and PAR dose not seem to be effective in induction of MAAs (Sinha *et al.*, 2002).

A terrestrial species *of Nostoc commune* when exposed to high solar radiation, was found to possess a UV-A/B absorbing pigment with absorption maxima at 312 and 325 nm (Scherer *et al.*, 1988). In this species, an oligosaccharide MAA (OS-MAA), has been shown to be located in the extracellular glycan sheath (Bohm *et al.*, 1995; Ehlin-Schulz *et al.*, 1997). It is the first mycosporine so far known to be covalently linked to the oligosaccharide and located outside the cell (Bohm *et al., 1995).* This pigment besides providing protection by absorbing UV radiation, may also provide additional protection by quenching (Dunlap and Yamamoto, 1995). In this species which is subjected to regular cycles of desiccation and rewetting, UV-B radiation is known to induce a cascade of physiological reactions where both passive and active protective mechanisms operate. Carotenoids provide a fast active SOS response whereas OS-MAAs provide passive protection against long term exposure to UV-B (Ehling-*Schulz et al.*, 1997)

Besides scytonemin and MAAs, occurrence of a few other UV-absorbing compounds, conferring resistance to UV radiation to the organisms, have been reported (Kumar *et al.*, 1996b). The shielding role of a brown-coloured pigment (having major absorption peak at 314 nm) from *Scytonema hofmanii* and a phycoerythrin rich pink extract fom *Nostoc spongiaeforme,* against UV-B-induced damage, has been reported (Kumar *et al.*, 1996b). A strain of *Nostoc* sp. showing resistance to high light intensity and UV radiation has been reported to produce three different UV-absorbing compounds having absorption maxima at 256, 314 and 400 nm (de Chazal and Smith. 1994).

During the last two decades vast information on UV-B absorbing/sun-screening pigments from diverse organisms have been gathered and this has led to the development of a database on photoprotective compounds (Groniger *et al.*, 2000). Readers may find detailed information on the absorption maxima and extinction coefficients of different compounds together with experimental protocols on the web site (see http://www. biologie.uni.erlangen.de/botanik I/index. html).

Repair

The damaging effects of UV-B radiation can be alleviated by UV-A or visible radiation (Quesada *et al.*, 1999) suggesting that in nature severity of the damaging effects of UV-B can be reduced by PAR and UV-A present in the solar radiation. The existence of efficient mechanisms to repair the UV-inducd lesions is well known in a variety of organisms, including cyanobacteria (Castenholz, 1997; Quesada *et al.*, 1999; Groniger *et al.*, 2000; Sinha *et al.*, 2001b; Sinha and Hader, 2002). Basically there are three known mechanisms that can repair the UV-induced DNA lesions. These are photoreactivation which consists of enzymatic monomerization of UV-induced dimers brought about by visible/UV-A radiation through the activity of photolyase (Sancar and Sancar, 1988), the dark or excision repair and the recombinational repair (Sinha *et al.*, 2001a). Photorepair of UV-C-induced DNA damage is well known in several cyanobacteria (Asata, 1972, Levine and Thiel, 1987) and the photolyase has been purified from *Anacystis nidulans* (Saito and Werbin, 1970; Eker *et al.*, 1990). The action spectra for photoreactivation of UV-C-induced DNA damage in different species of cyanobacteria show that UV-A and blue-light are most effective. Similar kind of studies involving repair of UV-B-induced DNA damage in cyanobacteria is lacking although Karentz *et al.* (1991a) have studied the cellular and molecular aspects of UV-B-induced DNA damage and its repair in Antarctic diatom communities. Buma *et al.* (1995) have also reported UV-B-induced thymine-dimer formation and its repair in a marine diatom *Cyclostella* sp.

Han *et al.* (2001) have reported photoreactivation of UV-B-induced inhibition of photosynthesis in *Anabaena* sp. The action spectrum for photoreactivation showed major peaks at 352.5 and 383 nm along with an additional peak at 411 nm suggesting that both UV-A and blue-light are effective in photoreactivation in this species. The unusually large number of DNA copies in cyanobacteria may confer an additional protection against long-term UV-B damage. The existence of polyploidy in cyanobacteria may be of additional importance since this may mask the effect of single mutations of a DNA molecule.

Conclusions

Increased intensity of ultraviolet-B (UV-B) radiation has now been unequivocally demonstrated to show detrimental effect on a number of vital physiological and biochemical processes of cyanobacteria. Such effects lead to reduced growth and survival. UV-B affects pigmentation, disintegrates phycobilisomes, severely inhibits photosynthetic processes and causes inactivation of nitrogenase. Major alterations in general protein profile of a number of cyanobacteria have also been reported. Formation of dimers between two adjacent pyrimidine bases, cis-syn cyclobutane dimers and pyrimidine (6-4) pyrimidone products following UV-B exposure in certain cyanobacteria has also been demonstrated. Unfortunately most of the studies pertaining to UV-B radiation effects on cyanobacteria are based on short term laboratory experiments employing high UV-B flux. Only a few short-term experiments that utilize solar UV-B radiation obtained by using various long-pass filters are available (Donkor *et al.*, 1993). Conclusions about the long-term ecological impacts of ambient natural UV-B radiation cannot be accurately forecast from results obtained through short-term incubation studies.

Since in nature, organisms are seldom affected by only a single stress factor, it is worth to consider that the effectiveness of UV-B radiation can be greatly modified by a number of other factors such as water-stress, increase in atmospheric CO_2, global mean temperature and presence of heavy metals like cadmium, mercury, copper, lead and nickel (Rai *et al.*, 1998). Hence other environmental and biological variables should also be taken into account while attempting a realistic assessment of the effects of enhanced levels of UV-B radiation. The existence of various protective/repair strategies provide a high level of natural defense against the potential effects of elevated UV-B radiation. Information related to the ecological significance of sun-screening substances such as scytonemin and MAAs and spatial distribution of MAAs within the cyanobacterial cells is still in infancy stage. Since UV-B damage is ameliorated by PAR and UV-A, UV-B/UV-A ratio has a major influence on the extent of UV-B damage. The existence and significance of emzymatic defense mechanisms and repair of UV-B-induced DNA damage in cyanobacterial populations deserve critical study. Because of differential sensitivity of organisms to UV-B radiation, changes in species composition are bound to occur which might affect biological diversity. Although during the last 25 years vast knowledge covering diverse aspects of UV-B effects on living organisms have accumulated but there are a number of gaps left in our understanding which need immediate attention.

Acknowledgements

Research works in the author's laboratory (A. Kumar) was supported by research grants received from the Department of Biotechnology, Government of India (BT/PR 1239/AGR/02/065/9S). We are thankful to Prof. D.P. Hader and Dr. R.P. Sinha, Erlangen, Germany, for sending their research papers and other relevant information related to UV-B effects.

References

Asata, Y., 1972. Isolation and characterization of ultra-violet light sensitive mutants of the blue-green *alga Anacystis nidulans*. *Journal of Bacteriology*, 110: 1058–1064.

Bebout, B.M. and Garcia-Pichei, F., 1995. UV-B induced vertical migrations of cyanobacteria in a microbial mat. *Applied and Enviromental Microbiology*, 61: 4215–4222.

Blakefield, M.K. and Harris, D.O., 1994. Delay of cell differentiation in *Anabaena aequalis* caused by UV-B radiation and the role of photoreactivation and excision repair. *Photochemistry and Photobiology*, 59: 204–208.

Bohm, G.A., Pfleiderer, W., Boger, P. and Scherer, S., 1995. Structure of a novel oligosaccharide-mycosporine-amino acid ultraviolet-A/B sunscreen pigment from the terrestrial cyanobacterium *Nostoc commune*. *Journal of Biological Chemistry*, 270: 8536–8539.

Brenowitz, S. and Castenholz, R.W., 1997. Long-term effects of UV and visible irradiance on natural populations of scytonemin-containing cyanobacterium (*Calothrix* sp.). *FEMS Microbiology Ecology*, 24: 343–352.

Britt, A.B., 1995. Repair of DNA damage induced by ultraviolet radiation. *Plant Physiology*, 108: 891–896.

Buma, A.G.J., Van Hannen, E.J., Roza, L., Veldhuis, M.J.W. and Gieskes, W.W.C., 1995. Monitoring ultraviolet-B induced DNA damage in individual diatom cells by immunofluorescent thymine dimer detection. *Journal of Phycology*, 51: 314–321.

Campbell, D., Eriksson, M.-J., Öquist, G., Gustafsson, P. and Clarke, A.K., 1998. The cyanobacterium *Synechococcus* resists UV-B by exchanging photosystem II reaction-centre D_1 proteins. *Proceedings of the National Academy of Sciences (USA)*, 95: 364–369.

Canini, A., Grilli Caiola, M., Civitareale, P. and Galiazzo, F., 1992. Superoxide dismutase in symbiotic, free-living and wild *Anabaena* and *Nostoc* (Nostocales, Cyanophyta). *Phycologia*, 31: 225–230.

Carreto, J.I., Carignan, M.O., Daleo, G. and De Marco, S.G., 1990. Occurrence of mycosporine-like amino acids in the red tide dinoflagellate *Alexandrium excavatum:* UV-protective compounds? *Journal of Plankton Research*, 12: 909–921.

Castenholz, R.W., 1997. Multiple strategies for UV tolerance in cyanobacteria. *The Spectrum*, 10: 10–16.

Cockell, C.S. and Knowland, J., 1999. Ultraviolet radiation screening compounds. *Biological Reviews*, 74: 311–345.

Cockell, C.S. and Rothschild, L.J., 1999. The effects of UV radiation A and B on diurnal variation in photosynthesis in three taxonomically and ecologically diverse microbial mats. *Photochemistry and Photobiology*, 69: 203–210.

Crutzen, P.J., 1992. Ultraviolet on the increase. *Nature*, 356: 104–105.

de Chazal, N.M. and Smith, G.D., 1994. Characterization of a brown *Nostoc* sp. from java that is resistant to high light intensity and UV. *Microbiology*, 140: 3183–3189.

Döhler, G., Biermann, I. and Zink, J., 1986. Impact of UV-B radiation on photosynthetic assimilation of ^{14}C-bicarbonate and inorganic ^{15}N-compounds by cyanobacteria. *Z Naturforsch*, 41: 426–432.

Donkor, V.A. and Hader, D.-P., 1991. Effects of solar and ultraviolet radiation on motility, photomovement and pigmentation in filamentous, gliding cyanobacteria. *FEME Microbiology Ecology*, 86: 159–168.

Donkor, V.A., Damian, H.A.K. and Häder, D.-P., 1993. Effects of tropical solar radiation on the motility of filamentous cyanobacteria. *FEMS Microbiology Ecology*, 12: 143–148.

Dunlap, W.C., Williams, D.M., Chalker, B.E. and Banaszak, A.T., 1989. Biochemical photoadaptations in vision: UV-absorbing pigments in fish eye tissues. *Comparative Biochemistry and Physiology*, 93: 601–607.

Dunlap, W.C. and Yamamoto, Y., 1995. Small-molecule antioxidants in marine organisms: Antioxidant activity of mycosporine-glycine. *Comparative Biochemistry and Physiology*, 112: 105–114.

Dunlap, W.C. and Shick, J.M., 1998. Ultraviolet radiation-absorbing mycosporine-like amino acids in coral reef organisms: A biochemical and environmental perspective. *Journal of Phycology*, 34: 418–430.

Ehling-Schulz, M., Bilger, W. and Scherer, S., 1997. UV-B-induced synthesis of photoprotective pigments and extracellular polysaccharides in the terrestrial cyanobacterium *Nostoc commune*. *Journal of Bacteriology*, 179: 1940–1945.

Eker, A.P., Kooiman, P., Hessels, J.K.C. and Yasui, A., 1990. DNA photoreactivating enzyme from the cyanobacterium *Anacystis nidulans*. *Journal of Biological Chemistry*, 14: 8009–8015.

Frederick, J.E., Snell, H.E. and Haywood, E.K., 1989. Solar ultraviolet radiation at the earth's surface. *Photochemistry and Photobiology*, 50: 443–450.

Friedberg, E.C., Walker, G.C. and Siede, W., 1995. *DNA Repair and Mutagenesis*. American Society of Microbiology Press, Washington, DC.

Garcia-Pichel, F. and Castenholz, R.W., 1991. Characterization and biological implications of scytonemin: A cyanobacterial sheath pigment. *Journal of Phycology*, 27: 395–409.

Garcia-Pichel, F. and Castenholz, R.W., 1993. Occurrence of UV-absorbing mycosporine-like compounds among cyanobacterial isolates and an estimation of their screening capacity. *Applied and Environmental Microbiology*, 59: 163–169.

Garcia-Pichel, F., Sherry, N.D. and Castenholz, R.W., 1992. Evidence for an ultraviolet sunscreen role of the extracellular pigment scytonemin in the terrestrial cyanobacterium *Chlorogloeopsis* sp. *Photochemistry and Photobiology*, 56: 17–23.

Garcia-Pichel, F., Wingard, C.E. and Castenholz, R.W., 1993. Evidence regarding the UV sunscreen role of a mycosporine-like compound in the cyanobacterium *Gloeocapsa* sp. *Applied and Environmental Microbiology*, 59: 170–176.

Groniger, A., Sinha, R.P., Klisch, M. and Häder, D.-P., 2000. Photoprotective compounds in cyanobacteria, phytoplankton and macroalgae: A database. *Journal of Photochemistry and Photobiology B: Biology*, 58: 115–122.

Häder, D.-P. and Worrest, R.C., 1991. Effects of enhanced solar radiation on aquatic ecosystems. *Photochemistry and Photobiology*, 53: 717–725.

Häder. D.-P., Kumar, H. D., Smith, R.C. and Worrest, R.C., 1998. Effects of solar ultraviolet radiation on aquatic ecosystems. *Journal of Photochemistry and Photobiology B: Biology*, 46: 53–68.

Han, T., Sinha, R.P. and Häder, D.-P., 2001. UV-A/blue light-induced reactivation of photosynthesis in UV-B irradiated cyanobacterium, *Anabaena* sp. *Journal of Plant Physiology*, 158: 1403–1413.

He, Y.Y. and Häder, D.-P., 2002. Involvement of reactive oxygen species in the UV-B damage of the cyanobacterium, *Anabaena* sp. *Photochemistry and Photobiology*, 66: 73–80.

Jurgens, U.J. and Weckesser, J., 1985. Carotenoid-containing outer membrane of *Synechocystis* sp. strain PCC6714. *Journal of Bacteriology*, 164: 384–389.

Karentz, D., Cleaver, J.E. and Mitchell, D.L., 1991a. Cell survival characteristics and molecular responses of Antarctic phytoplankton to ultraviolet-B radiation. *Journal of Phycology*, 27: 326–341.

Karentz, D., McEuen, F.S., Land, M.C. and Dunlap, W.C., 1991b. Survey of mycosporine-like amino acid compounds in Antarctic marine organism: Potential protection from ultraviolet exposure. *Marine Biology*, 108: 157–166.

Karsten, U. and Garcia-Pichel, F., 1996. Carotenoids and mycosporine-like amino acid compounds in members of the genus *Microcoleus* (cyanobacteria): A chemosystematic study. *Systematic and Applied Microbiology*, 19: 285–294.

Kerr, J.B. and McElroy, C.T., 1993. Evidence for large upward trends of ultraviolet-B radiation linked to ozone depletion. *Science*, 262: 1032–1034.

Kumar, A., Sinha, R.P. and Häder, D.-P., 1996a. Effect of UV-B on enzymes of nitrogen metabolism in the cyanobacterium *Nostoc* sp. *Journal of Plant Physiology*, 148: 86–91.

Kumar, A., Tyagi, M.B., Srinivas, G., Singh, N., Kumar, H. D., Sinha, R. P. and Häder, D.-P., 1996b. UV-B shielding role of $FeCl_3$ and certain cyanobacterial pigments. *Photochemistry and Photobiology*, 64: 321–325.

Kumar, A., Tyagi, M.B., Jha, P.N., Srinivas, G. and Singh, A., 2003. Inactivation of cyanobacterial nitrogenase after exposure to ultraviolet-B radiation. *Current Microbiology*, 46: 380–384.

Levine, E. and Thiel, T., 1987. UV-inducible DNA repair in the cyanobacteria *Anabaena* sp. *Journal of Bacteriology*, 169: 3988–3993.

Lindahl, T. and Wood, R.D., 2000. Quality control by DNA repair. *Science*, 286: 1897–1905.

Madronich, S., McKenzie, R.L., Bjorn, L.O. and Caldwell, M.M., 1998. Changes in biologically active ultraviolet radiation reaching the Earth's surface. *Journal of Photochemistry and Photobiology B: Biology*, 46: 5–19.

McClintock, J.B. and Karentz, D., 1997. Mycosporine-like amino acids in 38 species of subtidal marine organisms from McMurdo Sound, Antarctica. *Antarctican Science*, 9: 392–398.

Mitchell, D.L. and Karentz, D., 1993. The induction and repair of DNA photodamage in the environment. In: *Environmental UV Photobiology*, (Eds.) A.R. Young, L. Bjorn, J. Moan and W. Nultsch. Plenum Press, London, pp 345–377.

Molina, M.J. and Molina, L.T., 1992. Stratospheric ozone. In: *The Science of Global Change: The Impact of Human Activities on the Environment*, (Eds.) D. Dunnette and R.J. O'Brien. American Chemical Society. Washington, DC, pp. 24–35.

Newton, J.W., Tyler, D.D. and Slodki, M.E., 1979. Effect of ultraviolet-B (280–320 nm) radiation on blue-green algae (cyanobacteria): Possible biological indicators of stratospheric ozone depletion. *Applied and Environmental Microbiology*, 37: 1137–1141.

OREN, A., 1997. Mycosporine-like amino acids as osmotic solutes in a community of halophilic cyanobacteria. *Geomicrobiology Journal*, 14: 231–240.

Pierson, B.K., Mitchell, H.K. and Ruff-Roberts, A.L., 1993. *Chloroflexus aurantiacus* and ultraviolet radiation: Implications for archean shallow water stromatolites. *Origins of Life and Evolution of the Biosphere*, 23: 243–260.

Portwich, A. and Garcia-Pichel, F., 1999. Ultraviolet and osmotic stresses induce and regulate the synthesis of mycosporines in the cyanobacterium *Chlorogloeopsis* PCC 6912. *Archives of Microbiology*, 172: 187–192.

Portwich, A. and Garcia-Pichel, F., 2000. A novel prokaryotic UV-B photoreceptor in the cyanobacterium *Chlorogloeopsis* PCC 6912. *Photochemistry and Photobiology*, 71: 493–498.

Proteau, P.J., Gerwick, W.H., Garcia-Pichel, F. and Castenholz, R.W., 1993. The structure of scytonemin: An ultraviolet sunscreen pigments from the sheaths of cyanobacteria. *Experientia*, 49: 825–829.

Quesada, A. and Vincent, W.F., 1997. Strategies of adaptation by antarctic cyanobacteria to ultraviolet radiation. *European Journal of Phycology*, 32: 335–342.

Quesada, A., Vincent, W.F. and Lean-David, R.S., 1999. Community and pigment structure of arctic cyanobacterial assemblages: The occurrence and distribution of UV-absorbing compounds. *FEMS Microbiology Ecology*, 28: 315–323.

Rai, L.C., Tyagi, B., Rai, P.K. and Mallick, N., 1998. Interactive effects of UV-B and heavy metals (Cu and Pb) on nitrogen and phosphorus metabolism of a N_2-fixing cyanobacterium, *Anabaena doliolum*. *Environmental and Experimental Botany*, 39: 221–231.

Reynolds, C.S., Oliver, R.L. and Walsby, A.E., 1987. Cyanobacterial dominance: the role of buoyancy regulation in dynamic lake environments. *Newzealand Journal of Marine and Freshwater Research*, 21: 379–390.

Richter, P., Krywult, M., Sinha, R.P. and Häder, D.-P., 1999. Calcium signals from heterocysts of *Anabaena* sp. after UV irradiation. *Journal of Plant Physiology*, 154: 137–139.

Saito, N. and Werbin, H., 1970. Purification of a blue-green algal deoxyribonucleic acid photoreactivating enzyme: an enzyme requiring light as a physical cofactor to perform its catalytic function. *Biochemistry*, 9: 2610–2620.

Sancar, A. and Sancar, G.B., 1988. DNA repair enzymes. *Annual Review of Biochemistry*, 57: 29–67.

Sass, L., Spetea, C, Mate, Z., Nagy, F. and Vass, I., 1997. Repair of UV-B induced damage of photosystem II via *de novo* synthesis of the D_1 and D_2 reaction centre subunits in *Synechocystis* sp. PCC6803. *Photosynthetic Research*, 54: 55–62.

Scherer, S., Chen, T.W. and Boger, P., 1988. A new UV-A/B protecting pigment in the terrestrial cyanobacterium *Nostoc commune*. *Plant Physiology*, 88: 1055–1057.

Schopf, J.W., 1992. The oldest fossils and what they mean. In: *Major Events in the History of Life*, (Ed.) J.W. Scdhopf. Jones and Bartlett Publishers, U.K., pp. 29–64.

Shibata, H., Baba, K. and Ochiai, H., 1991. Near UV irradiation induces shock proteins in *Anacystis nidulans* R-2: Possible role of oxygen. *Plant and Cell Physiology*, 32: 771–776.

Singh, R.N., 1961. *Role of Blue-Green Algae in Nitrogen Economy of Indian Agriculture.* Indian Council of Agricultural Research, New Delhi, India.

Sinha, R.P. and Häder, D.-P., 1998. Effects of ultraviolet-B radiation in three rice field cyanobacteria. *Journal of Plant Physiology,* 153: 763–769.

Sinha, R.P. and Häder, D.-P., 2002. Life under solar UV radiation in aquatic organisms. *Advances in Space Research,* 30: 1547–1556.

Sinha, R.P., Kumar, H.D., Kumar, A. and Häder, D.-P., 1995a. Effects of UV-B irradiation on growth, survival, pigmentation and nitrogen metabolism enzymes in cyanobacteria. *Acta Protozoologica,* 34: 187–192.

Sinha, R.P., Lebert, M., Kumar, A., Kumar, H.D. and Häder, D.-P., 1995b. Disintegration of phycobilisomes in a rice field cyanobacterium *Nostoc* sp. following. UV irradiation. *Biochemistry and Molecular Biology International,* 37: 697–706.

Sinha, R.P., Singh, N., Kumar, A., Kumar, H.D., Häder, M. and Hader, D.-P., 1996. Effect of UV radiation on certain physiological and biochemical processes in cyanobacteria. *Journal of Photochemistry and Photobiology B: Biology,* 32: 107–113.

Sinha, R.P., Singh, N., Kumar, A., Kumar, H.D. and Häder, D.-P., 1997. Impacts of ultraviolet-B irradiation on nitrogen-fixing cyanobacteria of rice paddy fields. *Journal of Plant Physiology,* 150: 188–193.

Sinha, R.P., Klisch, M., Groniger, A. and Häder, D.-P., 1998. Ultraviolet-absorbing / screening substances in cyanobacteria, phytoplankton and macroalgae. *Journal of Photochemistry and Photobiology B: Biology,* 47: 83–94.

Sinha, R.P., Klisch, M. and Häder, D.-P., 1999. Induction of a mycosporine-like amino acid (MAA) in the rice-field cyanobacterium *Anabaena* sp. by UV irradiation. *Journal of Photochemistry and Photobiology B: Biology,* 52: 59–64.

Sinha, R.P., Dautz, M. and Häder, D.-P., 2001a. A simple and efficient method for the quantitative analysis of thymine dimers in cyanobacteria, phytoplankton and macroalgae. *Acta Protozoologica,* 40: 187–195.

Sinha, R.P., Klisch, M., Helbling, E.W. and Häder, D.-P., 2001b. Induction of mycosporine-like amino acids (MAAs) in cyanobacteria by solar ultraviolet-B radiation. *Journal of Photochemistry and Photobiology B: Biology,* 60: 129–135.

Sinha, R.P., Sinha, J.P., Gröniger, A. and Häder, D.-P., 2002. Polychromatic action spectrum for the induction of a mycosporine-like amino acid (MAA) in a rice-field cyanobacterium, *Anabaena* sp. *Journal of Photochemistry and Photobiology B: Biology,* 66: 47–53.

Sivalingam, P.M., Ikawa, T. and Nisizawa, K., 1976. Physiological roles of a substance 334 in algae. *Botanica Marina,* 19: 9–21.

Smith, R.C., Prezelin, B.B., Baker, K.S., Bidigare, R.R., Boucher, N.P., Coley. T., Karentz, D., Macintyre, S., Matlick, H.A., Menzies, D., Ondrusek, M., Wan, Z. and Waters, K.J., 1992. Ozone depletion: Ultraviolet radiation and phytoplankton biology in Antarctic waters. *Science,* 255: 952–959.

Stanier, R.Y. and Cohen-Bazire, 1977. Phototrophic prokryotes: the cyanobacteria. *Annual Review of Microbiology*, 31: 225–274.

Stewart, W.D.P., 1980. Some aspects of structure and function in N_2-fixing cyanobacteria. *Annual Review of Microbiology*, 34: 497–536.

Tel-Or, E., Huflejt, M.E. and Packer, L., 1986. Hydrogen peroxide metabolism in cyanobacteria. *Archives of Biochemistry and Biophysics*, 246: 396–402.

Tyagi, R., Srinivas, G., Vyas, D., Kumar, A. and Kumar, H.D., 1992. Differential effect of ultraviolet-B radiation on certain metabolic processes in a chromatically adapting *Nostoc*. *Photochemistry and Photobiology*, 55: 401–407.

Vincent, W.F. and Roy, S., 1993. Solar ultraviolet-B radiation and aquatic primary production: damage, protection and recovery. *Environmental Reviews*, 1: 1–12.

Vincent, W.F. and Quesada, A., 1994. Ultraviolet radiation effects on cyanobacteria: implications for Antarctic microbial ecosystems. In: *Ultraviolet Radiation in Antarctica: Measurements and Biological Effects*, (Eds.) C.S. Weiler and P.A. Penhale. Antarctic Research Series Vol. 2, American Geophysical Union, Washington, DC, pp. 111–124.

Wilson, M.I., Ghosh, S., Gerhardt, K.E., Holland, N., Sudhakar Babu, T., Edeeman, M., Dumbroff, E.B. and Greenberg, B.M., 1995. *In vivo* photomodification of ribulose-1,5-bisphosphate carboxylase/oxygenase holoenzyme by ultraviolet-B radiation. *Plant Physiology*, 109: 221–229.

Chapter 9

The Effects of Antibiotics on Blue-Green Algae

S.B. Padhi and P.K. Swain

Algal Research Laboratory,
Department of Botany, Berhampur University,
Orissa – 760 007

Antibiotic research is one of the most exciting field not only in chemotherapy, but also in relation to their action on microscopic forms of plant life. An antibiotic can be defined as a substance produced by a living organism, which inhibits the growth or activity of another living organism, studies on the utilisation of antibiotics is not new, rather it was started in 1929, when Dr. Alexander Flemming, found that a mould had contaminated his cultures of a pathogenic bacterium, *Staphylococcus aureus* and had killed bacteria in its immediate vicinity leaving a clear ring of lysed bacteria. The mould was a *Penicillium notatum*. Since then, investigations have been made to ascertain the effects of antibiotics on various organisms. Provasoli *et al.* (1948) were among the first investigation to show interest in the use of antibiotics to obtain pure culture of algae. In 1951, Provasoli *et al*, reported the effect of Penicillin, Streptomycin, Chlorotetracycline, Oxytetracyclins, Chloromphenicol, Bacitracin and Polymixin tolerated by representative species of algae and protozoa. They found that, the use of appropriate concentrations of the antibiotics resulted in the elimination of both gram-positive and gram- negative bacterial contaminants from the algal and protozoan cultures. Successful attempts to obtain bacteria-free cultures of algae were reported also by Pappas and Hoffman (1952). They succeeded in obtaining bacteria-free cultures of *Euglena gracilis* by the use of combinations of Penicillin, Dihydrostreptomycin and Chlorotetracycline.

From 1952 to 1958 there was a paucity of literature concerning the use of antibiotics as a means of obtaining pure cultures of algae. Effect of antibiotics on the growth of microalgae have been extensively investigated (Gerloff *et al.*, 1950; Provasoli *et al.*, 1951; Foter *et al.*, 1953; Zehnder and Hughes, 1958; Hunter and Mc Veigh, 1961; Lampen and Arnow, 1961; Perlman, 1964) and have revealed that algae are generally less sensitive towards antibiotics than the bacteria and fungi. A number of articles concerning the potential use of antibiotics as algicides in the purification of water supplies were published during 1952 to 1958. Foter *et al.* (1953) evaluated the antialgal properties of Polymixin-B, Chlorotetracycline, Bacitracin, Chloramphenicol, Neomycin, Penicillin, Streptomycin, Gliotoxin and Oxytetracycline on selected species of diatoms, blue-green algae and green algae. Two years later, the results of an investigation of the inhibitory effects of numerous inorganic and organic compounds including 5 antibiotics on representatives of the same groups of algae were reported by Palmer and Maloney (1955). Their result showed that actidione at concentrations resulting in complete inhibition of green algae and diatoms had little or no effect on blue-green algae. As reported by Foter *et al.*, (1953), Streptomycin sulphate was particularly inhibitory to blue green algae and diatoms but had little effects on green algae. Interest in the use of antibiotics as a means of obtaining pure cultures of algae was revived in 1958. Zehnder and Hughes (1958) reported that actidione, at concentrations of 50 ppm or less, inhibited the growth of members of the chlorophyceae, xanthophyceae and bacillariophyceae. Much higher concentrations had no inhibitory effect on the development of the myxophyceae. In 1959, Galloway and Krauss reported an investigation of the differential action of chemical agents on certain algae, bacteria and fungi. Polymixin-B sulphate was found to have a high degree of selectivity inhibiting the bacteria tested at concentrations which had little or no effect on *Scenedesmus obliquus* and *Chlorella vulgaris*. Early studies on the action of polymixins showed that among bacteria, gram negative organisms were much more sensitive than gram-positive one.

A good deal of work has been done on the taxonomy, morphology and electron microscopy of blue-green algae (Fritsch, 1945; Fogg, 1953; Kratz and Myers, 1955). Recently blue-green algae and bacteria are seen as a coherent group which characteristically lack membrane bound sub-cellular organelles such as mitochondria and chloroplast and which do not have their genetic material located on more than one chromosome in a membrane-bound nucleus (Stainer and Van Niel, 1962). The distinction between blue-green algae and eukaryotic algae now appears to be complete, in that no putative intermediate forms have withstood close examination (Delaney *et al.*, 1976) however, the only major similarity being the possession of oxygenic photosynthesis (Krogmann, 1973). The importance of blue-green algae in tropical rice fields (Singh, 1961) and its utilisation as a biofertilizer (Henriksson *et al.*, 1975; Rodgers *et al.*, 1979; Venketraman, 1981) is very well documented. Even though they are very important economically, agriculturally than bacteria, one aspect, that is their genetical studies has remained sadly neglected. The main reason for this has been the unavailability of a sufficient number of suitable mutant strains that could be used in genetical experiments.

A large number of antibiotic substances have been found to inhibit the growth of the blue-green algae (Table 9.1). Although some work has previously been carried out on the effect of streptomycin, pencillin and other antibiotics on the growth of a few species of blue-green algae (Galloway and Krauss, 1959; Hunter and Mc Veigh, 1961), the production of strains resistant to such agents have not been reported for many members of blue-green algae. The first report of gene transfer among these organisms came from Professor Fogg's laboratory, when Kumar (1962) reported the isolation of doubly-resistant recombinant strains of *Anacystis nidulans*. However, this report has been criticized by Pikalek (1967) with the ground that, under the conditions employed, one of the antibiotics used (Penicillin) was unstable. With the report of recombination in a filamentous blue-green alga, *Cylindrospermum majus* (Singh and Sinha, 1965), these studies gained much interest. Singh and Sinha (1965) grew a spore-forming, penicillin resistant strain with a non-sporulating, streptomycin resistant strain and isolated from the mixed culture a doubly resistant recombinant strain that produced spores at low frequency. Bazin (1968) isolated stable clones of *Anacystis nidulans* resistant to both polymixin-B and streptomycin from a mixed culture of singly resistant strains. Gene transfer was also reported in the nitrogen-fixing alga *Nostoc muscorum* by Stewart and Singh (1975), who crossed Str^{+} nif'' strain with Str^{s} nif^{+} wild type and got recombinant clones which were both streptomycin resistant and able to fix nitrogen. Antibiotic and drug resistant mutants have also helped variously in studying mutagenesis and genetic recombination in prokaryotes (Van Baalen, 1973). Drug-resistant mutants of blue-green algae have been used as markers for studying genetic interchange and biochemical reactions since last two decades by several workers (Kumar, 1962, 1964,1968, Singh and Sinha, 1965, Pikalek, 1967, Bazin, 1968, Shestakov and Khyen, 1970, Srivastava, 1970; see Ladha and Kumar, 1978). However, possible biochemical alterations conferring resistance, such as the question of possible penicillinase production by blue-green algae remains unanswered.

Exposure to antibiotics behave in two different ways–either it inhibits growth without killing the alga, growth resume after removal of antibiotic (algistatic) or in other cases, that is the algicidal way, where it kills the organism and growth does not resume by removal of antibiotics

Antibiotics Disc Study

Antibiotics disc study is one of the most efficient methods to determine the sensitivity of microorganisms towards antibiotics. It was introduced in 1940s (Morley, 1945; Bondi *et al.*, 1947). Vance (1966) determined the sensitivity of blue green algae, especially *Microcystis aeruginosa* to 32 antibiotics by antibiotics disc method. The principle behind method is that the applied organism, if sensitive, its growth is prevented in a circular zone around the disc. Cooper (1955) proposed a model to explain zone formation and to show how zone size depends on several factors. Rate of diffusion of the drug is largely dependent on the molecular weight of the agent (Garrod, 1958).

Table 9.1: Summary of the Reports on Effect of Antibiotics on Cyanobacteria

Organism Studied	*Antibiotic Tested*	*References*
Anabaena variabilis	Bacitracin	Krauss (1962)
Anabaena variabilis	Bacitracin	Kumar (1968)
Anabaena variabilis	Chloramphenicol	Galloway and Krauss (1959)
Anacystis nidulans spp.	Chloramphenicol Dihydrostreptomycin (and others)	Kumar (1964c) Vance (1966)
Anacystis nidulans	Mitomycin C	Kumar, 1968
Anacystis nidulans	Neomycin	Kumar, 1968
Anacystis nidulans	Penicillin	Kumar, (1964b)
Phormidium muscicola	Penicillin	Srivastava (1970)
Various	Pimafucin	Wieringa(1968)
Various	Polymixin B	Galloway and Krauss (1959)
Anacystis nidulans	Polymixin B	Whitton (1967)
Chlorogloea fritschii	Polymixin B	Whitton (1967)
Anacystis montana	Rifarnycins	Rodriguez-Lopez *et al.* (1971)
Anacystis nidulans	Streptomycin	Kumar *et al,* 1967
Phormidium muscicola	Streptomycin	Srivastava, 1970.
Various	Various	Tchan and Gould, 1961
Anacystis nidulans	Various (including bacitracin tyrocidine)	Whitton, 1968
Microcystis aeruginosa	Bacitracin	Vance (1966)
Anabaena variabilis	Chloromycetin	Krauss(1962)
Nostoc sp.	Neomycin	Krauss(1962)
Phormidium sp.	Neomycin	Krauss (1962)
Microcystis aeruginosa	Neomycin	Vance (1966)
Anabaena variabilis	Penicillin	Krauss(1962)
Microcystis aeruginosa	Penicillin	Krauss(1962)
Microcystis aeruginosa	Penicillin	Vance (1966)
Anacystis nidulans	Penicillin	Kumar, 1964
Phormidium muscicola	Penicillin	Srivastava, 1969
Nostoc sp.	Polymixin A	Krauss (1962)
Anabaena variabilis	Polymixin B	Krauss (1962)
Cylindrospermum licheniforme	Polymixin B	Krauss (1962)
Microcystis aeruginosa	Polymixin B	Krauss (1962)
Microcystis aeruginosa	Polymixin B	Vance (1966)
Anacystis nidulans	Polymixin B	Whitton (1967b)
Chlorogloea fritschii	Polymixin B	Whitton (1967b)
Anabaena variabilis	Streptomycin	Krauss (1962)

Contd...

Table 9.1–Contd...

Organism Studied	Antibiotic Tested	References
Cylindrospermum licheniformis	Streptomycin	Krauss (1962)
Microcystis aeruginosa	Streptomycin	Krauss (1962)
Microcystis aeruginosa	Streptomycin	Vance (1966)
Nostoc sp.	Streptomycin	Krauss (1962)
Phormidium sp.	Streptomycin	Krauss (1962)
Scytonema schrnidlei,	Penicillin	B. Nanda, P.C. Choudhury,
Anabaena sp.		B.K. Routroy and S. Padhi (1989)
Scytonema schmidlei *Anabaena sp.*	Penicillin	R.K. Moharana, S.B. Padhi, S. Padhi and H. Patnaik(1986)
Anabaena sp. and *Westiellopsis prolifica* Janet.	Chloramphenicol	S.B. Padhi, R.K. Moharana and S. Padhi (1987)
Anabaena sp. and *Westiellopsis prolifica* Janet.	Polymixin-B	S.B. Padhi, S. Mahapatra and S. Padhi (1989)
Westiellopsis prolifica	Streptomycin	S. Padhi, S.B. Padhi and P.I. Das (1988)

In this disc study method, one can know the effect of several factors that affects this type of study. Among the various factors, pH of the medium is one which effects the activity of many agents like Tetracycline and Novobiocin (Sabath *et al.*, 1968) were as Polymixin and Penicillin are affected very little by such variations (Brown and Blowers, 1978). Minerals and salts can considerably affect the activity of novobiocin against staphylococci and that of Penicillin against *Proteus* sp. (Garrod *et al.* 1973). Free divalent cations chelate tetracycline zone sizes (Ericssion and Sherris, 1971, Byenner and Sherris, 1972) and reduced the activity of novobiocin against the gram-negative bacteria (Brock, 1956). Magnesium and Calcium ions reduce the activity of polymixin (Newton, 1953). Binding of antibiotics to agar has been reported (Hances *et al*, 1967, Kunin and Edmondson, 1968; El-Nakeeb and Yousef, 1970). Acid and sulphate groups of the agar bind peptide antibiotics *e.g.* polymixin or antibiotics with basic groups like Kanamycin and Neomycins (Ericssion and Sherris, 1971).

Mutants of Blue-green Algae

Mutation may be defined as a heritable change and a mutant is the changed organism that results. This process constitutes the most important single factor on which prokaryotic systems which are devoid of sexuality depend for their evolution. Although it is a rare event which may occur spontaneously as a result of mistakes in either replication, repair or recombination (the three 'R' process) the ability of the systems to repair the defects during DNA replication may well have a significant role in evolution. Repair process, on the other hand, would be advantageous in enabling species to stabilize its genetic material, in an environment, that could cause of mutation at high rate. The spontaneous mutation rate can be increased several fold by treating the organisms with physical and chemical mutagenes and mutations resulting from such treatments are called induced mutations.

Development of methodology for producing mutants is essential for investigation of blue-green algal genetics. Most of the accumulated informations on blue-green algae pertains to their growth and nutritional characteristics. Little is known of their biochemistry and even less of their genetics (Ladha and Kumar, 1978).

Different types of mutants can be expected to aid in investigating biochemical problems, in finding recombination and transformation in various blue-green algae. Many mutagens have been used including various radiations, base analogues, alkalytic agents and nitrosoamines (Van Baalen, 1973). Shestakov and Jevner (1968) commented on the efficiency of the nitrosoarnins, mtrosomethyl urea (NMU) and N-methyl-N′-nitro-N-nitrosoguanidine (MNNG) in inducing mutations in blue-green algae.

Antibiotic Resistant Mutants

Antibiotic resistant mutants have mainly been employed to detect recombination and transformation in blue green algae. Biochemists and physiologists derive information concerning intermediate steps in metabolism and the mode of action of drugs by experimenting with susceptible and resistant strains of microorganisms. Penicillin resistant forms have been reported by Kumar (1964b). Susceptibility to Penicillin vary widely between different species of blue-green algae. In many cases Penicillin resistant results from the micro-organism having acquired ability to make penicllinase and thus enzyme activity has been marked in other antibiotic resistant strains also (Gupta, Talpasayi, 1981) Streptomycin-resistant mutants have been reported by Kumar (1964b), Singh and Sinha (1965). Pikalek (1967) obtained strains of *A. nidulans* resistant to Streptomycin and Erythromycin. Bazin (1968) obtained polymixin-B resistant mutants after treatment of *A. nidulans* with 8-azaguanine.

The isolation of the above strains resistant to antibiotics have been made in certain unicellular and filamentous blue-green algae both after spontaneous screening and after exposure to mutagens (Table 9.1) Zhevner, Glazer and Shestakov (1973) studied several types of filamentous mutants of *A.nidulans* with reference to colony and cell morphology, population heterogenisity, DNA content and resistant to different antibiotics and mutagens (Table 9.1). In gram and Van Baalen (1970) and Ingram and Thurston (1970) have isolated two types of filamentous mutants in another coccoid blue-green algae, *Agmenellum quadruplicatum* Brebission. The mutation of a long filamentous form to one with shorter filaments has been obtained in *Plectonema boryanum* gomont by Padan and Shilo (1969) without concomitant growth impairment. Attempts have been also made inheterocystous forms. Singh and Tiwari (1969) have isolated a mutants of *Nostoc linckia* (Roth). Bornet et Born et Flah, after ultraviolet irradiation which produced chains of 2-8-heterocysts. Ladha and Kumar (1977a) reported two phenotypic classes of morphological mutants in *N. linckia* after nitrosoguamdine mutagenesis. Currier, Haury and Wok (1977) reported mutants of *Anabaena variabilis.* Mutants of *Anabaena raciborskii* and *Anabaena flos-aquae* lacking gas vacuoles have been isolated by Das and Singh (1976) and Walsby (1977) respectively. Mutants of several blue-green algae, particularly *A. nidulans* have been reported which are altered in pigment ratio (Stevens and Myers, 1976). Stoletov *et al.* (1965), Van Baalen (1965), Asato and Folsome (1969) and Stevens and Van Baalen

(1970) have all commented on the isolation of pigment mutants. Stevens and Van Baalen (1976) have also isolated two classes of pigment mutants, namely, yellow green (YG) and blue (BL) of *A. nidulans* with nitrosoguanidine mutagenesis.

Stevens, Stevans and Myers (1975) have isolated a uracil requiring mutant of *A, nidulans.* Herdman, Delaney and Carr (1973) reported a mutant of *A. nidulans* which requires a combination of pyruvate or acetate plus malate or succinate or fumarate for growth. Currier *et al.* (1977) have isolated mutants of *Anabaena variabilis* requiring uracil, adenine, biotin and nicotinic acid.

Many attempts have been also done to isolate mutants of nitrogen-fixing blue-green algae. Ladha and Kumar (1977b) and Singh *et al,* (1977) have isolated two classes of non-nitrogen-fixing mutants of *Nostoc linckia* and *N. muscorum* and an additional class of nitrogen-fixing mutants of *N. linckia* producing multiple heterocysts. Several types of mutants affecting heterocysts formation, nitrogen fixation and the capacity to reduce nitrate to nitrite have been reported. Nitrate reductase is a substrate inducible enzyme (Hattory 1962; Carr, 1967). Its activity has been little explored in blue green algae. Nitrate reductase deficient mutants of blue-green algae have been reported by relatively few workers (Singh and Singh, 1968; Stevens and Van Baalen, 1969; 1970; Sinha and Kumar, 1973; Singh and Sonie, 1977).

These reports suggests that much more study is needed to emphasize the enormous amount of work still required in this area. As different groups of blue green algae differ considerably towards their action with antibiotics, mutagens, such type of studies can be successfully used for classifying the blue-green algae in a more systematic way.

References

Asato, Y. and Folsome, C.E., 1969. *Mutation Res.,* 8: 531–536.

Bazin, M.J., 1968. *Nature. London,* 218: 282–283.

Bharati, S.G. and Bongale, V.D., 1976. *Phykos.,* 15: 75–79.

Bondi, A., Sapulding, E.H., Smith, D.E. and Dietz, C.C., 1947. *Ann. J Med. Sci.,* 213: 221–225.

Brock, T.D., 1956. *J. Bacteriol.,* 72: 320–323.

Brown, D. and Blowers, R., 1978. In: *Laboratory Methods in Antimicrobial Chemotherapy,* (Eds.) D.S. Reeves, I. Phillips, J.D. Williams and R. Wise. Churchil Livingstone, p. 8–30.

Byenner, V.C. and Sherris, J.C., 1972. *Antimicrobial Agents and Chemotherapy,* 1: 116–122.

Carr, N.G., 1967. *Abstr. 4th meet, Fedrn. Europ. Biochem. Soc,* p. 89, University off., Oslo, Norway.

Cooper, K.E., 1955. *Nature, London,* 176: 510–511.

Currier, T.C., Haury, J.F. and Wolk, C.P., 1977. *J. Bacteriol.,* 129: 1556–1562.

Dash, B. and Singh, P.K., 1976. *Arch. Microbiol.,* 111: 195–196.

Delaney, S.F., Herdmann, M. and Carr, N.G., 1976. In: *The Genetics of Algae*, (Ed.) R.A. Lewin. Blackwell Scientific Publication, Oxford, p. 7–28.

El Nakeeb, M.A. and Yousef, R.T., 1970. *T. Pharmacy*, 32: 6–9.

Ericsson, H. and Sheerıs, J.C., 1971. *Acta Pathologia et Microbiologia Scandinavia, Sec. B.* Suppl. No. 217.

Fogg, G.E., 1953. *The Metabolism of Algae*. Methuen Co., London, p. 149.

Fogg, G.E., Stewart, W.D.V., Fay, P. and Walsby, A.E., 1973. *The Blue-Green Algae*. Academic Press, New York, London.

Foter, M.J., Palmer, CM. and Maloney, T.E., 1953. *Antibiot. Chemther*, 3: 505.

Fritsch, F.E., 1945. *The Structure and Reproduction of Algae*, Vol. II. Cambridge University Press, England, 939 p.

Galloway, R.A. and Krauss, R.W., 1959. *Am. J. Bot.*, 46: 40–49.

Garrod, L.P., 1958. *Proc. 3rd Internat. Congress. Clinical Pathology, Brussels*, 1957, (Eds.) M. Welsch, P. Dustm and J. Dagnelie, Press Academiques Europeennes, Brussels, p. 129–139.

Gerloff, G.C., Fitzgerald, G.P. and Skoog, F., 1950. *Am. J. Bot.*, 37: 216–218.

Gupta, R.K. and Talpasayi, E.R.S., 1981. *Curr. Sci.*, 50(23): 1039–1040.

Hances, F.J., Sands, J.G. and Bennett., E.O., 1967. *Appl. Microbiol.*, 15: 31–34.

Hattori, A., 1962. *Pl. Cell. Physiol. (Tokyo)*, 3: 371–377.

Henriksson, E., Henriksson, L.E. and De Silva, E.J., 1975. In: *Nitrogen Fixation by Free Living Microorganisms*, (Ed.) W.D.P. Stewart. Cambridge Univ. Press, Cambridge.

Herdman, M., Delaney, S.F. and Carr, N.G., 1973. *J. Gen. Microbiol.*, 79: 233–237.

Hunter, E.O. and Miveigh, I., 1961. *Am. J. Bot.* 48: 179.

Ingram, L.O. and Thurston, R.L., 1970. *Protoplasma*, 71: 55–75.

Ingram, L.O. and Van, Baaleon, C., 1970. *J. Bacteriol.*, 102: 784–789.

Janet, M., 1941. p. 167–170.

Krauss, R.V., 1962. In: *Physiology and Biochemistry of Algae*, (Ed.) R.A. Lewin. Academic Press, New York and London, p. 673–685.

Kratz, W.A. and Myers, J., 1955. *Am. J. Bot.*, 42: 282–287.

Krogmann, D.W., 1973. In: *Biology of Blue-Green Algae*, (Eds.) N.G.Carr. and B.A. Whitton. Blackwell Scientific Publication, Oxford, p. 80–98.

Kumar, H.D., 1962. *Nature, London*, 196: 1121–1122.

Kumar, H.D., 1964a. *Ann. Bot. N.S.*, 28: 555–564.

Kumar, H.D., 1964b. *J. Expt. Bot.*, 15: 232–250.

Kumar, H.D., 1964c. *Pl. Cell. Physiol. (Tokyo)*, 5: 465–472.

Kumar, H.D., 1968a. *Beit. Biol. Pflanzen.*, 45: 161–170.

Kumar, H.D., Singh, H.N. and Prakash, G., 1967. *Cell Physiology (Tokyo)*, 8: 171–179.

Kunin, C.M. and Edmondson, W.P., 1968. *Proc. Soc. Expt. Biol. Med.*, 129: 118–122.

Ladha, J.K. and Kumar, H.D., 1977a. *Zft. Allg. Microbiol.*, 17: 513–519.

Ladha, J.K. and Kumar, H.D., 1977b. In: *Proc. L.C. Dunn. and T. Dobzhansky Symp. Genetics*, Univ. Mysore.

Ladha, J.K. and Kumar, H.D., 1978. *Biol. Rev.*, 53: 355–386.

Lampen, J.O. and Arnow, P., 1961. *J. Bacteriol.*, 82: 247.

Maharana, R.K., Padhi, S.B., Padhi, S. and Patnaik, H., 1986. *Phykos*, 25: 12–15.

Murali, D.C., 1945. *J. Pathol. Bacteriol.*, 57: 379–382.

Nanda, B., Choudhury, P.C., Routray, B.K. and Padhy, S., 1989. *IBC*, 6: 71–72.

Newton, B.A., 1953. *J. Gen. Bacteriol.*, 57: 379–382.

Newton, B.A., 1956. *Bact. Rev.*, 20: 14–27.

Paden, E. and Shilo, M., 1969. *T. Bacteriol.*, 97: 975–976.

Padhi, S., Padhi, S.B. and Dash, P.I., 1988. *J. (Curr.) Bio. Sci.*, 4: 125–127.

Padhi, S.B., Maharana, R.K. and Padhi, S., 1987. *Comp. Physiol and Ecol.*, 13: 13–16.

Padhi, S.B., Mahapatra, S. and Padhi, S., 1989. *Comp. Physiol and Ecol.*, 14(3): 155–158.

Palmer, C.N. and Maloney, T.E., 1955. *Ohio. J. Sci.*, 55: 1–8.

Pappas, G.D. and Hoffman, H., 1952. *Ohio. J. Sci.*, 52: 102–105.

Perlman, D., 1964. In: *Antimicrobial Agents and Chemotherapy*, p. 114–119.

Pikalek, P., 1967. *Nature, London*, p. 215–666.

Provasoli, L., Hunter, S.H. and Schatz, A., 1948. *Proc. Soc. Exp. Biol. Ad. Med.*, 69: 279–282.

Provasoli, L., Pinter, I. and Packer, L., 1951. *Proc. Am. Soc. Protozol.*, 2: 6.

Rodgers, G.A., Bergman, B., Heriksson, E. and Udris, M., 1979. *Plant and Soil*, 52: 99-107.

Rodgriguez-Lopez, M., Calvo, M.L.M. and Gomez-Acebo, J., 1971. *J. Ultrastruc. Res.*, 36: 595–602.

Rosen, W.G., and Gawlik, S.P., 1961. *T. Protozool.*, 8: 90.

Sabth, L.D.., Gerstein, D.A. and Finland, M., 1968. *Antimicrobial Agents and Chemotherapy*, p. 398–404.

Shestakov, S.V and Khyen, N.T., 1970. *Mol. Gen. Genetics*, 107: 372–375.

Shestakov, S.V. and Jevner, V.D., 1968. *Proc. XIII. Internat. Congr. Genetics*, 1: 84.

Singh, H.N. and Sonie, K.C., 1971. *Mutation, Res.*, 43: 205–212.

Singh, R.N., 1961. *The Role of Blue-Green Algae in Nitrogen Economy of Indian Agriculture.* ICAR, New Delhi, India, p. 78.

Singh, R.N. and Singh, P.K., 1968. In: *Abs. India. Conf. Microbiol.*, Baroda, p.17.

Singh, R.N. and Sinha, K., 1965. *Nature, London*, 207: 782–783.

Singh, R.N. and Tiwari, D.N., 1969. *Nature, London*, 221: 62–64.

Sinha, B.D. and Kumar, H.D., 1973. *Ann. Bot.*, 37: 673–679.

Srivastava, B.S., 1969. *Arch. Microbiol.*, 66: 234–238.

Srivastava, B.S., 1970. *Phycologia*, 9: 205–208

Stainer, R.Y. and Van Niel, C.B., 1962. *Arch. Microbiol.*, 42: 17–35.

Stevens, C.L.R. and Stevans, S.E. (Jr.) and Myers, J., 1975. *J. Bacteriol.*, 124: 247–251.

Stevens, S.E. and Van Baalen, C., 1969. *J. Phycol.*, 5: 136–139.

Stevens, S.E. and Van Baalen, C., 1970. *Arch. Microbiol.*, 72: 1–8.

Stewart, W.D.P. and Singh, H.N., 1975. *Biophys. Biochem. Res. Comm.*, 62: 62–69.

Stoletov, V.N., Zhevner, V.D., Garibyan, D.V. and Shestakov, S.V., 1965. *Genetika*, 6: 77–83.

Tchan, Y.T. and Gould, J., 1961. *Nature, London*, 192: 1276–1277.

Van Baalen, C., 1973. In: *The Biology of Blue-Green Algae*, (Eds.) N.G. Carr. and B.A. Whitton. Blackwell Scientific Publications, Oxford, p. 201–213,

Vance, B.D., 1966. *J. Phycol.*, 2: 125–128.

Venkataraman, G.S., 1981. *Curr Sci.*, 50(6): 253–256.

Walsby, A.E., 1977. *Arch.Microbiol.*, 114: 160–167.

Whitton, B.A., 1967. *Can. J. Microbiol.*, 13: 987–993.

Whitton, B.A., 1968. *Pl. Cell. Physiol. (Tokyo)*, 9: 23–26

Wieringa, K.T., 1968. *Ant. Leeuwenhoek*, 34: 54–56.

Zehnder, A. and Hughes, E.O., 1958. *Can. J. Microbiol.*, 4: 399–408.

Zhevner, V.D., Glazev, V.M. and Shestakov, S.V., 1973. *Mikrobiologiya*, 42: 290–297.

Chapter 10

Effect of Diethyl Sulphonate on Growth and Development of *Chara vulgaris*

Kalpana Kund and Lakshman Labh*

University Department of Botany,
J.P. University, Chapra, Bihar

Introduction

A number of studies are on record reporting mitotic disorders in response to various chemical mutagens in different species of *Chara* and *Nitella* (Chatterjee and Sharma, 1972; Delay and Carpentier, 1955; Moutschen, *et al.*, 1956; Ramjee, 1968; Sharma and Tripathi, 1976a, b, Turner,1970) but their post-treatment effects on the morphological variability have not been worked out so far. Since, departure from the normal morphological features may be used as an early indicator of the effectiveness of particular mutagen, such studies deserve due attention. Consequently a maiden effort has been made to study the effects of Diethyl Sulphonate (dES) on *Chara vulgaris* var. *vulgaris* f. *contraria* with special reference to induced morphological variability. Besides the karyological effects of the aforesaid mutagens, recovery of normal mitotic divisions have been studied. An attempt has also been made to ascertain the recovery time of normal mitotic division which is not yet worked out for the alga at hand.

Heritability of induced morphological variability has also been studied after raising the M_1 plants from the oospores formed after treatments.

---

* Retired Professor & Head.

Materials and Methods

Clonal population of locally collected *Chara valgaris* var. *vulgaris* f. *contraria* was obtained from a bulbil and was used as the experimental material for the proposed study. Since the present study also deals with the post treatment morphologial variations, it is desirable to enumerate its diagnostic morphological feature. It is a monoecious and corticated form having upper one or two segments ecorticated with primary cortical cells relatively more prominent, diplostephanous with the stipulodes in both tiers equally developed and dactyls unicellular with acute tip and coronal base.

25 young twigs of almost same age and length were treated with 0.00625 per cent, 0.0125 per cent, 0.025 per cent, 0.05 per cent, 0.1 per cent and 0.15 per cent of dES for nine hours at a temperature 20±2°C and pH 7. The volume ratio of plants and mutagen solution was maintained as 1: 5 respectively for each set of treatment. Desired concentrations of dES were prepared by dissolving it in double glass distilled water. The treated filaments were thoroughly washed and were planted in different glass jars containing 1/4 sterile sand and 3/4 (1) Hope's medium (Hope and Walker, 1975). The planted jars were allowed to grow under the laboratory conditions at daily cycle of 16 hours (3300 Lux) and 8 hours of darkness.

Cultures of untreated filaments of almost same age and length were also grown similarly for reference.

The spectrum and frequency of the dES induced morphological variations were calculated out of the observations made on one thousand segments. The observations on the stability of the morphological variants were made after raising the populations from oospores Plate 10.4, Figure 4.3 developed on treated filaments (M_1). The obtained M_1 populations were successively propagated through cuttings to ascertain the stability of the morphological variants. Vegetatively propagated plants were named VM-generations and the successive generations were numbered after the times it was propagated.

Persistent effects of the mutagen on the mitotic events were studied in the fixations made after 24 hours and 48 hours of the treatment. The same was also studied in the fixations of the filaments of different VM generations for ascertaining recovery of induced mitotic disorders, if any.

Fixation of treated and untreated filaments (3-4 nodes long) bearing sex organs were made between 8-10 A.M. in aceto alcohol (1 : 3). Iron-alum-acetocarmine method (Godword, 1948) was followed for cytological preparations. Since the antheridia are free from stored food as well as any pigment and exhibit a high degree of synchronisation of mitotic division the proposed cytological studies were made on it. The frequency of mitotic events was calculated after examining 1000 antheridia.

The plants after each treatment along with the control were allowed to grow for observing the persistent effects of the mutagen.

The following three parameters were taken into consideration for evaluating the effectiveness of the mutagens:

Growth Performance

It was ascertained in terms of (*i*) linear growth *i.e.* the total increase in the length of the main axis as well as the lateral branches of unlimited growth; (*ii*) total differentiation of nodes and (*iii*) total emergence of branches after treatments. The observations were made at an interval of one week and the data were scored as the mean of five replicates.

Morphological Variability

The treated plants were vegetatively propagated to raise the population. The plants so obtained were screened for morphological variability, if any, under the effects of the treatment. The spectrum and frequency of anomalies observed were calculated out of 1,000 segments. Scoring of spectrum and its frequency was performed right from heading to maturity.

Mitotic Events

Three major but interlocking parameters which were considered to evaluate the effects of mutagens on the mitotic events are as follows:

1. *The frequency of division figures*: Changes in the proportion of nuclei undergoing division may be marked as a partial index of the degree and kind of effect of a given treatment. Hence it has been calculated after scoring 1,000 cells of the antheridial filaments.
2. *The relative frequency of different stages of mitosis*: The most readily detectable effects of any chemical on mitoses are a more or less marked change in the relative frequency of prophase, metaphase and post-metaphase. Hence, the frequency of different mitotic stages among the total dividing nuclei was scored for each treatment as well as control.
3. *Departure from the normal organization and movement of the mitotic chromosomes*: Observations on the heritability of the morphologial variants once induced under the effects of mutagens, were made after raising the M_1 population from oospores developed after treatment. The M_1 plants, so obtained, were vegetatively propagated through cutting to see the stability of the variants. The different generations of vegetatively propagated plants were named as VM-generations for ready reference and were numbered after the times it has been propagated. Similarly, the period for the recovery of mitotic disorders, caused due to the mutagens were studied by propagating the treated filaments through cutting and the successive generations were numbered as V_1, V_2 and so on indicating the number of times it has been propogated.

Observations on Growth Performance

Alga survived after treatments with all the concentrations except 0.15 per cent which proved lethal for the alga. The growth performance amongst the surviving treated plants was recorded, which may be discussed under the following heads:

Linear Growth

The linear growth responded basically to the same pattern to all the six concentrations of dES except the degree of increase or decrease against the control (Table 10.1).

Table 10.1: Per cent Increase or Decrease in the Linear Growth Over Control After Treatments with dES

Concentrations	*Observations After One Week*	*Observations After Two Weeks*	*Observations After Three Weeks*
0.00625%	–2.88%	+14.78%	+13.44%
0.0125%	–2.88%	+18.26%	+16.80%
0.025%	–3.77%	+60.0%	+59.66%
0.05%	–3.77%	+13.91%	+11.76%
0.1%	–2.88%	+5.21%	+2.52%

As shown in the aforesaid table and figure, the linear growth in the concentration 0.00625–0.1 per cent was initially inhibited during first week and then was abruptly enhanced during second week. This increasing trend was also maintained during third week but comparatively at slow rate. It is further evident that the rate of enhancement regularly increased with increase in the concentration unto 0.0125 per cent and thereafter it progressively decreased in the higher concentrations.

Differentiation of Nodes

Treatments with different concentrations of dES yielded consistent result on the differentiation of nodes except 0.1 per cent which witnessed a different result. The details of results have been summarized in the Table 10.2.

Table 10.2: Per cent Increasing or Decrease in the Frequency of Nodes Compared to Control After Treatment with dES

Concentrations	*Observations After One Week*	*Observations After Two Weeks*	*Observations After Three Weeks*
0.00625%	– 3.85%	+33.03%	+31.86%
0.0125%	– 3.85%	+37.61%	+41.59%
0.025%	– 1.92%	+37.61%	+39.82%
0.05%	– 1.92%	+14.68%	+13.27%
0.1%	– 3.85%	– 8.26%	– 10.62%

As indicated in the Table 10.2, the rate of the differentiation of nodes was initially inhibited during first week in the concentration 0.00625 per cent–0.05 per cent, during second week it was abruptly increased and then slightly decreased in the concentrations 0.00625 per cent and 0.05 per cent during the third week. In the concentrations 0.0125 per cent and 0.25 per cent the same rate of increase (higher than others) was recorded during second week and this increasing trend was also continued in both the concentration during third week, but it was a bit higher in 0.0125 per cent, 0.1 per cent of dES acted adversely on the differentiation of nodes.

Emergence of Branches

No branches were seen emerging either in control or in any of the six concentrations during first week but its emergence was recorded during second week in all the concentrations except 0.1 per cent in which the branches were yet to be induced like control. Increase in the frequency of branches over control was also noticed during third week in the concentrations from 0.00625 per cent to 0.05 per cent but comparatively at slow rate. In the concentration 0.1 per cent, the branches emerged during third week but its frequency was recorded below the control level.

Of all the concentrations highest frequency of branches was recorded in 0.0125 per cent. The details of data computed for different treatment may be seen in the Table 10.3.

Table 10.3: Growth Performance in Relation to Control After Treatment with dES

Concentrations	*Observations After One Week*	*Observations After Two Weeks*	*Observations After Three Weeks*
0.00625%	–00	+63	+61.54
0.0125%	–00	+70	+73.08
0.025%	–00	+70	+69.23
0.05%	–00	+45	+42.31
0.1%	–00	–00	–2.88

Observations on Morphological Variabilities

Persistent effects of dES on the alga at hand were observed in the form of various types of morphological variations which may be classified under five heads *viz.* changes in cortication, stipulods, dactyls and reduction in sex-organs, Besides, production of abnormal branches were also seen. The above noted variants were seen in the branches as well as in the segments developed after treatments.

The details of morphological variability obtained after treatments with dES may be discussed categorically as follows.

Change in Cortical Cells

Following three types of changes were observed.

1. Shifting of prominence from primary to secondary series of cortical cells was observed in all the treatments except 0.00625 per cent. The frequency of such changes was noted dose dependent.
2. *Elimination of Cortical cells*: Basal as well as intercalary segments became ecorticated (Plate 10.1, Figure 1.2) in addition to the upper most 1-3 segments. This was observed in all the five concentration and its frequency was scored (cf. Table 10.4). In certain instances, incomplete ecortication of the segment (Plate 10.1, Figure 1.1) was also observed but only in 0.1 per cent and its

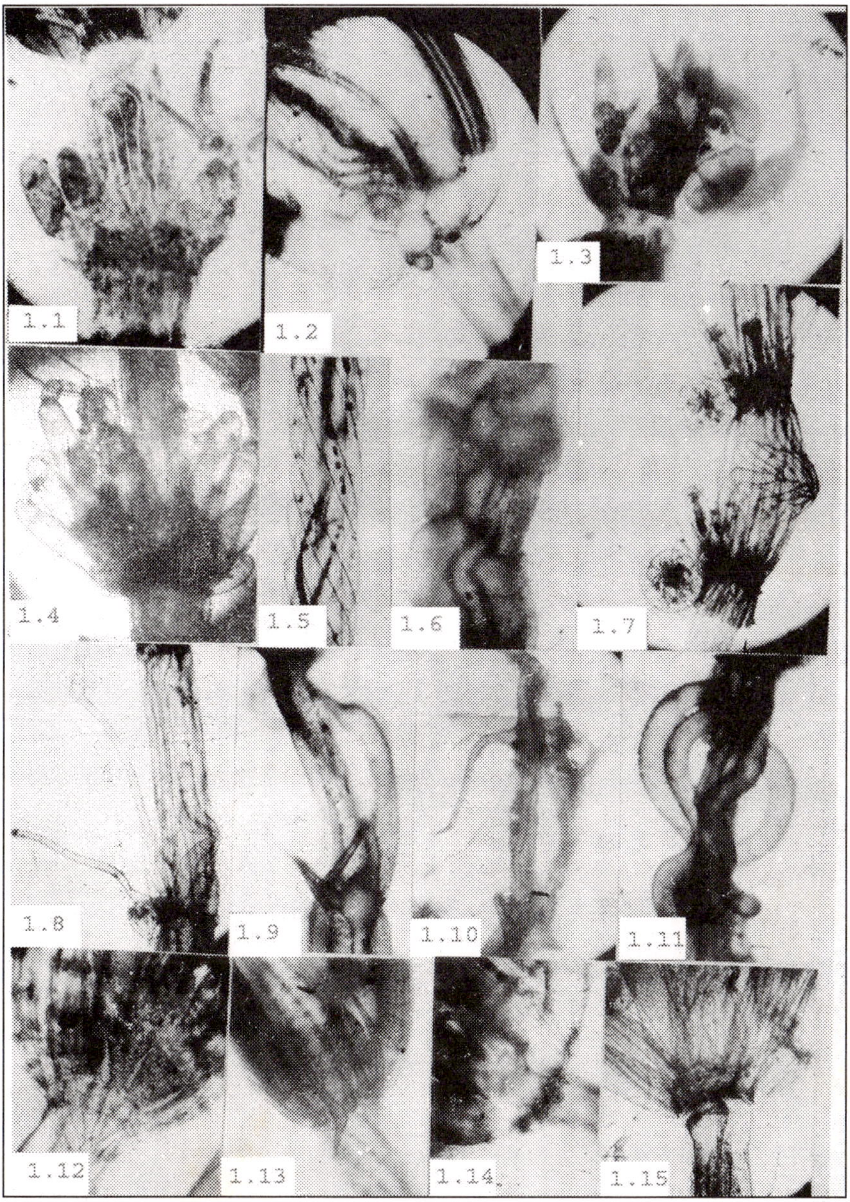

Plate 10.1: Morphological Variations After Treatment with dES in *Chara vulgaris*.

Figures 1.1: Half ecorticated segment; 1.2: Ecorticated basal segment; 1.3 and 1.4: Stunted branchlets with completely ecorticated segment; 1.5: Spiral cortication; 1.6: Convoluted cortication; 1.7: Hyperirophic cortication showing two nodes bearing two oogonia instead of one; 1.8 and 1.10: Divergent cortical cells; 1.9 and 1.11: Cortical cells forming loop; 1.12: Reduced number of stipulodes in upper tier; 1.13: Reduced number of stipulodes in lower tier; 1.14: Stipulodes reduced to knob like structure in both tiers; 1.15: Complete elimination of stipulodes in both tiers (Magnifications–Figures 1.1–1.4 and 1.7–1.11: X50; Figures 1.5 and 1.: –X150; Figures 1.12–1.15: X100).

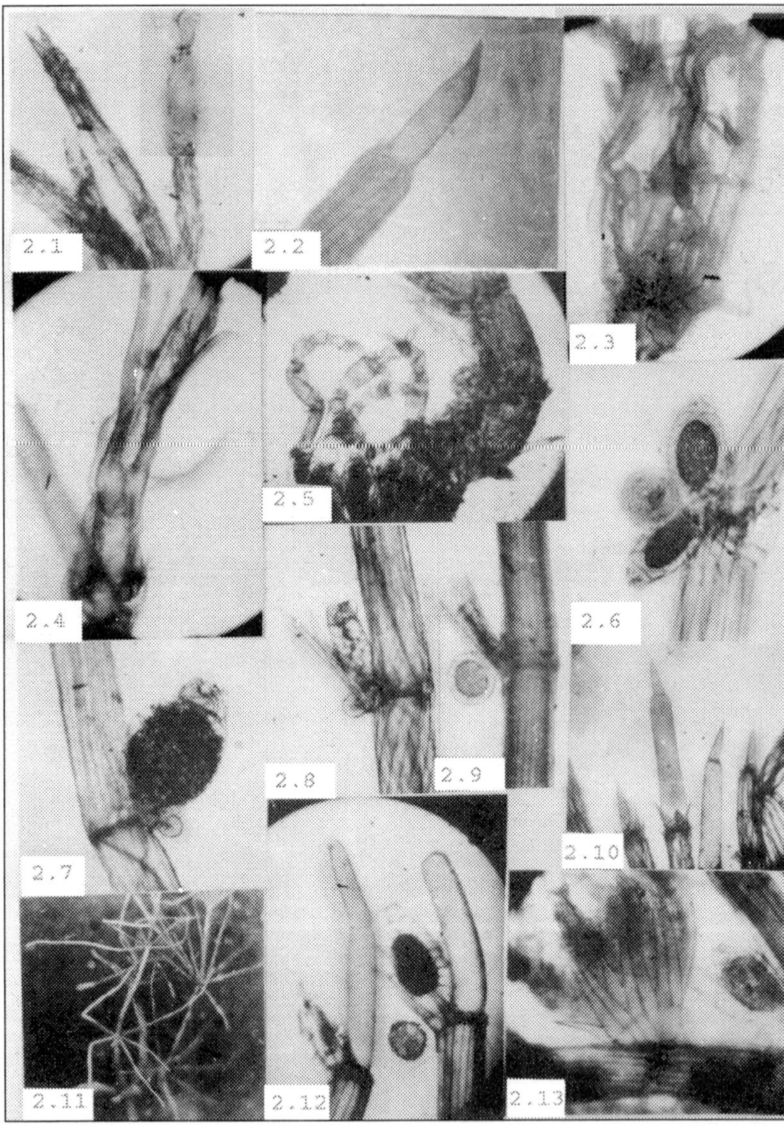

Plate 10.2: Morphological Variations After Treatment with dES in *Chara vulgaris*.

Figures 2.1: Branchlet tip with two dactyls and Bicelled dactyls; 2.2: Dactyl; 2.3 and 2.4: Stunted branchlets with completely ecorticated segment without coronal base; 2.3: Clumping and Curling of branchlets; 2.4: Deformed branchlets showing stunted growth and fusion with axis; 2.5: Moniliform branchlets; 2.6: Node bearing two oogonia and one antheridium; 2.7: Node bearing rudimentary antheridium below matured oogonium; 2.8: Node with normal young oogonium and rudimentary antheridium; 2.9: Node bearing normal antheridium and rudimentary oogonium; 2.10: Normal dactyl with coronal base and bicelled dactyl without corona; 2.11: Chlorophyll deficient branchlets; 2.12: Branchlets bearing normal sex organs but tip without dactyl; 2.13: Node with normal stipulodes in both tiers. (Magnification: Figure 2.2: X100; Figure 2.11: X0.1 and rest all X50).

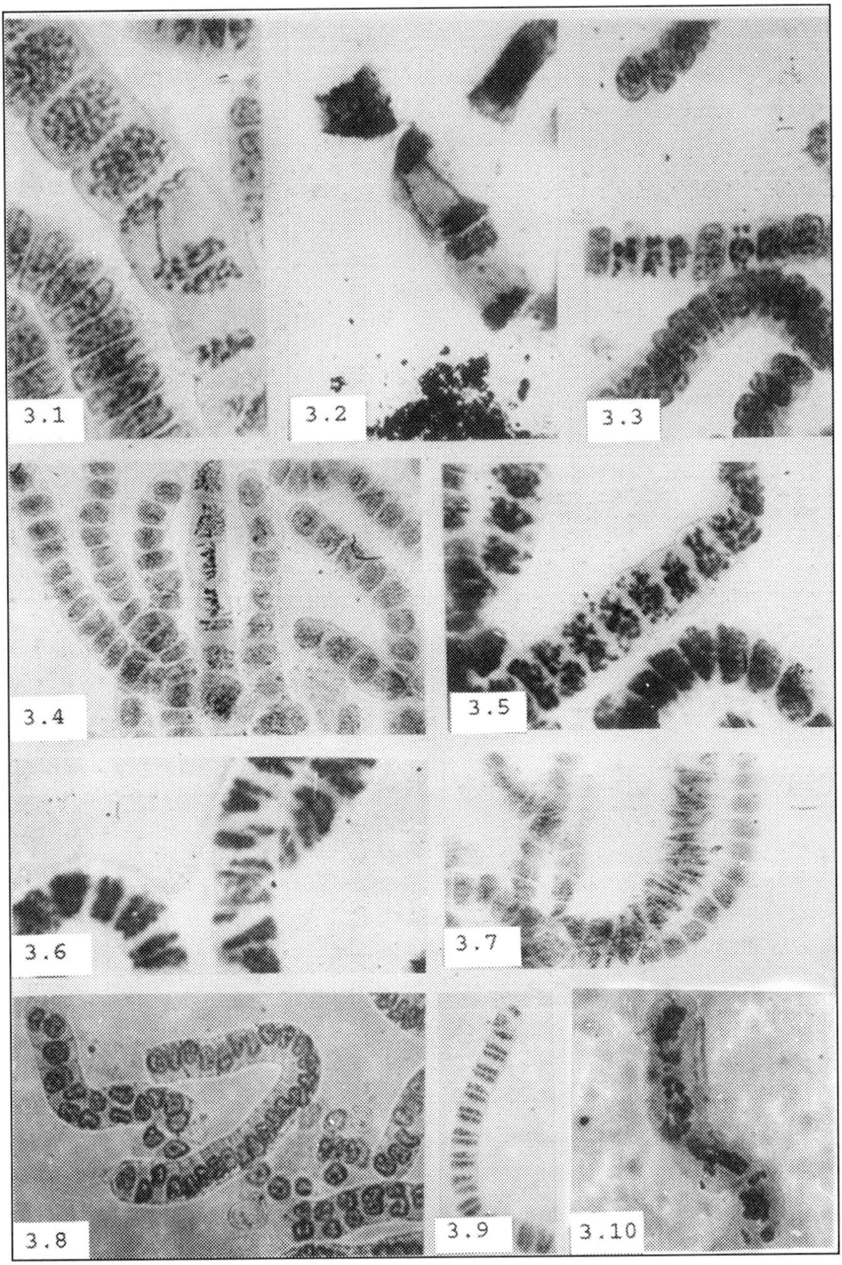

Plate 10.3: Mitotic Variations After Treatment with dES in *Chara vulgaris*

Figures 3.1 and 3.2: Anaphasic bridge; 3.3: Clotting of chromosomes; 3.4: Contraction and stickiness of chromosomes; 3.5: Smashed metaphase chromosomes; 3.6: Unequal distribution of chromosomes at anaphase; 3.7: Unequal distribution of chromosomes at anaphase; 3.8: Binucleate antheridial cells; 3.9: Tripolar spindle; 3.10: Clotting of chromosomes forming single lump (Magnification: Figures 3.1–3.4, 3.7 and 3.8: X900; Figures 3.5 and 3.6: X 1500; Figure 3.9: X500 and Figure 3.10: X3000).

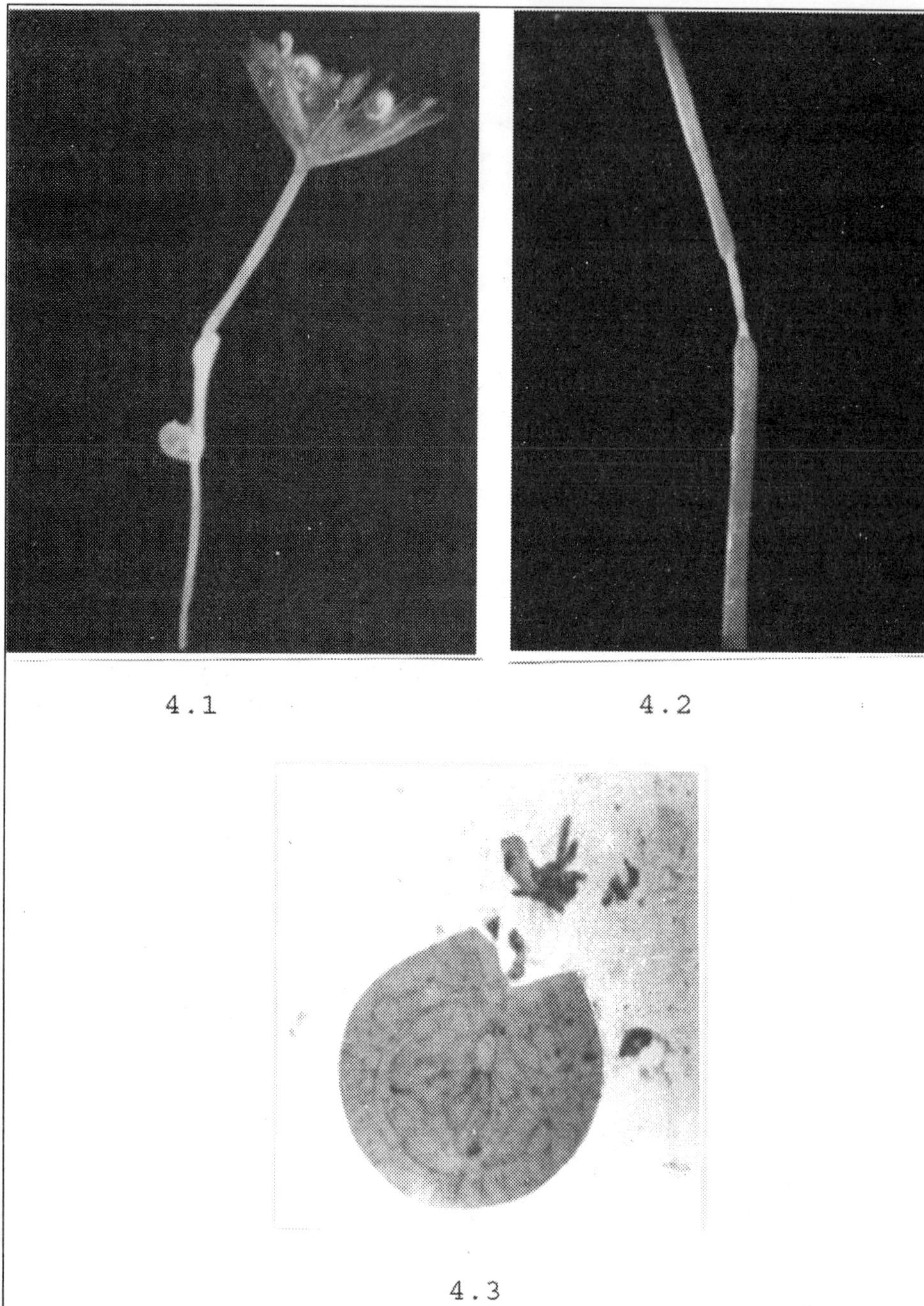

Plate 10.4

Figures 4.1: Germinating oospore of *Chara vulgaris* (X900); 4.2: Portion of branchlet showing chlorophyll deficient band after treatment with dES (X1600); 4.3: Disarticulated antheridium showing rudimentary antheridial filaments. Just above the punctured antheridium seen capitula with few celled rudimentary antheridial filaments (X3000).

frequency was computed 0.6 per cent. Apart from the segmental ecortication, ecortication of entire branches (Plate 10.1, Figures 1.3 and 1.4) was also seen in the concentrations from 0.025–0.1 per cent. Its frequency was computed dose dependent (cf. Table 10.4).

Table 10.4: Ecortication of the Segment Following dES Treatments

Parameters	*0.00625%*	*0.0125%*	*0.025%*	*0.05%*	*0.1%*
Ecortication of the Basal and/or intercalary segment	1.1%	4.9%	13.6%	6.5%	4.0%
Incompletely corticated segment	—	—	—	—	0.6%
Entirely ecorticated branches	—	—	0.63%	5.25%	5.96%
Overall Frequency	1.1%	4.9%	14.23%	11.75%	10.56%

3. *Changes in cortication pattern*: Departure from the usual cortication *i.e.* cortical cells running straight way parallel to each other between the nodes was observed in the form of spiral (Plate 10.1, Figure 1.5) convoluted (Plate 10.1, Figure 1.6) and hypertrophic (Plate 10.1, Figure 1.7) organization of the cortical cells, while the spiral cortication was noted only in 0.0125 per cent, 0.05 per cent and 0.1 per cent, rest two were seen consistently in all the five concentrations. The cortical cells in 0.05 per cent were also seen forming loops of various sizes and shapes after being diverged (Plate 10.1, Figure 1.10) from the central cells (Plate 10.1, Figures 1.8 and 1.11). The frequency of above abnormalities as in the different treatments may be seen in the Table 10.5.

Table 10.5: Spectrum and Frequency of Changes in Cortical Cells after Treatment with dES

Parameters	*0.00625%*	*0.0125%*	*0.025%*	*0.05%*	*0.1%*
Spiral Cortication	—	1.1	—	1.9	1.3
Convoluted Cortication	1.1	1.3	2.2	7.2	5.2
Hypertrophic Cortication	0.7	0.6	0.6	1.9	1.4
Cortical Cells forming loop	—	—	—	1.2	—
Overall Frequency	1.8	3.0	2.8	12.2	6.9

Change in Stipulodes

It was observed either in the form of reduction in the number of stipulodes from either of the two tiers (Plate 10.1, Figure 1.12) or both tiers, (Plate 10.1, Figure 1.13) or complete elimination of stipulodes (Plate 10.1, Figure 1.15). Changes of first category was found as a common feature of almost all the treatments but that of second was observed only in the treatment with 0.05 per cent. The frequency of above noted changes was computed for each treatment (Table 10.6).

Table 10.6: Spectrum and Frequency of Changes in Stipulodes Following the Treatment with dES

Type of Changes	*0.00625%*	*0.0125%*	*0.025%*	*0.05%*	*0.1%*
Reduced number of stipulodes in upper tier	7.1	—	33.3	8.7	6.7
Reduced number of stipulodes in lower tier	35.7	20.0	11.1	4.3	20.7
Reduced number of stipulodes in both tiers	—	30.0	11.1	17.4	3.4
Complete elimination of stipulodes in lower tier	—	—	—	26.0	—
Complete elimination of stipulodes in both tiers	—	—	—	8.7	—
Overall Frequency	42.8	50.0	55.5	65.1	30.8

Changes in Dactyls

It was observed in the form of elimination of coronal outgrowth from the base (Plate 10.2, Figure 2.2), elimination of dactyl cell as a whole (Plate 10.2, Figure 2.11). While the change of first type was noted as common feature of all the treatments, second was observed only in 0.05 per cent and 0.1 per cent. In case of 0.05 per cent, also the bicelled dactyls were observed (Plate 10.2, Figures 2.1 and 2.10). Frequency of above noted changes was computed in each case (Table 10.7)

Table 10.7: Spectrum and Frequency (Per cent) of the Changes in Dactyls Following the Treatments with dES

Type of Changes	*0.00625%*	*0.0125%*	*0.025%*	*0.05%*	*0.1%*
Elimination of dactyl	—	—	—	26.6	5.3
Dactyl without coronal base	—	9.7	10.7	15.6	6.7
Bicelled dactyl	—	—	—	2.3	—
Overall Frequency	—	9.7	10.7	44.5	12.0

Production of Abnormal Branches

It was observed only in the concentration 0.05 per cent. The abnormal branches were of the following two types.

Stunted

The branches with abnormally stunted growth (Plate 10.1, Figures 1.3, 1.4 and Plate 10.2, FIgure 2.4) and their short laterals being highly curled and clumped (Plate 10.2, Figure 2.3). Its frequency was scored 21.6 per cent.

Moniliform

The branches having highly condensed and swollen segments (Plate 10.2, Figure 2.5). Occurrence of such abnormal branches was observed in 5.25 per cent cases.

Observation on Sex Organs

The concentration 0.0625 per cent and 0.0125 per cent could not bring any effects on sex organs but the concentrations above to it acted adversely on both male as well as female sex organs. They were either reduced in size (Plate 10.2, Figures 2.7–2.9) or were seen carrying rudimentary spermatogenous filaments (Plate 10.4, Figure 4.3) and rudimentary egg (Plate 10.2, Figure 2.9). Reduction in size of the gametangia was recorded as a common feature of the treatments with 0.25–0.1 per cent but the production of rudimentary sex organs was observed in the concentrations 0.05 per cent and 0.1 per cent only. Reductions of both sex organs were never seen simultaneously on the same node. In very few instances emergence of two antheridial and/or two oogonia at a node was also observed (Plate 10.1, Figure 1.7 and Plate 10.2, Figure 2.6). Categorical frequency of the above variation may be seen in the Table 10.8.

Table 10.8: Spectrum and Frequency of the Changes in Sex Organs

Spectrum	*0.00625%*	*0.0125%*	*0.025%*	*0.05%*	*0.1%*
Antheridia reduced in size	—	—	2.4	3.6	5.8
Antheridia carrying either rudimentary or no spermatogenous filaments	—	—	—	4.0	10.2
Oogonia reduced in size	—	—	4.0	8.0	11.0
Oogonia carrying rudimentary egg	—	—	—	4.0	13.8
Overall Frequency	—	—	6.4	19.6	40.8

Inheritance of Induded Morphological Variations

M_1- Plants could be obtained only in the concentration 0.05 per cent and hence observation on the inheritance of morphological abnormalities is restricted only to this concentration.

Of all the abnormalities enumerated above only those of cortical cells as well as of stipulodes inherited, true in M_1–generation and rest all were eliminated. Spectrum and frequency of the abnormalities recorded in M_1-population may be seen in the Table 10.9.

Apparently the sex organs on M_1-Plants were normal with respect to its size and organizations but their emergence was noted only on 7 per cent of the total branches.

In addition to above morphological abnormalities M_1-Plants were seen chlorophyll mutant of virina albscens (Transescens) type as recognised by Gustafsson (1940). As soon the short laterals attained a length of 8-10 nodes, their tip turned white (Chlorosis), later it spreaded downward with age and ultimately the entire plant became white within a fortnight (Plate 10.2, Figure 2.11). In certain instances the short laterals witnessed chlorosis in 2-3 intercalary segments imparting white banded appearance (Plate 10.4, Figure 4.2). These albinic branches did not bore any sex organs. However, a good number of young branches were seen emerging in succession from the subterranean portion of the dying plants to meet the same fate.

Table 10.9: Spectrum and Frequency (Per cent) of dES-induced Morphological Variations Recorded Among M_1 Populations

Parameters	*Frequency in Per cent*
Changes in Cortication	
(a) Spiral	71.8
(b) Convoluted	1.2
(c) Hypertrophic	0.3
(d) Ecorticated	1.47
Changes in Stipulodes	
(a) Reduction in the number of stipulodes in both tiers.	54.5
(b) Nodes without stipulode	18.2

Stability of the morphological variants appeared in M_1-generation was traced unto VM_{10}–generations.

Observation on Mitotic Events

The effects of dES on the mitotic events may be described under the following three heads:

Frequency of Division Figures:

No mitotic figure could be obtained in any of the concentrations till 24 hours after treatment.

The mitotic activities were restored after 48 hours in all the concentrations but consistently at the frequency below the control level (Table 10.10). It was further noted that the degree of depression increased with the concentrations.

Table 10.10: Increase or Decrease in the Frequency of Dividing Nuclei Against Control Following dES-Treatments (Figures in per cent)

Concentrations	*After 48 hours*	VM_1	VM_2
0.00625%	–2.06	+28.05	+24.5
0.0125%	–3.71	+29.13	+26.82
0.025%	–11.57	+30.93	+27.48
0.05%	–26.03	+6.47	+6.95
0.1%	–45.42	–17.98	–19.86

The frequency of dividing nuclei was computed, though high in all the concentrations of both VM_1 and VM_2 it was comparatively lower in VM_2 generations (Table 10.10). It is further apparent that the frequency of diving nuclei gradually increased unto 0.025 per cent and thereafter it decreased remaining above control level in the concentration 0.05 per cent and finally fell down below the control level in 0.1 per cent.

Relative Frequency of Mitotic Stages

The relative frequency of prophase, metaphase and post-metaphase was scored after 48 hours as well as in VM_1 and VM_2 generations of all the five concentrations along with control. Increase or decrease in the relative frequency of mitotic stages against control, as ascertained, may be seen in the Table 10.11.

Table 10.11: Increase or Decrease in the Relative Frequency of Mitotic Stages Against Control

Fixation	*Concentration (%)*	*Frequency of Different Mitotic Stages*		
		Prophase (%)	*Metaphase (%)*	*Post-metaphase (%)*
Observation after	0.00625	–9.31	–41.47	+191.70
48 hours treatment	0.0125	–18.39	–55.17	+262.59
	0.025	–44.92	–65.13	+338.08
	0.05	–100.00	–100.00	+557.00
	0.1	–100.00	–100.00	+162.14
V_1	0.00625	–9.50	–9.50	+51.97
	0.0125	–16.45	–8.79	+58.06
	0.025	–33.34	–12.81	+96.88
	0.05	–38.42	–53.55	+272.58
	0.1	–58.03	–60.87	+329.81
V_2	0.00625	–9.42	–4.92	+17.49
	0.0125	–6.43	–5.59	+25.86
	0.025	–18.89	–7.09	+43.90
	0.05	–27.80	–23.93	+110.98
	0.1	–32.20	–38.38	+165.06

In the specimens fixed after 48 hours, the frequency of prophase and metaphase was found decreasing with increasing in the concentrations unto 0.025 per cent and beyond this completely checked. In contrast with prophase and metaphase, the frequency of post-metaphase increased with concentrations (Table 10.11)

It is further evident from the table under reference that the frequency of prophase, metaphase and post metaphase in VM_1 and VM_2 generations followed the same course of decrease increase in all the five concentrations discussed above.

Mitotic Deviates:

The karyological observations made 48 hours after treatment revealed various types of mitotic irregularities. Such observations were also made on VM_1 and VM_2 generations to ascertain the recovery of normal mitotic figures. The details of spectrum

Table 10.12: The Spectrum and Frequency (Per cent) of the Aberrant Mitotic Figures in Three Successive Fixations

Concentrations	*Unequal Anaphasic Separation*	*Failure of Anaphasic Separation*	*Lagging*	*Bridge*	*Tripolar Anaphase*	*Clotting of Chromosome*	*Smashed Metaphase*	*Micronuclei*	*Binucleate Cell*	*Total Anomaly*
After 48 hours										
0.00625%	–	–	–	0.5	0.97	–	–	–	–	1.47%
0.0125%	1.09	–	–	–	–	–	–	0.9	–	1.99%
0.025%	2.81	–	1.02	–	–	1.08	–	1.05	1.05	7.46%
0.05%	8.62	0.98	2.94	1.15	–	1.47	0.98	2.39	–	18.53%
0.1%	18.72	0.49	8.73	–	–	3.74	5.98	4.22	–	41.88%
V_1										
0.00625%	–	–	–	–	–	–	–	–	–	–
0.0125%	–	–	–	–	–	–	–	–	–	–
0.025%	1	–	–	–	–	–	–	–	–	1.00%
0.05%	1.5	–	0.49	–	–	0.3	–	0.8	–	3.09%
0.1%	8.05	–	2.21	–	–	0.76	0.5	1.54	–	13.06%
V_2										
0.00625%	–	–	–	–	–	–	–	–	–	–
0.0125%	–	–	–	–	–	–	–	–	–	–
0.025%	–	–	–	–	–	–	–	–	–	–
0.05%	–	–	–	–	–	–	–	–	–	–
0.1%	4.52	–	1.24	–	–	0.52	–	–	–	6.28%

and frequency of mitotic irregularities, as ascertained in the aforesaid three successive fixations may be seen in the Table 10.12.

As indicated in the aforesaid table of all the nine types of aberrant mitotic figures seen in the first fixation *i.e.* after 48 hours, the formation of tripolar spindle (Plate 10.3, Figure 3.9) and binucleate cells (Plate 10.3, Figure 3.8) were of rare occurrence being recorded at very low frequency only in the concentrations 0.00625 per cent and 0.0125 per cent respectively. The most frequently observed mitotic irregularities were the unequal anaphasic separation (Plate 10.3, Figures 3.6 and 3.7) and the formation of micronuclei. Among the cells witnessing unequal anaphasic separation, the diagonal orientation of anaphasic axis was also seen very frequently.

The formation of laggards (Plate 10.3, Figures 3.1 and 3.2)and the clotting of chromosomes (Plate 10.3, Figures 3.3 and 3.10) were also among the most frequently observed mitotic irregularities but only in three higher concentrations *i.e.* from 0.025 per cent to 0.1 per cent. Surprisingly, the laggard fragments of the chromosome were frequently seen migrating to either of the poles of anaphase.

The bridges (Plate 10.3, Figures 3.1 and 3.2) were found in the concentrations 0.00625 per cent and 0.05 per cent. Failure of anaphasic separation and smashed metaphase with severely broken chromosomes (Plate 10.3, Figure 3.5) were recorded only in the two higher concentrations *viz.* 0.05 per cent and 0.1 per cent.

Overall frequency of mitotic irregularities was found increasing with the concentrations to reach as high as 41.88 per cent in the concentration 0.1 per cent. Spectrum of the aberrant mitotic figures also increased accordingly but the bridges, which were recorded in 0.05 per cent were not found in the concentration 0.1 per cent.

Apart from the above noted mitotic irregularities, the differential staining of interphase nuclei was recorded for the treatment with 0.025 per cent. In the concentration 0.05 per cent the doubling of the chromosome number was seen in the restricted cases.

In the second fixation of VM_1 generation, failure of anaphasic separation, bridges, tripolar spindle and binucleate cells were not seen in any of the five concentrations. The remaining five types of irregularities were recorded in the concentrations 0.05 per cent and 0.1 per cent but comparatively at very low frequency.

In VM_2 generation, all the irregularities were eliminated except unequal anaphasic separation, laggards and clotting of chromosomes which persisted in 0.1 per cent. However, the frequency of last two types of irregularities, was very low (Table 10.12).

Observations on M_1 Generation

M_1 plants could be obtained only in 0.05 per cent and hence the accounts on the inheritance of morphological variants are restricted only to this concentration.

Of all the categories of morphological variations, only those of cortical cells as well as stipulodes were inherited in the M_1 generations while remaining other types of variations were eliminated. The frequency of different morphological variants with respect to the changes in the cortical cells and stipulods may be seen in the

Table 10.13. Normal sex organs were observed, but at late stage of growth and only in 7 per cent of branches.

Table 10.13: Spectrum and Frequency (Per cent) of dES-induced Morphological Variations Recorded among M_1-Population

Parameters	*Frequency in Per cent*
Change in cortication	
(a) Spiral	71.8
(b) Convoluted	1.2
(c) Hypertrophic	0.3
(d) Ecorticated	1.47
Change in stipulodes	
(a) Reduction in the number of stipulodes in both tiers	54.5
(b) Nodes without stipulode	18.2

In addition to above variations M_1 plants were seen chlorophyll mutant of *Virina-albscens* (Transescens) type (Gustafsson, 1940). As the filaments achieved a length of 8-10 nodes, first the tip of short laterals turned white after losing chlorophyll. Some of the short laterals were also seen with 2-3 alternating bands of green and non-green segments. Late, the whitening of the segments spreaded in the remaining portion and finally the entire plant turned white within a fortnight after its first appearance leaving only the young apical portion green. In a few cases, the antheridia also turned white. The *Virina-albscens* filaments did not bore any new branches but a good number of plants were seen emerging from the subterranean portion of the dying filaments (VM_2). These plants of VM_2 generations were screened for the morphological variations and were found true for all categories of variants observed on M_1 plant except minor variations of their respective frequency. These VM_2 plants also met the same fate of *Virina-albscens* (Transcens) type of chlorophyll mutation.

Discussion

The overall growth of the alga was evaluated after considering three parameters *i.e.* linear growth, differentiation of nodes and emergence of branches. All the three parameters responded consistently on the same pattern to the action of dES. Initially, the growth was inhibited during the first week, it was considerably enhanced during the second week and then roughly became static during the third week after the treatment with all the concentration of dES.

The mitotic activities in the present study also responded to the treatments with all the concentrations of dES on the pattern of growth being initially inhibited, at least unto 48 hours, enhanced in VM_1 generation and became roughly static in VM_2 generation.

Since linear relationship has been observed between the growth performance and mitotic activities following the treatments, it is quite likely that the biosynthesis of phytohormones, responsible for the growth, might have been influenced at the pre-synthesis level (DNA-RNA level).

The initial inhibition of growth as well as the mitotic activities may be attributed to the temporary physiological imbalance caused due to the treatments with dES. The stimulatory effects of the treatments were manifested only after escaping from the said imbalance through the adaptation of protoplasm against mutagen reaction and then became static after reaching the peak of its stimulatory effects.

The treatments with dES yielded positive results on the growth and mitotic activities of the alga at hand unto the concentrations 0.05 per cent. The magnitude of the positive effects has been noted increasing with the concentrations unto 0.0125 per cent in case of growth and up to 0.025 per cent in case of mitotic activities, thereafter it gradually decreased remaining above control unto 0.05 per cent and finally fell down below it in the concentration 0.1 per cent. Alga could not survive after treatment with 0.15 per cent, Therefore, 0.1 per cent of dES may be taken as an adverse dose for the growth and 0.15 per cent as the lethal. The adverse effects of higher concentrations of dES have been attributed to inhibition and/or destruction of auxin synthesis, change of nutrition levels, inhibition of mitosis and chromosomal damage associated with secondary physiological changes (Skoog, 1953, Went and Thiman, 1957). The depression in mitotic activity by the higher concentrations of dES may be attributed to the chemical changes brought about in the cell (Lea, 1955), production acceleration of antimitotic substances (Rubin and Metlisky, 1958 and Kuzin, 1963), and blockage of DNA synthesis (Heiner, 1971).

Since the highest mitotic induction has been recorded after treatment with 0.025 per cent, this may be taken as the most effective dose for inducing mitosis.

Of the three parameters of growth, the differentiation of nodes and the emergence of branches have been induced maximum through 0.0125 per cent. The linear growth, though, has been induced maximum through 0.025 per cent, the differentiation of nodes and emergence of branches has been noted comparatively lower than 0.0125 per cent. Thus, 0.0125 per cent may be considered as effective does for inducing overall growth in the alga under study.

Comparative effects of dES on three parameters of growth further reveals that the stimulatory effects of the treatment was more apparent on the emergence of branches and least on the linear growth. It appears, therefore, that the dES preferentially induces the development of axial branches. This is consistent with the observations of Avadhani and Rao (1968) who also observed stimulatory effects of dES unto the concentration 0.0304M in *Arachis hpogea L.*

The treatments with increasing concentrations of dES have been witnessed with gradual decrease in the frequency of prophase and metaphase accompanied by the increasing frequency of post-metaphase figures. The frequency of post-metaphase, however, decreased abruptly in the concentration 0.1 per cent. The relative frequency of different stages in the treated VM_1 and VM_2 generations has also been observed on the above described pattern.

dES is a well-known alkylating agent with two reactive groups that produce intra and inter steridcross links in the DNA. Alkylation of DNA may be induced in different ways *viz.* by alkylation of phosphate groups, breaking the backbone of DNA by some of the bases giving rise to unstable quaternary nitrogen as a result of alkylation and by deprivation of DNA (Freese, 1963), Alkylation of bases G,A,C ant T occurs at 7 N, 3N and 7 N, 3 N and 3 N position respectively (Banks, 1971).

The negative response of the prophase and metaphase to the increasing concentration appears to be due to the helix disruption and/or helix distortion followed by interference of the dES with normal DNA synthesis cycle *i.e.* S-phase. This assumption gets support from the breaking of the chromosomes at metaphase and anaphase as well as the clotting of the chromosomes at metaphase observed frequently after treatments and also its frequency has been found increasing along the concentration. Production of aberrant mitotic figures after the treatment with chemical mutagens as well as ionizing radiations has been reported by Natarajan and Upadhyay (1964), Mikaelsen *et al.* (1968), Bose and Bose (1971), and Chowdhary and Dnyanasagar (1980).

The frequency of post-metaphase has been noted dose dependent. This may be due to the adverse effects of dES on spindle mechanism. This assumption rests with the fact that the anaphasic disturbances in the form of the unequal anaphasic separation was observed most frequently and its frequency was noted increasing along with the concentration.Suppression of the anaphasic separation was also observed in higher concentrations. Therefore, the increased frequency of post-metaphase might have been caused due to the accumulation of the dividing nuclei at anaphase. Earlier, the anaphasic disturbances following the treatments with dES have been reported by Heiner (1971) and Chowdhary and Dnyansagar (1980).

Various types of morphological variants with respect to the elimination of cortical cells, reduction in stipulodes, changes in dactyls and reduction in sex organs have been observed by the author after treatments with dES and their frequency has been found dose dependent as also recorded for the aberrant mitotic figures. Therefore, all the morphological variability may be presumed to be the manifestation of mitotic irregularities (cf. Sparrow, 1951) Wada 1952; Awadhani and Rao, 1968; Olatunde and Chheda 1971 and Tarar and Dnyansagar, 1980). This assumption may also be supported form the fact that in the higher concentration where the mitotic irregularities were high, completely deformed branches was recorded. The changes in the prominence of cortical cells, observed after treatments, were eliminated in the successive vegetatively propagated generation and hence may be presumed simply on level of physiological interference of the mutagen used.

It is worth mentioning here that all the observed changes on the level of cortical cells, stipulodes, dactyls and sex organs are either of additive nature or of reductive. Consequently the unequal distribution of the chromosomes at anaphase and/or fragmentation of the chromosomes, which have been observed as the dominating irregularities, might have played an important role in producing morphological variants.

Further, it is to be mentioned that the elimination of either upper tier or lower tier has been observed very frequently in addition to the elimination of both tiers. The partial reduction in the number of stipulodes has also been recorded very frequently. This trend of elimination and reduction has also been inherited with great fidelity in M_1 generation and retained in successive vegetatively propagated generations. Although, inheritance of any variants in M generation cannot be taken as an evidence to predict point mutation, being of primary nature, it would not be out of way to say that if at all it is a case of point mutation or a case of simple chromosomal aberrations, the expression of two tiers as well as all the component stipulodes even in a single tier is not controlled by a single gene but by a pull of genes. Similarly, the elimination of dactyls and coronal base has been noted independently besides in combination (vide infra.) Thus, dactyl and coronal base appear to be controlled by two different set of genes.

The reduction of both antheridia and oogonia has never been observed on the same branches. Therefore, the expression of ♂ and ♀ sex organs at least in *chara*, appears to be controlled by two different genes and probably located either distantly in cis-position or in trans position to conceive their independent movement easy following breaking or unequal distribution.

The lagging of very high number of acentric fragments were seen at anaphase following treatments and also their poleward movements. The movement of these acentric fragments may be attributed to the fact that they might have been transferred to the pole by the movement of the healthy chromosomes due to the shortening of chromosomal fibers as reported earlier in *Tradescantia* by Wadá (1952).

The clotting of the chromosome has not been observed in the lower concentration but has been observed in the higher concentrations from 0.025 per cent onwards Tanake and Sato (1952) reported its presence in *Tradescantia* treated with 0.600 ppm. Streptomycin for 3 hours. According to them, this phenomenon seems to occur in a critical stage of cell following severe treatments and to be considered as a stage due to a sort of cogulation. Therefore, it will have scarcely any genetical significance.

The formation of micronuclei which has been observed along with unequal anaphasic separation and/or lagging of chromosomes may be presumed to be clearly ramnants originating from accentric fragments produced in the preceding mitosis.

The unequal distribution of chromosomes at anaphase, which was observed by the author as one of the dominating mitotic irregularities may by presumed due to stickiness of the chromosomes which was also observed frequently after treatments. This is consistent with the results of Wada (1952) who also reported unequal distribution of daughter chromosomes in the sister cells due to stickiness of the chromosomes caused by the tropon compounds in *Tradescantia*.

Binucleate cells observed in 0.05 per cent treated samples might have been induced through cell plate inhibition as reported earlier by Sarma and Tripathi (1976) in *C. globularis*, *C. braunii* and *N. flagelliformis* following treatments with the bromine and caffeine.

References

Avadhani, K.K. and Rao, B.V.R., 1968. Studies on the effects of Diethylsulphonate on T.M.V. 2 Groundnut (*Arachis hypogaea* L.). *Indian J. Sci. and Indust.*, 2(2): 77–82.

Banks, G. R., 1971. Mutagenesis: A review of some molecular aspects. *Science Progress*, 59(236): 457–503.

Bose, S. and Bose, M. S., 1971. Morphological and cytological effects of pre- and post-irrandiation treatments with different chemicals in Tomato. *The Nucleus*, 14(1): 9–14

Chatterjee, P. and Sharma, A.K., 1972. Effect of Phenols on nuclear division in *Chara zeylanica*. *The Nucleus*, 15(3): 214–218

Chowdhary, A.D. and Dnyansagar, V.R., 1980. Induced chromosomal aberrations in garlic. *Cytol. Genet.*, 15: 58–60.

Delay, C. and Carpentier, S., 1955. Action de la colchicine sur *chara vulgaris* L. 1. Action Sur les filaments spermatogenes. *Rev. Cytol et Biol. Veg*, 16: 415.

Freese, E., 1963. Molecular mechanism of mutations. In: *Molecular Biology, Part I*, (Ed.) Herbert Taylur. Academic Press, New York.

Godward, M. B. E., 1948. The iron alum acetocaramine method for algae. *Nature*, 161: 203.

Gustafsson, A., 1940. The mutation systems of the chlorophyll apparatus Lunds. *Univ. Arsskr.*, 23: 1–40.

Heiner, R.E., 1971. Alterations in the nuclear cycle, mitotic index and chromosomes of *Vicia* as affected by diethylsulphonate. *Mut. Res.*, 12: 249–254.

Hope, A.B. and Walker, N.A., 1975. *The Physiology of Giant Algal Cells*. Camb. Univ. Press, Cambridge.

Kuzin, A.M., 1963. The theoretical basis of pre-sowing irradiation of seeds. In: *Pre-Sowing Irradiation of Seeds of Agricultural Crops*. USSR Acad. Sci. Press, Moscow.

Lea, D.E., 1955. *Action of Radiation in Living Cells*, 2nd edn. Camb. Univ. Press, Cambridge.

Mikaelsen, K., Ahnstrom, G. and Li, W.C., 1968. Genetic effects of alkylating agents in barley: Influence of post-storage, metabolic state and the pH of mutagen solution. *Hereditas, Lund.*, 59: 353–74.

Moutschen, J.M., Dahmen, M. and Gillet, C., 1956. Sur les modifications induties per les hydrazide malique et isonicotinique dans les antheridies de *Chara vulgaris* L. La *Cellude*, 58: 65.

Natarajan, A.T. and Upadhyay, M.D., 1964. Localized chromosome breakage induced by ethylmethane sulphonate and hydroxylamine in *Vicia faba*. *Chromosoma*, 15: 156–169.

Olatunde, A.O. and Chheda, H.R., 1971. Mitotic events and other disorders induced in cowpeas, *Vigna unquiculata* (L.) Walp. by ionizing radiation. *Radiat. Bot.*, 11: 375–381.

Ramjee, 1968. Contributions to the systematics, cytology and cytotaxonomy of Indian Charophyta. *Ph.D. Thesis*, BHU, Varanasi.

Rubin, B A. and Metlisky, L.V., 1958. A Study of the action of ionizing radiation on the metabolism of potato tubers in relation to their problem of all year-round storage. In: *Proc 2nd Intn. Conf. Peaceful Use of Atomic Energy.*

Sarma, Y.S.R.K. and Tripathi, S.N., 1976a. Effects of chemicals on some members of Indian Charophyta–I. *Caryologia*, 29(3): 247– 262.

Sarma, Y.S.R.K. and Tripathi S.N., 1976b. Effects of chemicals on some members of Indian Charophyta–II. *Caryologia*, 29(3): 263–276.

Skoog, F., 1953. Substances involved in normal growth and differentiation of plants. *Brook Haven Symposia in Biology*, 6: 1–21.

Sparrow, A.H., 1951. Some cytological and morphological changes induced in plants by ionizing radiation. *Science*, 114–48.

Tanaka, N. and Sato, Sy., 1952. Effects of streptomycin on the mitotic cells of *Tradescantia paludosa*. *Cytologia*, 17(2): 124–133.

Tarar, J.L. and Dnyansagar, V.R., 1980. Mutagen-induced variation in flowering and floral morphology in plants of *Turnera ulmifolia* Linn. *J. Indian Bot. Soc.*, 59: 164–169.

Turner, F.R., 1970. The effect of colchicine on spermatogenesis in *Nitella*. *J. Cell. Biol.*, 46: 220–234.

Wada, B., 1952. The effect of chemicals on mitosis studied in *Tradescantia* cells *in vivo*. I. Acetylaminotropolone. *Cytologia*, 17(1): 14–34.

Went, F.W. and Thiman, K. V., 1957. *Phytohormones*. The McMillan Co., New York.

Chapter 11

Field Study of Pesticides Residue in Nepal

Surendra Narayan Das

Co-ordinator,
Museum-cum-Library and Documentation Center,
Institute of Forestry, Hetauda, Makwanpur, Nepal
A Constituent Unit of Tribhuvan University, Kathmandu (Nepal)

Introduction

Rice is one of the important staple foods for more than one and half of the world population. It constitutes about 80 per cent human food need in Asian Countries (Way, 1976). This important crop is prone to more than 70 species of insect pests amounting to 40 per cent loss in its annual production (Gruzdyev, 1983). One of the well known means of protecting the rice plant from these pests is to apply organo chlorine insecticides amongst which use of HCH is most common all over the world since 1940s. The insecticidal properties of HCH are due to its stable sterio isomers of 1, 2, 3, 4, 5, and 6. The use of HCH was completely banned in many developed countries on account of its long persistent side effects. It is however, continued to be used in the large quantity in India, Nepal and other developing Asian Countries.

Persistence of Lindane (γHCH) residue in harvested rice grain and straw, bran has been reported (Das *et al.*, 1975). Survey carried out in different states of India by the researchers has revealed contamination of paddy crops with lindane in general. In Nepal where rice is predominant crop and lindane is widely used in its cultivation, such survey has not been made so far. In the present endeavour an attempt has been made to survey the every component of paddy cultivation in Nepal for ascertaining the residue of organo chlorine pesticides in them to avoid related health hazard.

Only a small fraction of pesticides used in the agricultural field are absorbed by the target organism and rest fall on the field soil from where they are drifted to the low lying aquatic ecosystem. Persistence of the pesticides in aquatic environment is of paramount significance. Algae, the predominating primary producer in aquatic system, are adversely affected which in turn affect the whole ecosystem through interference with food chain and energy flow. Therefore attempt has been made to study the effect of the pesticides, if present in the river or canal water the sole source of irrigation in Nepal, on the algae.

Methodology

Pesticides Residue Analysis of Environmental Samples Sampling

Samples of soil, paddy plants, rice grains, and water (used for irrigation) were collected from different sites at Jagannathpur, Parwanipur, Birganj, and Barainia, Fhentaphulwaria farms of Parsa and Bara districts respectively of Central Nepal during 1999-2000. All the samples were collected from the fields where pesticides have been sprayed by the farmers. Samples were carefully brought to laboratory in a closed polythene bag after proper tagging.

Rice grain samples were also collected from the market of Kathmandu and Birganj. Roots, straw and grains were cut apart from the plant and kept separately after proper drying in open sunlight to avoid any fungal or microbial infection.

Soil samples collected from four corners of the paddy fields to be studied were screened for visible debris and mixed thoroughly. The samples were air dried, sieved through 2 mm size mesh and stored in defreezer along with other samples before analysis.

Extraction and Clean Up of Water Samples

A water sample of 500 ml was taken in one liter separatory funnel and extracted with n-hexane (100 ml) by shaking following the methods of Karanth *et al.* (1983). The extraction was repeated three times for 30 minutes each. The extracts were pooled and collected each time in conical flask and the volume was evaporated to 5 ml under rotary evaporator. The concentrate was passed through florisil column; (2.0 cm in diameter and 30 cm long) containing bilayer of anhydrous Sodium sulphate sand wiched between a thin layer of florisil over a small pad of glass wool at bottom. The hexane eluted was collected and further reduced to 2ml.

Extraction and Clean Up of Soil Samples

A representative of 100g air-dried soil sample was taken in 250 ml of conical flask and extracted with n-hexane on magnetic stirrer putting magnetic bar in the flask for 6 min.The extraction was repeated three times (Karanth 1982,1983). The extracts were pooled, filtered and concentrated to 5ml. The concentrate was passed through florisil column and volume reduced to 2ml for GLC analysis as mentioned above.

Extraction and Clean Up of Grains, Straw and Root

Rice grains, straw and root samples collected from different field stations and markets during year 1999, 2000 were powdered separately in Booty mixer grinder.

The powdered grains, straw and roots each weighing 10g were extracted with a mixture of acetonitrile: water (65: 35 v/v) in separate conical flask on magnetic stirrer following the methods of Bertuzzi *et al.* (1967). All the samples were extracted 3 times. The residues were taken in n-hexane and clean up was performed as per Verirov and Aharonson (1978). The eluted samples were concentrated to 2ml volume and used for GLC analysis.

Toxicity and Interaction of Gamma HCH to Blue Green and Green Algae

Growth and Maintenance of Experimental Algal Culture

Axenic cultures of green alga *Oocystis borgei* and nitrogen fixing blue green alga *Nostoc linckia* isolated, purified and maintained in the laboratory were used throughout the experiment. *Oocystis borgei* was sub cultured Hughes's medium supplemented with nitrogen in the form of Nitrate and that of *Nostoc linckia* in H^- nitrogen free Hughes's medium and were grown in cotton stoppered conical flask maintained under continuous fluorescent light (Philips)200 mole m-2s-1 at 25±5°C.(EPA 1975).

Incorporation of Sand with γ-HCH

Pure acid washed calcinated sea sand (E.Merck) with particle size of 0.1-0.3 mm was used as simulated sediment by incorporating known amount of desired organics for environmental dynamics studies. Acid washed sand was further washed with n-hexane and evaporated to remove any traces of pesticide. The sand was coated homogeneously with gamma HCH (500 µg/g) by continuous shaking with desired amount in n-hexane (Mishra *et al.*, 1996). The gamma HCH level was up to the saturation limit.The solvent was evaporated to dryness under controlled condition (temperature and pressure) to avoid vaporization of gamma HCH.

Incorporation of Algal Test Organism with γ-HCH Coated Sand

In experimental system, 14g loosely packed γ-HCH coated sand (2cm height) were taken in six 100ml conical flasks in triplicate set for each of the 2 sets of test organism for *Oocystis borgei* and *Nostoc linckia*. Similarly, four 100ml conical flasks containing 14g pure sand free from γ-HCH were performed in duplicate for each of the two test organisms (*Oocystis* and *Nostoc*) as control. All the 10 sets of conical flask were divided into 2 groups for both the test organisms each having 5 sets (three experimental and two control). In first 5 sets of conical flask,50ml of nitrate containing medium (H^+) inoculated with test organism *Oocystis borgei* at the rate of OD=0.05 was added in both experimental and control flask. 50 ml nitrogen free Hughes's medium was inoculated with the culture of *Nostoc linckia* having the OD=0.05 in another 5 sets of conical flasks containing sand coated with γ-HCH and without γ-HCH. All the flasks (experimental and control) were incubated for 21 days under fluorescent light intensity 200 mole $1^{-2}5'$ at 25±0.5° C.

The algal culture were harvested and separated from sediment at the end of the incubation (21 days) and test organisms were separated from culture medium by centrifugation at 3500 rpm for 10 minutes. The pellets were washed twice with 5 ml of distilled water recentrifuged and biomass was determined as per the method of Lillelund and Wehrmann (1974).

The pellet homogenized in Potter elvehjehm type homogenizer under chilled conditions and 2.5 per cent homogenate was prepared in prechilled distilled water. I ml homogenate was extracted with 95 per cent chilled acetone and kept for 20 hours in darkness in refrigerator. The homozenate was centrifuged and chlorophyll was determined from supernatant following the method of Arnon (1949) using spectronic-21 (Bausch and Lomb).

The protein contents in experimental and control samples were determined after precipitation of homogenate with 10 per cent TCA following the method of Lowry *et al.* (1951). The TCA precipitate was washed twice with acetone in view of interference of pigment during color development with folin ciocalteu reagent.

Extraction and Estimation of HCH in Algae

The uptake of gamma HCH by algae from experimental flasks and extraction of HCH from test organism were done following the method of Diercking and Hansen (1974) with certain modifications.

One ml (1.0 ml) homozenate was centrifuged after adding 5 ml of distilled water. The supernatant pooled and pellet was resuspended in 1.0 ml n-hexane and immediately centrifuged for 2 minutes. The hexane wash was repeated twice more and all the supernatant combined and made to known volume. The hexane wash was cleaned up by passing through florisil column pre-soaked with n-hexane as shown in the flow chart. The final volume of elute was made up to 5 ml and used for GLC analysis.

Estimation of HCH in Medium

Estimation of HCH in the medium was determined as per the method described by Misra *et al.* (1996). 50 ml samples were decanted from each flask and extracted with 15 ml of AR grade n-hexane in separatory funnel. The extraction was done three times and all the extracts were pooled, dried over anhydrous sodium sulphate and filtered through glass wool. The filtered extracts were evaporated by rotatory evaporator and concentrated to 3-4.0 ml under control conditions to avoid loss of gamma HCH by vaporisation. Finally, the volume was made to 5ml for analysis.

Estimation of HCH from Sand

The separated sand mass in each flask were further extracted with AR grade n-hexane using magnetic bar over magnetic stirrer for one hour interval. The extraction was made thrice and the extracts were pooled each time in separate flask. The extracts were dried over anhydrous sodium sulphate granules and finally passed through florisil column. The hexane extracts were concentrated finally to known volume for gas chromatographic analysis.

Gas Chromatographic Analysis

The extracts were subjected to a gas chromatographic analysis on a Varian Vista 6000 equipped with a 63 Nielectron capture detector was used for quantification of the residues. A 1.8 m x 4 mm column packed with Chrom WHP mesh size 80/100 coated with 1.5 per cent OV-17 and 1.95 per cent OV-210 was used. Conditions (all

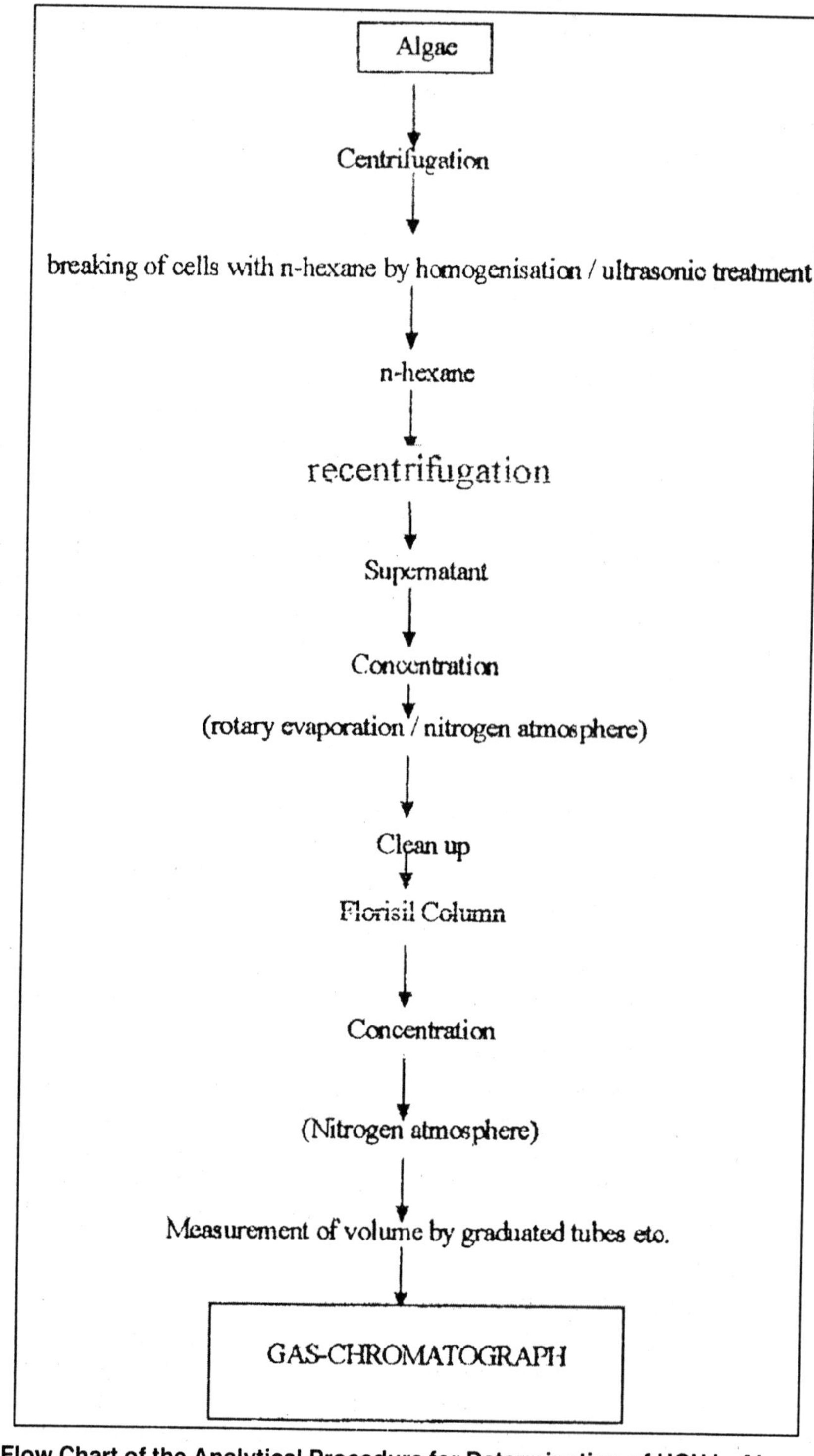

Flow Chart of the Analytical Procedure for Determination of HCH in Algae

temperature in °C): injector 250, column 180, detector 250, carrier gas nitrogen 60 ml/min.

Observations

Exploratory studies of pesticides residue conducted in seven varieties of paddy shown in the paddy fields of three districts of Nepal have been summarized in the Tables 11.1 to 11.8. As shown in the tables under reference, the samples of paddy grain, straw, root and field soil were found to contain varying amount of five organochlorine pesticides among which HCH with its four isomers was observed more frequently and it was followed by Aldrin and Endosulphan. PP-DDE and OP-DDT were detected but in restricted samples.

Table 11.1: Organochlorine Insecticides Residue Analysis of Soil Samples and Paddy Crops from Parwanipur, Bara District (Year 2000) Station–II (Type Mansuli)

Parameters	αHCH	γHCH	βHCH	δHCH	Aldrin	Endo Sulf-1	PP-DDE	OP-DDT
Grains	0.002	ND	ND	0.264	0.020	ND	ND	ND
Straw	0.208	ND	1.377	ND	0.015	ND	ND	ND
Roots	0.816	ND	2.017	0.055	ND	ND	ND	ND
Soils	0.003	ND	0.057	0.006	ND	ND	ND	ND

Table 11.2: Organochlorine Insecticides Residue Analysis of Soil Samples and Paddy Crops from Fentaphulwaria, Bara District (Year 2000) Station–I (Type Basmati)

Parameters	αHCH	γHCH	βHCH	δHCH	Aldrin	Endo Sulf-1	PP-DDE	OP-DDT
Grains	0.025	ND	0.146	0.040	ND	ND	ND	ND
Straw	0.019	ND	0.217	ND	0.005	ND	ND	ND
Roots	0.046	ND	0.656	ND	0.095	ND	ND	ND
Soils	0.148	ND	0.451	ND	0.001	ND	ND	ND

Table 11.3: Organochlorine Insecticides Residue Analysis of Soil Samples and Paddy Crops from Brahmpuri, Parsa District (Year 2000) Station–III (Type Kasturi)

Parameters	αHCH	γHCH	βHCH	δHCH	Aldrin	Endo Sulf-1	PP-DDE	OP-DDT
Grains	0.003	0.010	0.052	ND	0.053	0.004	ND	ND
Straw	0.015	0.006	0.015	ND	0.039	ND	ND	ND
Roots	0.03	0.067	2.01	ND	0.218	ND	ND	ND
Soils	0.006	0.002	0.023	ND	0.014	ND	ND	ND

Data detected in µg/g.

ND: Not detected

Table 11.4: Organochlorine Insecticides Residue Analysis of Soil Samples and Paddy Crops from Jagannathpur, Parsa District (Year 2000) Station–I (Type-Sabitri)

Parameters	αHCH	γHCH	βHCH	δHCH	Aldrin	Endo Sulf-1	PP-DDE	OP-DDT
Grains	0.019	0.006	0.008	0.001	0.003	ND	ND	ND
Straw	0.034	0.012	0.031	ND	0.007	ND	ND	ND
Roots	0.032	ND	0.249	ND	ND	0.185	ND	ND
Soils	0.0123	0.0041	0.0303	ND	0.0053	ND	ND	ND

Table 11.5: Organochlorine Insecticides Residue Analysis of Soil Samples and Paddy Crops from Fentaphulwaria, Bara District (Year 2000) Station–II (Type-Sabitri)

Parameters	αHCH	γHCH	βHCH	δHCH	Aldrin	Endo Sulf-1	PP-DDE	OP-DDT
Grains	0.0067	0.002	0.004	0.002	ND	0.064	ND	ND
Straw	0.038	0.009	ND	ND	ND	ND	ND	0.222
Roots	ND	ND	ND	ND	ND	ND	ND	ND
Soils	0.025	0.017	0.039	0.01	0.012	ND	0.01	ND

Table 11.6: Organochlorine Insecticides Residue Analysis of Soil Samples and Paddy Crops from Parwanipur, Bara District (Year 2000) Station–III (Type-Chaite)

Parameters	αHCH	γHCH	βHCH	δHCH	Aldrin	Endo Sulf-1	PP-DDE	OP-DDT
Grains	0.015	0.016	0.141	ND	ND	0.237	ND	0.081
Straw	0.068	0.019	0.061	0.003	ND	ND	ND	ND
Roots	0.086	0.036	0.244	ND	0.023	ND	0.155	ND
Soils	0.014	0.005	0.014	0.003	ND	ND	ND	ND

Table 11.7: Organochlorine Insecticides Residue Analysis of Soil Samples and Paddy Crops from Parwanipur, Bara District (Year 2000) Station–IV (Type-Radha–17)

Parameters	αHCH	γHCH	βHCH	δHCH	Aldrin	Endo Sulf-1	PP-DDE	OP-DDT
Grains	0.012	0.004	0.01	0.001	ND	0.009	0.007	ND
Straw	0.097	0.024	0.034	0.011	ND	0.077	ND	ND
Roots	0.037	0.01	0.004	ND	ND	ND	ND	ND
Soils	0.015	0.002	0.042	0.001	ND	ND	ND	ND

Data detected in µg/g.

ND: Not detected

Table 11.8: Organochlorine Insecticides Residue Analysis of Soil Samples and Paddy Crops from Parwanipur, Bara District (Year 2000) Station–V (Type-Pakistani-17)

Parameters	*αHCH*	*γHCH*	*βHCH*	*δHCH*	*Aldrin*	*Endo Sulf-1*	*PP-DDE*	*OP-DDT*
Grains	ND	ND	ND	ND	ND	ND	ND	ND
Straw	0.064	0.012	0.032	ND	ND	0.038	ND	ND
Roots	ND	ND	ND	ND	ND	ND	ND	ND
Soils	0.001	0.004	ND	ND	ND	ND	ND	ND

Data detected in μg/g.

ND: Not detected

Nine verities of rice grains procured from the market of Birganj and Kathmandu were also analysed and were found carrying residues of the same five pesticides (Tables 11.9 and 11.10)

Table 11.9: Organochlorine Insecticides Residue Analysis of Rice Grain Samples Collected from Birganj Market, Parsa District (Year 2000)

Parameters	*αHCH*	*γHCH*	*βHCH*	*δHCH*	*Aldrin*	*Endo Sulf-1*	*PP-DDE*	*OP-DDT*
Taichun	0.0018	0.0012	0.0256	0.0095	0.0061	ND	0.0151	ND
Pokhreli	0.0032	0.0013	ND	0.0644	0.0045	0.1137	ND	ND
Kasturi	0.0094	0.0116	0.1787	0.0408	0.0309	ND	0.0631	ND
Ekhattar	0.0098	0.0163	ND	0.0665	0.006	ND	ND	ND
Mansuli	0.0096	0.0079	0.0321	0.0223	0.0067	ND	0.0067	ND

Table 11.10: Organochlorine Insecticides Residue Analysis of Rice Grain Samples Collected from Kathmandu Market (Year 2000)

Parameters	*αHCH*	*γHCH*	*βHCH*	*δHCH*	*Aldrin*	*Endo Sulf-1*	*PP-DDE*	*OP-DDT*
Sabitri	0.014	0.002	0.004	0.007	ND	0.039	ND	3.465
Katika	0.013	0.004	0.006	0.005	ND	ND	ND	ND
Philips	0.012	0.002	ND	ND	0.006	ND	ND	ND
Mansuli	0.027	0.013	0.014	ND	0.012	0.048	ND	ND
Basmati	0.019	0.006	0.005	0.002	ND	0.022	ND	ND

Data detected in μg/g.

ND: Not detected

A random study of water samples collected from three rivers, which are sole source of irrigation in the paddy fields also revealed presence of HCH isomers and in one place Aldrin and PP-DDE (Table 11.11)

Table 11.11: Organochlorine Insecticides Residue Analysis of Water Samples Collected from Rivers Used for Irrigation (Year 2000)

Locality and Name of District	αHCH	γHCH	βHCH	δHCH	Aldrin	Endo Sulf-1	PP-DDE	OP-DDT
Dorwa (Parsa district)	0.014	0.003	ND	0.003	ND	ND	ND	ND
Gandak (Parsa district)	0.017	0.004	ND	0.008	ND	ND	ND	ND
Bangari (Bara district)	0.018	0.002	0.063	0.011	0.015	ND	0.057	ND

Data detected in µg/g.

ND: Not detected

Table 11.12: Over All Accumulation of Organochlorine Pesticides in Paddy and Paddy Field Soil

Sl.No.	Variety of Paddy	Source	Grain	Straw	Root	Soil
1.	Taichun	Birganj market	0.0593	–	–	–
2.	Pokhreli	Kathmando market	0.1871	–	–	–
3.	Kasturi	Birganj market	0.3345	–	–	–
		Paddy Field (Brahmapuri)	0.122	0.075	2.325	0.045
4.	Ekhattar	Birganj market	0.0932	–	–	–
5.	Mansuli	Birganj market	0.0853	–	–	–
		Kathmandu market	0.114	–	–	–
		Paddy Field (Parwanipur)	0.286	1.600	2.888	0.066
6.	Sabitri	Kathmandu market	3.531	–	–	–
		Paddy Field (Jagannathpur)	0.037	0.084	0.466	0.052
		Paddy Field (Fentaphulwaria)	0.787	0.269	–	0.113
7.	Katika	Kathmandu market	0.028	–	–	–
8.	Philips	Kathmandu market	0.020	–	–	–
9.	Basmati	Kathmandu market	0.54	–	–	–
		Paddy Field (Fentaphulwaria)	0.211	0.241	0.702	0.600
10.	Chaita	Paddy Field (Parwanipur)	0.490	0.151	0.544	0.036
11.	Radha–17	Paddy Field (Parwanipur)	0.043	0.243	0.051	0.060
12.	Pakistani	Paddy Field (Parwanipur)	Nil	0.146	Nil	0.005

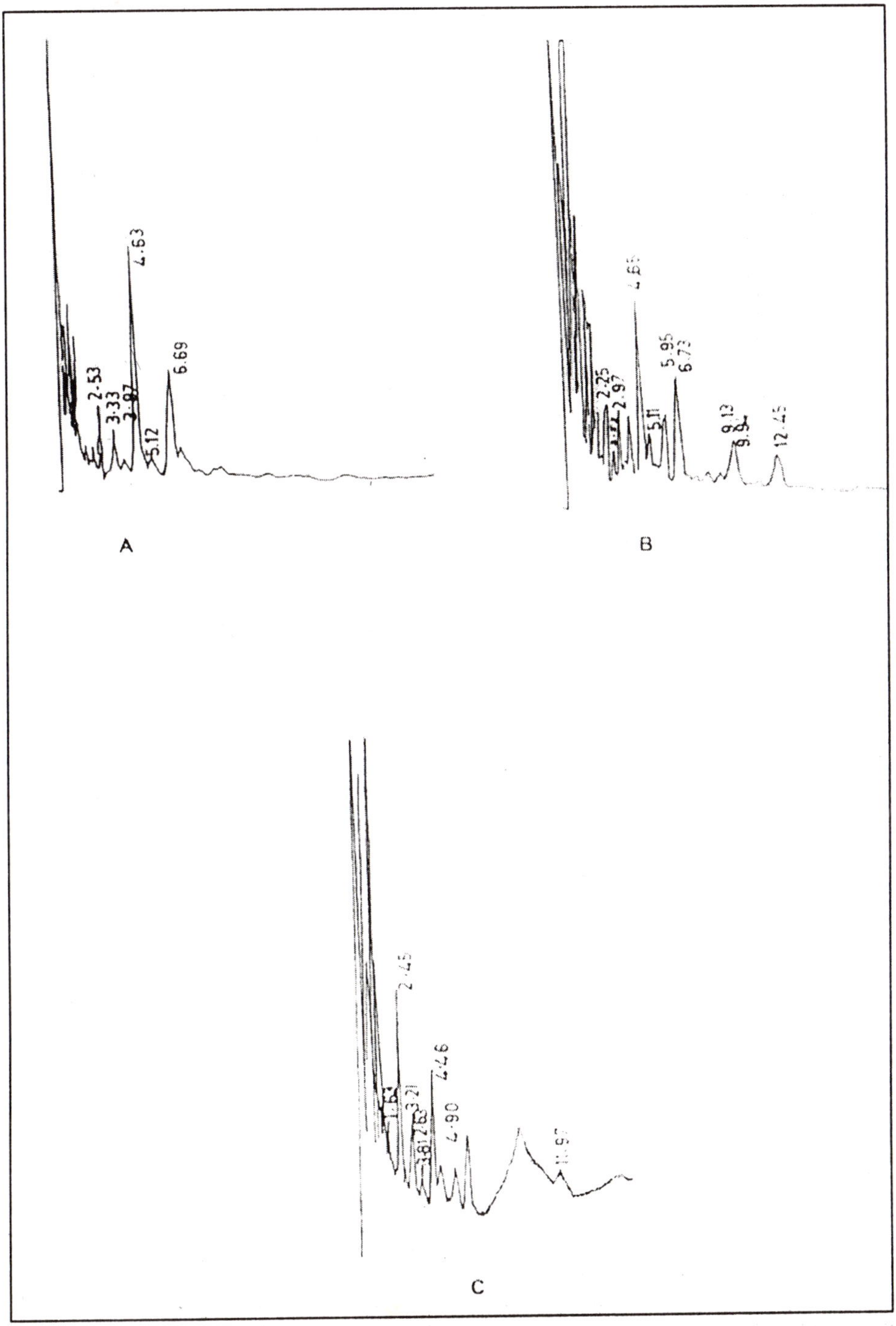

Figure 11.1: Gas Liquid Chromatograph Profile of Pesticide Residues in Paddy Crops. A: Kasturi (Rice grain); B: Kasturi (Straw) and C: Mansuli (Rice grains).

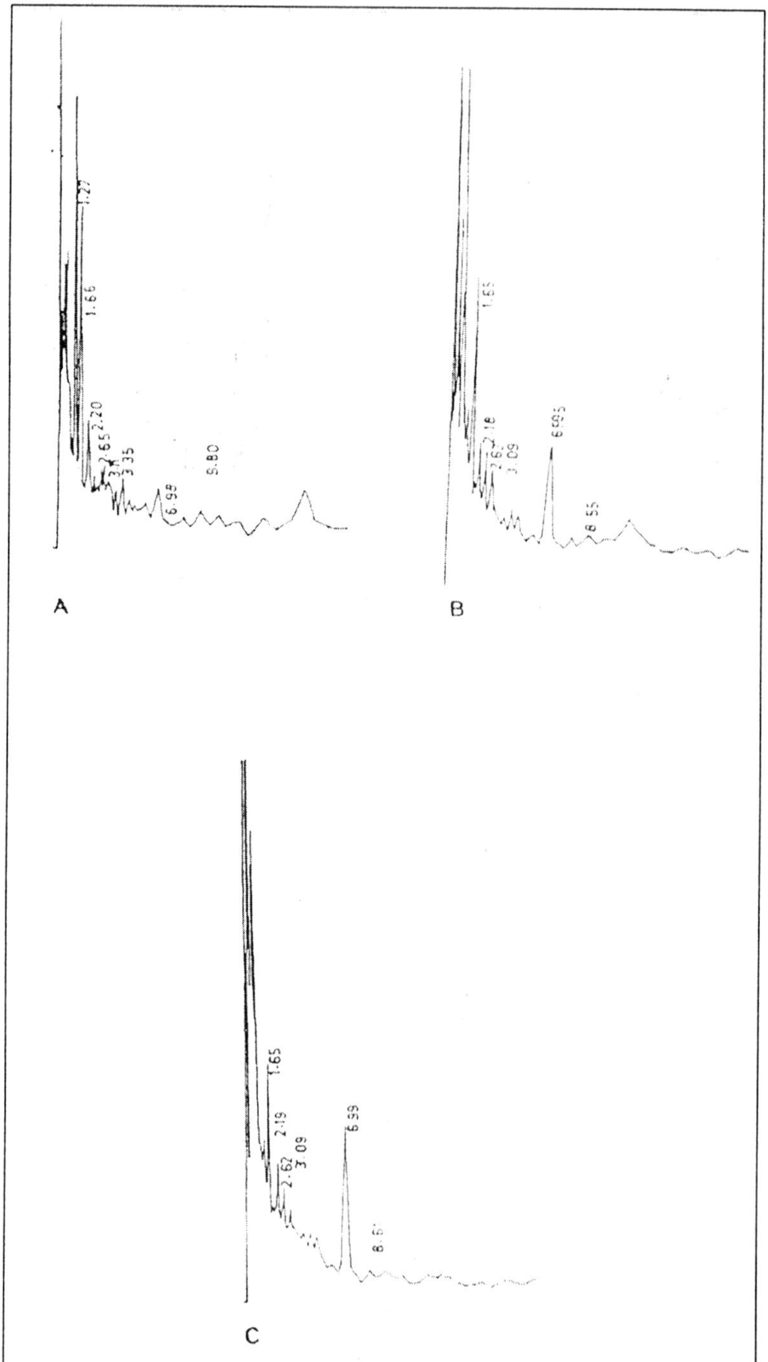

Figure 11.2: Gas Liquid Chromatograph Profile of Pesticide Residues in Rice Grains Collected from Paddy Fields.

A: Basmati; B: Chaite-4 and C: Radha-17.

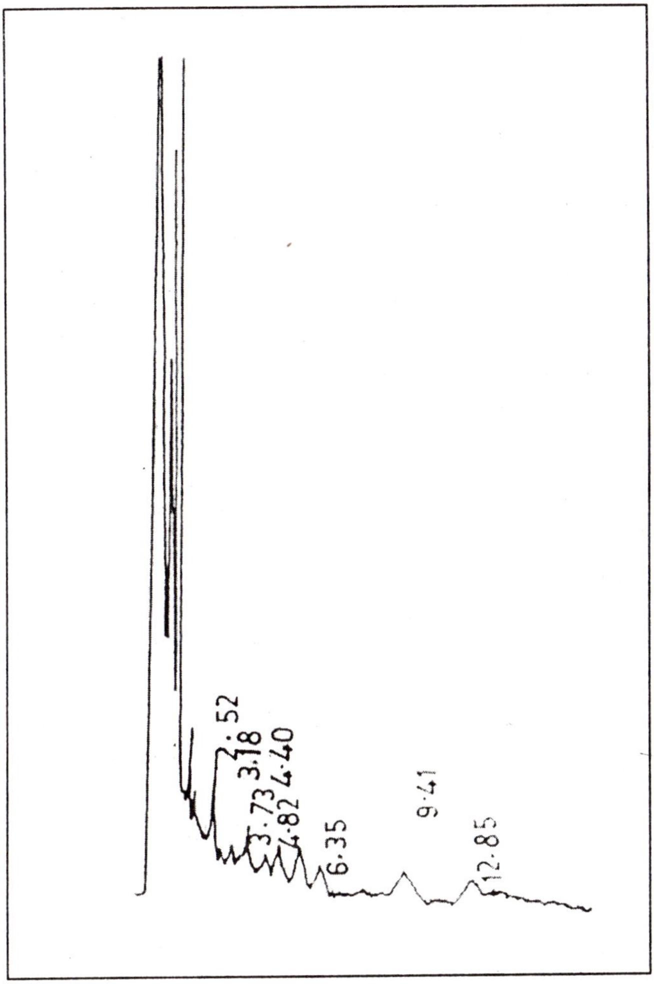

Figure 11.3: Gas Liquid Chromatograph Profile of Pesticide Residues Level in Water Sample of Gandak Canal

GLC profiles of organochlorine pesticide residues in paddy corps as well as rice grains procured from the market and gandak canal water are given in Figures 11.1–11.3. In all the samples a and β isomers of HCH could be detected and in many cases γ and δ isomers also. Aldrin residues were also generally present with a few cases of PP-DDE and Endosulphan.

Sum total of the organochlorine pesticide residues detected in different parts of paddy crops studied in fields, respective paddy field soil as well as the rice grains procured from the markets of Birganj and Kathmandu have been computed (Table 11.12) on the basis of the fractional data presented in the Tables from 11.1 to 11.10.

The presence of pesticides in the river as well as canal water led to observe its effect on the algae which are extremely important components of aquatic ecosystem.

It has been estimated that they account for some 60 per cent of annual CO_2 fixation and they form the base of most aquatic food waves (Odum-1974). Therefore the effect of γ HCH, which is most common organochlorine insecticide and also detected in river as well as canal water both, was observed in a N_2-fixer cyanobacterium *Nostoc linkia* and a green algae *Oocystis borgei* at 500μg/g sediment. Post treatment observations. Reference to their biomass, total chlorophyll, total proteins and total starch has been summarized in the Table 11.13. As shown in the table under reference biomass of *N. linckia* was reduced to less than half in 21 days (P<.01) where as no statistically significant effect was observed on *Oocystis borgei* (Table 11.13). Chlorophyll and Protein tended to decrease in both taxa (P<0.1) starch increased by 40 per cent in *N. linkia* (P<0.01) but in *O.borgei*, there was a decrease.

During the study, pesticide got released into the medium as seen from the lindane content of the culture filtrates in both algae (Table 11.14). The pesticide taken up by the algae were also significant. There was no bioconcentration in *N. linckia* but *O. borgei* over two-fold higher value (in term of weight) than the sediment. The GLC profile of γ HCH residues recovery in biota-medium (water)–sediment (sand) are shown in Figures 11.4 and 11.5.

Table 11.13: Effects of γ-HCH on Biochemical Parameters of Green and Blue Green Algae

Test Organisms	*Nostoc linckia*		*Oocystis borgei*	
Parameters	*Control*	*Treated*	*Control*	*Treated*
Biomass (gm/L)	2.240±0.005	1.000±0.003	0.960±0.006	1.220±0.005
Total chlorophyll (mg/g)	0.395±0.123	0.235±0.053	0.283±0.043	0.218±0.020
Total proteins (mg/g)	15.777±0.791	14.937±3.190	17.225±2.15	12.992±2.233
Total starch (μg/ml)	1.485±0.000	2.005±0.034	1.771±0.222	1.578±0.05

Value are represented in A.M.±SD of 2 determinations

Concentration of gamma – HCH 500 μg/g of sand

Table 11.14: γ-HCH Residue Recovery in Biota, Sand and Medium (Culture Filtrates)

Test Organisms	*Nostoc linckia*		*Oocystis borgei*	
Parameters	*Control*	*γ-HCH Treated (500 μg/g) Sand*	*Control*	*γ-HCH Treated (500 μg/g) Sand*
Sand	0.088±0.00	10.92±0.25	0.030±0.01	11.30±1.70
Algal mass	0.106±0.02	12.601±3.83	0.089±0.02	30.38±5.71
Culture filtrate	0.005±0.00	2.070±0.75	0.001±0.00	2.40±0.03

Value are represented in A.M.±SD of 2 determinations

Concentration of gamma – HCH 500 μg/g of sand

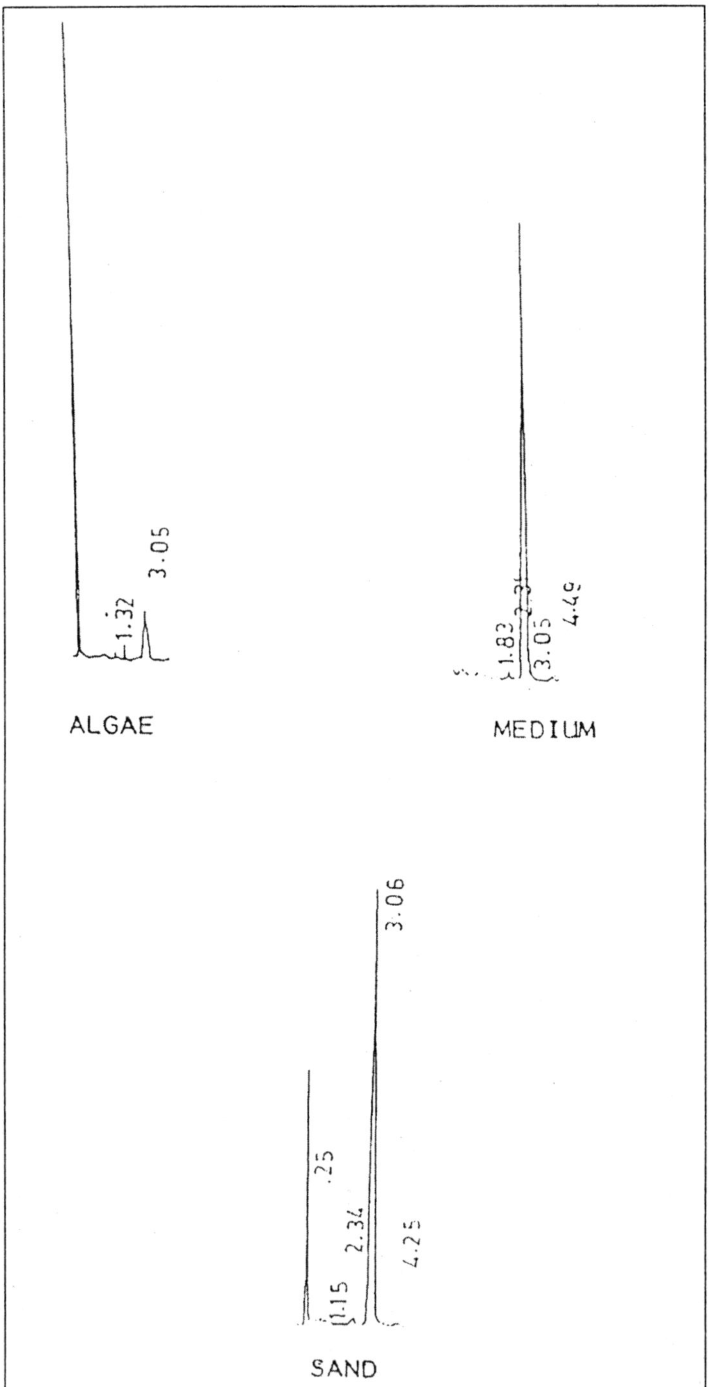

Figure 11.4: Gas Liquid Chromatograph Profile of Lindane Residues in Blue-Green Algae–*Nostoc linckia* (Biota), Medium (Water) and Sand (Sediment)

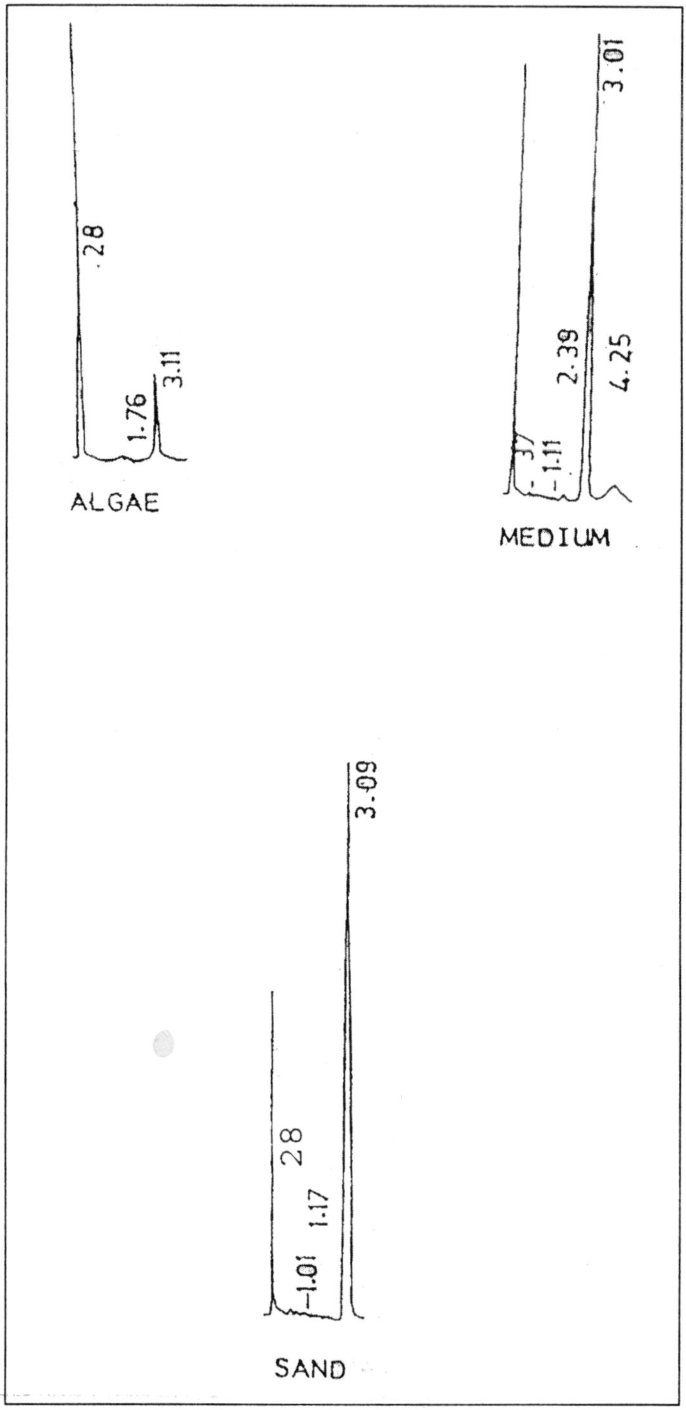

Figure 11.5: GLC Profile for Lindane Residues in Green Algae *Oocystis borgei* (Biota), Medium (Water) and Sand (Sediment)

Discussion

Detectable amount of five organochlorine pesticides residue has been recorded in the grains, straw and root of paddy crop as well as in the field soil (Tables 11.1 to 11.8). Among the recorded five pesticides, HCH along with its four isomers was observed most frequently. It was followed by aldrin and endosulufan. OP-DDT and PP-DDE were detected in very restricted samples. The same pesticides were also detected in nine varieties of rice grains procured from the market of Birganj and Kathmandu (Tables 11.9 and 11.10 and Figures 11.1 and 11.2). This indicates that the incidence of pesticides residue is widely distributed in Nepal. The presence of pesticide residue in the canal water (Figure 11.3) and in the water of three rivers (Table 11.11) is also of utmost importance being only source of irrigation. Accumulation of β HCH in aquatic system has been also reported in India by Ramesh *et al.* (1990) and in other countries of the world (Tatsukava *et al.*, 1972; Huat *et al.*, 1991; and Chessels *et al.*, 1988). Chessels *et al.* (1988) have reported the accumulation of β HCH in agricultural soils. Persistence of the organic chemicals in the aquatic system is of paramount importance not only in view of its biological magnification but also because it reduces the biodiversity rendering the ecosystem unstable. Predominance of HCH discussed above is mainly due to its β isomer. Which has low solubility in water, high potential for accumulation because of its low vapor pressure as well as low Henery's law constant. The chemical structures of this isomer in which all the chlorine atoms are present in equatorial position also increases its stability and resistance against microbial degradation. This is carcinogenic in nature and by virtue of its stable chemical properties it is liable to be bioconcentrated and biomagnified and could also biotransformed to some other toxic compound. The presence of PP-DDE is also significant due to its presence and well known anti-androgenic and estrogenic effects (Adler Creutz *et al.*, 1982; Katzenellen Oboger, 1995).

The levels of organochlorine pesticide residue detected in the present study are comparable with that in India (Das *et al.*, 1975; Pillai and Aggrawal, 1979; Jain *et al.*, 1980; Singh *et al.*, 1981) and many other countries. It is further evident from the analytical data (Tables 11.1–11.11 and Figures 11.1–11.3), that the level of pesticides residue in no case detected beyond prescribed acceptable delay intake limit. But the use of pesticides in Nepal is bound to increase in the coming years due to intensive agriculture and disease vector control programmes. Therefore the problem of persistence of recalcitrant toxic residues could emerge. Regular exploratory studies are essential for having baseline data to develop suitable interventional strategy.

Overall accumulation of organochlorine pesticides in the grains of 12 varieties of paddy projected in the Table 11.12 apparently points to preferential response of paddy varieties. While the grains of Pakistani variety were free from pesticide residue, it was detected as low as 0.02/0.028µg/g in Philips and Katika respectively followed by Radha-17 (0.043µg/g) and Taichun (0.0593µg/g). On the other hand accumulation of pesticide was detected as high as 3.531µg/g in Sabitri. In between these to extremes level of pesticides residue in the grains of remaining eight varieties ranged from 0.043ug/g to 0.3345µg/g with graded variation (Table 11.12). Thus the threat of pesticides residue can be minimized by encouraging the cultivation of paddy variety having low potency of accumulating pesticides.

It is further evident from the Table 11.12 that the pesticides residue has been detected consistently in the field soil and paddy root, straw and grains. Its level in the paddy has been ascertained highest in its root, followed by straw and grain in general showing thereby that the pesticides from soil is absorbed through the root from where it is transported to grain. However this assumption is applied to the four isomers of HCH and aldrin as the bulk of pesticides in the soil is contributed only by them (Tables 11.1 to 11.8). Endosulfan, PP-DDE and OP-DDT were never detected in the soil except in one case (Table 11.5). Their presence in the tissues of paddy crop may be attributed to their foliar spray.

Presence of pesticide residue in grain is of direct concern but in straw is of utmost importance. Straw is most frequently used as the fodder of domestic animals and hence pesticide residue present there in is biologically magnified along the grazing chain.

Effect study of Lindane (γHCH) showed its apparent dose and time dependent adverse affect on *Nostoc linckia* and *Oocystis borgei* both but the biomass reduction was more severe in first case. Evidence was also obtained to show the release of HCH from the sediment and uptake by algae. The mechanistic effect of pesticide on algae is not known in detail but the present result show that agricultural run off and aerosol setting could affect algal biodiversity rendering the ecosystem unstable, Bioconcentration of pesticides in the green alga is of much concern due to its biomagnifications along the grazing chain. Adverse effect of pesticide on biomass production especially of N_2 fixer Cynobacterium *Nostoc linckia* leave a big question mark over the increasing popularity of *Anabaena-Azolla* system used as biofertiliger in wetland agriculture. Care has to be taken for production, selection and application in such situation.

The present investigation has thus attempted to understand the overall impact of pesticides on paddy and algal diversity, soil and water pollution rather than mechanistic in depth study on one facet only. Besides this, it will help to identify the lines of promoting action in Nepal regarding assessment and abatement of future pollution and its effects. The test protocols and models developed in the present investigation will help in ecotoxicity evaluation and the observed no adverse effect level in prescribing water quality standards in Nepal.

References

Adlercreutz, H., Heikkinen, R., Woods, M., Fotsis, T., Dwyer, J.T., Goldin, B.R. and Gorbach, S.L., 1982. Excretion of the ligans enterolactone and enterodiol and of equal in omnivorous and vegetarian post menopausal women and in women with breast cancer. *Lancet*, pp. 1295.

Arnon, D.I., 1949. Copper enzyme in isolated chloroplasts. I. Polyphenol oxidase in *Beta vulgarica. Plant Physiol.*, 24: 1–16.

Bertuzzi, P.F., Laverne, K, Miles, C.I. and Burke, J.A., 1967. Extraction of chlorinated pesticide residues form non-fatty chlorinated samples of low moisture content. *J. Assoc. Off. Anal. Chem.*, 150(3): 623–627.

Chessels, M.J., D.W. Hawkar, D.W. Connel and I.A. Papajcsik (1988. Factor influencing the distribution of lindane and Isomers in Soil of an agricultural environment; *Chemosphere*, 17: 1741–1749.

Das, G.K., Thayumanavan, B., Kulshrestha, J.P. and Dani, R.C., 1975. Mass spectrometric method for the estimation of lindane and carbonyl residues in straw, rice and bran. *Rees Ind.*, 20: 85–88.

Diercking, R. and Hansen, P.D., 1974. Aufarbeitung von einzelligen algen zur gaschromatographischen bestimmung des Lindans. *Ber. dt. Wiss. Kommn. Meerefosh.*, 23: 326.

EPA, 1975. *Test Methods for Assessing the Effect of Chemicals on Plants*, (Eds.) R. Rubinstein and J. Smith. EPA–560–17–75–008. Final Report, U.S. Environmental Protection Agency, Washington DC.

Gruzdyer, 1983. *The Chemical Protection of Plants.* Mir. Publ., Moscow.

Huat T.G., Hock, G.S. and Chumy, K. Vijayalet, 1991. Analysis of pesticide residues in Peninsular Malaysian water ways. *Environ. Monit. Assessment*, 19: 496–497.

Jain, H.K, Pandey, S. Y., Agnihotri, N.P. and Srivastava, K.P., 1980. Residues of insecticides in rice crops. *Indian J. Entomol.*, 42(4): 675–679.

Karanth, N.G.K., Jayram, M., Majumdar, S.K., 1982. Insecticidal residues in vegetables obtained from soil treated with hexachlorocyclohexane *J. Food Sci. Technol.*, 19: 14–19

Karanth, N.G.M., Jayram, M. and Majumdar, S. K., 1983. Observation of the growth in vegetable plants residue of hexachlorocyclohexane in treated soil. *Comp. Physiol. Ecol.*, 8(4): 357–361.

Katzenellenboboger, J.A., 1995. The structure pervasiveness of esterogenic activities. *Environ. Health. Perspective*, 103: 99–101

Lillelund, K., and Wehrmann, L., 1974. Akkumulations versuche von Lindan in einer kultur von *Dunaliella* spec. Zur standardisierung lines algen test. *Ber dt. Wiss kommn Meeresforsch.* 23: 317.

Lowry. O.H., Rosebrough, Randall, N.J. and Farr, A., 1951. Protein determination by the folinphenol reagents. *J. Biol. Chem.*, 193: 265–275.

Misra, V., Pandey, S. D. and Viswanathan, P.N., 1996. Binding of gamma hexachlorocycle hexane with humic acid and chlorophyll in aquatic environment in relation to ecotoxicity. *Intern. J. Environ. Anal. Chem.*, 64: 257–264.

Odum, E.P.U., 1971. *Fundamentals of Ecology*, 3rd edn. W.B. Saunders Comp., Philadelphia.

Pillai, M.K.K. and Agarwal, H.C., 1979. *DDT Residue and its Degradation in Soil, Water and Few Aquatic Animals of Yamuna in Delhi.* Prof. Rep. Dept. Sci. and Technol, New Delhi.

Ramesh, A., Tanabe, S., Iwata, H., Tatsukawa, R., Subramanian, A.N., Mohan, D. and Venugopalan, V.K., 1990. Seasonal variation of persistent organochlorine insecticide residues in Veller river water in Tamil Nadu, South India. *Environ. Poll.*, 67: 289–304.

Singh, K.P., Pandey, S.Y. and Kapadia, G.D., 1981. Note on carbaryl and lindane residues in and on rice. *Indian J. Agri. Sci.*, 51: 824–826.

Tatsukawa, R.T. Wakimoto and Ogawa, T., 1972. BHC residues in the environment. In: *Environmental Toxicology of Pesticides*, (Eds.) F. Matsumura, C.M. Boush and T. Misato. Academic Press, New York, pp. 229–238.

Verirov, D. and Aharonson, N., 1978. Simplified fat extraction with sulphuric acid as clean up procedure for residue determination of chlorinated hydrocarbons in butter. *J. Assoc. Off. Anal. Chem.*, 61: 253–260

Way, M.J., 1976. Entomology and the world food situation. *Bull. Soc. Amer.*, 22(2): 125–129.

Chapter 12

Karyotype Analysis in the Genus *Nitella* (Ag.) L.

***Kiran Bala*[1] *and M.P. Singh*[2]**

[1]*Department of Botany, T.R.M. Campus, Birganj, Constituent Unit of Tribhuvan University, Kathamandu, Nepal*
[2]*Professor and Head, Department of Botany, R.P.S. College, Jaintpur, Constituent Unit of B.R.A. Bihar University, Muzaffarpur, Bihar*

Introduction

Nitella Ag. Leonh is among one of the widely distributed genus of the living Charophyta. Taxonomically it has emerged as most fascinating genus following the amalgamation of long existing distinct taxa by wood and Imahori (1965). Karyotype study in this genus has emerged of immense value for taxonomic treatment of this genus (Khan and Sarma, 1967, Sarma and Ramjee, 1967a, b, '69 and 71 and Verma, 1974). In the present investigation karyotype of five taxa belonging to this genus *viz. N. furcata* sub. sp. *mucronata* var. and f. *mucronata* R.D.W., *N. furcata* sub. sp. *flagelliformis* Braun. R.D.W., *N. pseudoflabata* sub sp. *pseudoflabata* var. *elegans* f. *sejuncta* (Imahori) R.D.W., *N. pseudoflabata* sub. Sp. *pseudoflabata* var. and f. *mucosa* (Nordst) Bailey, R.D.W. and *N. hyalina* (Decond.) Ag. R.D.W. The presently ascertained chromosome number in all the five taxa under reference has been reported earlier (Sinha and Verma, 1970, Imahori and kato, 1961, Hotchkiss, 1965 and Sato, 1959) but their karyotypic details are wanting

Materials and Methods

The five taxa under investigations were collected from three districts of North

Bihar *viz.* Darbhanga, Gopalganj and East Champaran. They were growing either in waterlogged paddy fields or in temporary pools.

Fixation of the filaments, carrying young antheridia, for cytological preparations was made on the collection spot during the morning hours between 10-11 A.M. Cytological slides for chromosome study was made from young antheridial filaments following the Iron-Allum-Acetocarmine method devised by Godward (1948).

Total chromatin length, coefficient of variations among the chromosome complements and TF per cent of each karyotype were computed as per Verma (1980).

Observations

Nuclear organization in all the five taxa was observed on the same pattern being almost spherical, bounded by a distinct nuclear membrane and rich in chromatin material. The diameter of interphase nuclei among the investigated five taxa ranged between 6.6 and 14.3 µm (Table 12.1). Presence of a single nucleolus in the nuclei has been noted as a common feature of all the taxa and its diameter has been measured from 2.2 to 4.4 µm (Table 12.1).

Table 12.1: Karyological Measurements and Chromosome Number

Sl.No.	*Name of the Taxon*	*Diameter Nucleus (µm)*	*Diameter Nucleolus (µm)*	*Chromosome Number*	*Chromosome Length (µm)*	*Centromere*	
						m	*Sm*
1.	*N furcata* sub. sp. *mucronata* var. and f. *mucronata*	8.8–14.3	2.2–3.3	n = 12	3.3-4.9	5	7
2.	*N. hyalina*	8.8–13.2	2.2–4.4	n = 12	2.6–4.0	5	7
3.	*N. pseudoflabata* sub. sp. *pseudoflabata* var. *elegans* f. *sejuncta*	6.6–9.9	2.2–3.3	n = 18	2.0–5.3	5	13
4.	*N. pseudoflabata* sub. sp. *pseudoflabata* var. and f. *mucosa*	6.6–12.1	2.2–6.6	n = 18	2.2–4.4	12	6
5.	*N. furcata* sub. sp. *flagelliformis*	6.6–12.1	2.2–3.3	n = 24	2.7–5.3	5	19

Events of unclear division have been observed on typical mitotic pattern. At metaphase, the highly condensed, short and thick chromosomes become spreaded at equatorial plate. At this stage, count of n=12 (Plate 12.1 , Figures 1.1 and 1.2) and n = 18 (Plate 12.1 , Figures 1.3 and 1.4) has been ascertained for two taxa each and n = 24 (Plate 12.1 , Figures 1.5) for one (Table 12.1). Length of the chromosomes among five investigated taxa ranged between 2.0 µm and 5.3 µm. Position of centromere in each chromosome has ascertained after measuring its two arms (Table 12.1) and the idiogram of each karyotype has been made out (Plate 12.2 , Figures 2.1–2.5).

Total chromatin length, coefficient of variation (C.V.) and TF per cent computed for each of the five worked out karyotypes may be seen in the Table 12.2.

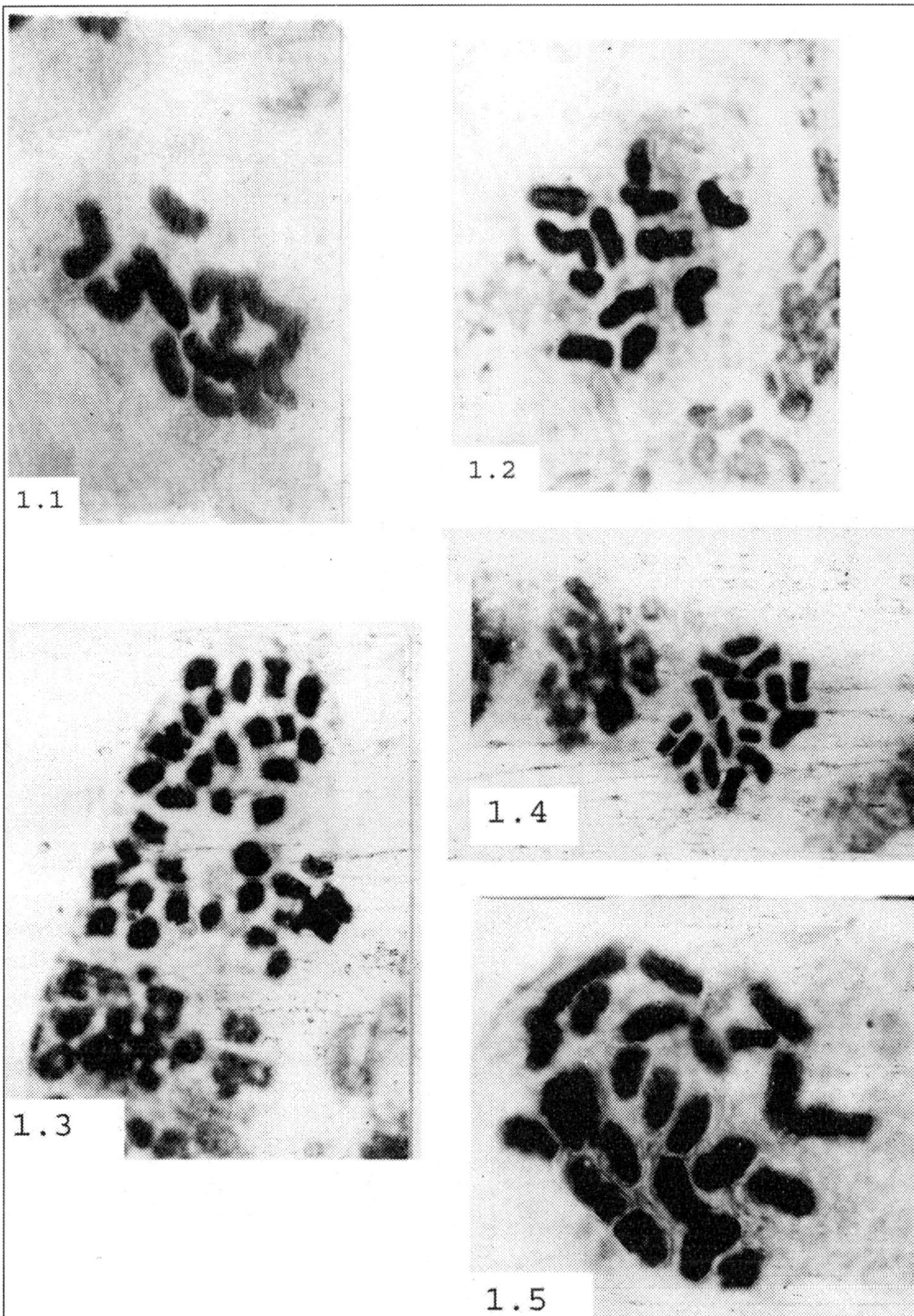

Plate 12.1: Metaphase Showing Chromosome Number in *Nitella*

Figures 1.1: *N. furcata* sub. sp. *mucronata* var. and f. *mucronata* n = 12 (X5200); 1.2: *N. hyalina*. N=12 (X 5200); 1.3: *N- pseudoflabellata* sub. sp. *pseudoflabellata* var. *elegans* f. *sejuncta,* n = 18 (X5000); 1.4: *N- pseudoflabellata* sub. sp. *pseudoflabellata* var. and *F. mucosa.* n = 18 (X5000); 1.5: *N. furcata* sub. sp. *flagelliformis. n= 24* (X6500)

Plate 12.2: Ideogram

Figures 12.1: *N. furcata* sub. sp. *mucronata* var. and f. *mucronata*; 2.2: *N. hyalina*; 2.3: *N- pseudoflabellata* sub. sp. *pseudoflabellata* var. *elegans* f. *sejuncta*; 2.4: *N- pseudoflabellata* sub. sp. *pseudoflabellata* var. and *F. mucosa;* 2.5: *N. furcata* sub. sp. *flagelliformis.*

Table 12.2: Total Chromatin Length, C.V. and TF %

Sl.No.	Name of the Taxon	Total Chromatin Length (μm)	C.V. (%)	TF %
1.	*N. furcata* sub. sp. *mucronata* var. and f. *mucronata*	50.8	11.6	45.4
2.	*N. hyalina*	39.3	15.8	42.7
3.	*N. pseudoflabata* sub. sp. *pseudoflabata* var. *elegans* f. *sejuncta*	66.1	24.3	40.8
4.	*N. pseudoflabata* sub. sp. *pseudoflabata* var. and f. *mucosa*	57.2	30.5	44.3
5.	*N. furcata* sub. sp. *flagelliformis*	88.4	22.4	39.7

Discussion

Karyological observations of five taxa at hand have revealed no difference as regards the nuclear organisation and events of mitoses and hence bear no taxonomic importance.

The diameter of nuclei and nucleoli for five taxa (Table 12.1) revealed no significant and conclusive variation in regard to the chromosome number.

Presence of single nucleolus in the nuclei has been noted as common feature of all the five taxa as reported earlier by (Lindenbein, (1927), Hotchkiss, (1965), Khan and Sarma, (1967), Sinha and Verma, (1970) and Mukherjee and Noor, 1973)

Chromosomes observed in all the five investigated taxa are equally big with distinct localized centromere. These two features show quite close resemblance of the Charophyta with Cladophorales and Oedogoniales on one hand and higher plants on the other.

Length of the chromosomes among the taxa bearing the mitotic counts n = 12, 18 and 24 has been measured 2.6–4.9 μm, 2.0–5.3 μm and 2.7–5.3 μm respectively (Table 12.1). Sarma and Ramjee (1971) considered an inverse relationship between the length of the chromosomes and extent of ploidy.

Counts of n = 12 (Plate 12.1, Figures 1.1 and 1.2) and n = 18 (Plate 12.1, Figures 1.3 and 1.4) have been ascertained for two taxa each and n = 24 (Plate 12.1, Figures 1.5) for one taxon (Cf. Table 12.1). As indicated in the Table 12.3, the present counts confirm some of the previous reports and differ from some others. Altogether 32-taxa excluding 5-worked out in the present investigation, have been studied so far and the euploid counts ascertained among them are n = 6, 9, 15, 21, 27 and 36 in addition to n = 12, 18 and 24 ascertained by the authors (Strasburger, 1908, Gillet, 1959, Imahori and Kato, 1961, Hotchkiss, 1965, Sarma and Ramjee, 1967a, b, 1969 and 1971, Khan and Sarma, 1967, Mukherjee and Noor, 1973 and Noor and Mukherjee, 1977). The frequency of different counts has been computed and on this basis it is concluded that n-18 seems to be the most common number in the genus *Nitella* being reported in 64.9 per cent cases. Next highest frequency comes 16.2 per cent which has been scored for the counts n = 9 and 36.

Review of chromosome number in the genus *Nitella* apparently reveals the following facts:

A single taxon frequently exists at different ploidy levels in nature. This appears to be a successful means of speciation. Since, interspecific numerical variations of chromosome are also not uncommon, speciation within the genus also appears to be encouraged through euploidy. The annual habit and frequent vegetative multiplication of the Charophytes are supposed to be the remarkable assets to establish the new polyploids.

Sharing of a single count by a number of morphologically distinct taxa is a common feature of the genus (Table 12.3). This indicates that the polyploidy is not the sole means of speciation rather the structural alteration of chromosomes involving gene mutation and/or rearrangement of genes might have played significant role in speciation. Observed variations as regards the C.V. and TF per cent among five investigated taxa (Table 12.2) further support the above assumption.

Table 12.3: Chromosome Number in Five Investigated Taxa Ascertained in the Present Study and Reported Earlier

Sl.No.	Name of the Taxon	Present Count	Previous	Reports
1.	*N. furcata* sub. sp. *mucronata* var. and f. *mucronata*	12 (Plate 12.1, Figure 1.1)	n = 12	Noor (1969–70), Sinha and Verma (1970)
			N = 18	Gillet (1959), Sarma and Ramjee (1969)
			N = 36	Sarma and Ramjee (1969)
2.	*N. hyalina*	12 (Plate 12.1 Figure 1.2)	n = 12	Sato (1959), Sinha and Verma (1970)
			n = 12	Stewart (1937), Sato (1959)
			n = 16	Gillet (1959)
			n = 18	Guerlesquin (1961), Hotchkiss (1965), Khan and Sarma (1967)
3.	*N. pseudoflabata* sub. sp. *pseudoflabata* var. *elegans* f. *sejuncta*	18 (Plate 12.1 Figure 1.3)	n = 18	Imahori and Kato (1961), Noor (1969 and 70), Sinha and Verma (1970)
4.	*N. pseudoflabata* sub. sp. *pseudoflabata* var. and f. *mucosa*	18 (Plate 12.1 Figure 1.4)	n = 18 n = 21	Hotchkiss (1965), Noor and Mukherjee (1977)
5.	*N. furcata* sub. sp. *flagelliformis*	24 (Plate 12.1 Figure 1.5)	n = 9	Khan and Sarma (1967),
			n = 24	Sinha and Verma (1970)

From the foregoing discussion it is further apparent that the alone chromosome number is of least importance for characterising any of taxon within the genus. However, the genus may be characterized by the basic number "3" as reported by Sarma and Khan (1964). Presently ascertained counts 12, 18 and 24 are in multiple of "3" and further strengthen the basic number suggested by Sarma and Khan (loc.cit.)

The genus *Nitella* seems no way less than *Chara* in achieving the polyploids. Adhering to the basic number "3", *N. furcata* has been observed existing at 16x–ploidy level in nature (Imahori and Kato, 1961) as against the 10x, highest polyploids reported so far in the genus *Chara i.e.* in *C. zeylanica* willd. (Hotchkiss, 1963).

The karyotypes of five worked out taxa seem closely related in regard to gross morphology of the chromosomes (Plate 12.2, Figures 2.1 to 2.5). The sex-chromosomal heteromorphism and N.O. chromosome are altogether absent. The basic number has been conservatively retained by all of them. All these facts go in support of the derivation of one karyotype from the other. It is apparent from the Table 12.2 that the karyotypes with the chromosome number n = 18 have comparatively higher C.V. and lower TF per cent than those with n = 12. It is, therefore, concluded that the karyotypes with n = 12 are primitive than the karyotypes with n = 18. The C.V. of the karyotype having n = 24 is apparently advanced than those bearing n = 12 but not n = 18. However, its TF per cent is pretty lower than any of the four worked out karyotypes (Table 12.2) claiming its advancement over rest. Increasing asymmetry of the karyotypes with increase in the chromosome number has also been reported by Verma (1970) among the 18-taxa of this genus. The relative advancement discussed above dose not essentially implies that their derivation has taken place along a single evolutionary line. Sarma, *et.al.* (1970) have traced variously bifurcating evolutionary lines within the genus on cytological basis.

Summary and Conlcusions

Five taxa belonging to the genus *Nitella viz., N. furcata* sub. sp. *mucronata* var. and f. *mucronata*, *N. hyalina*, *N. pseudoflabata* sub. sp. *pseudoflabata* var. var. *elegans* f. *sejuncta*, *N. pseudoflabata* sub. sp. *pseudoflabata* var. and f. *mucosa* and *N. furcata* sub. sp *flagelliformis* have been collected from three districts of North Bihar and their karyotypes have been worked out.

The chromosome number n = 12 has been ascertained for first two named taxa, n = 18 for next two named taxa and n = 24 for last named taxon. The frequency of different counts reported so far in the genus has been calculated and it has been included that n = 18 is the most common number for the genus being reported in 64.9 per cent cases. It has been further concluded that the euploidy as well as structural alteration of chromosomes involving gene mutation and/or rearrangement of genes are the two signification events underlying the speciation within the genus. The annual habit and frequent vegetative multiplication of this genus help to establish two now polyploids.

Of the five worked out karyotypes those with the chromosome number n = 12 have been regarded most primitive and that having n = 24 as most advanced on the basis of its symmetry as ascertained after computing total chromatin length, C.V. and TF per cent. The karyotypes with n = 18 occupy intermediate position between the two.

References

Gillet, C., 1959. Numbress chromosomique de plusiers especes de Charophyceae (Generes *Nitella* et *Chara*). *Rev. Cytol. Biol. Veg.*, 20: 229–234.

Godward, M.B.E., 1948. The iron-alum-acetocarmine method for algae. *Nature*, 161: 203.

Guerlesquin, M., 1961. Contribution a letude chromosomique des Charophyces d' Europe occidentale et d' Afrique du Nord. *Rev. Gen. Bot.*, 68: 306–372.

Hotchkiss, A.T., 1963. A first report of chromosome numbers in Charophyte genus *Lychnothamnus* (Rupr.) Leonh. and comparison with the other Charophyte genera. *Proc. Linn. Soc., New South Wales*, 138: 368–372.

Hotchkiss, A.T., 1965. Chromosome numbers in Characeae from the South Pacific. *Science*, 19: 31–37.

Imahori, K. and Kato, T., 1961. Notes on chromosome numbers of Charophytes of Fukui Prefecture Japan (1) Sci. Rep. No. 10. Osaka Univ., Japan. 39–48.

Khan, M. and Sarma, Y.S.R.K., 1967a. Studies on Cytotaxonomy of Indian Charophyta. I. *Chara. Phykos.*, 6 (1 and 2): 36–47.

Khan, M. and Sarma, Y.S.R.K., 1967b. Studies on Cytotaxonomy of Indian Chrophyte. II. *Nitella. Ibid*, 6: 48–61.

Lindenbein, W., 1927. Beitrag zur Cytologie der Charales. *Planta*, 4: 436–466

Mukherjee, S. and Noor, M.N., 1973. A first report of karyological studies in *Nitella* wattii Groves from India. *Science and Culture*, 39: 459–460

Noor, M.N., 1965. Cytotaxonomic and cultural studies of some members of Chlorophyceae of Chhotanagpur. *Ph.D. Thesis*, Ranchi University, Ranchi (Bihar).

Noor, M.N., 1969–70. A Preliminary report on chromosome members of some Indian characeae. *J. Ranchi Univ.*, Vol. VI and VII: 232–239.

Noor, M.N. and Mukherjee, S., 1977. Some new records of chromosome numbers in Indian Charophyta. *Cytologia*, 42: 227–232.

Stewart, L.M., 1937. Studies in the life-history of *Nitella hyalina* Ag. *J. Elisha Mitchell Sci. Soc.*, 53: 173–190.

Sarma, Y.S.R.K. and Khan, M., 1964. Chromosome numbers in some Indian species of *Nitella. Chromosoma (Berl.)*, 15: 246–247.

Sarma, Y.S.R.K., Khan, M. and Ramjee, 1970. Cytological approach to phylogeny, inter-relationship and evolution in Charophyta. *Indian Biologist*, 2: 11–19.

Sarma, Y.S.R.K., and Ramjee, 1967. A new chromosome number in *Nitella tenuissima* (Desr.) Kutz. em. R.D.W. *Caryologia*, 20: 61–63.

Sarma, Y.S.R.K., and Ramjee, 1969. Chromosome numbers in three taxa of Indian Charophyta. *J. Cytol. Genet.*, 4: 32–36.

Sarma, Y.S.R.K., and Ramjee, 1971. Significance of chromosome numbers in Charophyta: A discussion. *Caryologia*, 24: 391–401.

Sato, D., 1959. The karyotype and phylogeny in plants. *Sci. Pap. Cell. Gen. Ed. Univ. Tokyo*, 9: 303–327.

Sinha, J.P. and Verma, B.N., 1970. Cytological analysis of the charophytes of Bihar. *Phykos*, 9(2): 92–99.

Strasburger, E., 1908. Einiges uber Characeen and Amitoses. Linsbauer, Wiesner Festchr. *Viena*, p. 24–47.

Verma, B.N., 1980. Karyotype analysis in three species of *Rhizoclonium* Kutz. *Cytologia*, 45: 433– 440.

Verma, S., 1974. *PhD. Thesis,* Ranchi University, Ranchi, Bihar (India).

Wood, R.D. and Imahori, K., 1965. A revision of the Characeae. Monograph and Iconograph. Verlag von J. Cramer-Publ.

Chapter 13

Karyomorphological Studies in *Rhizoclonium hieroglyphicum* (Ag.) Kütz. Complex

Prem Kumar Prasad

Principal,
B.M. College, Rahika, Madhubani, Bihar

Introduction

Rhizoclonium hieroglyphicum (Ag.) Kutz. has nine varieties underneath (De Toni, 1989). From the available literature it appears, while the type species has been repeatedly investigated in the past (Peterschilka, 1923; Geitler, 1935; Sinha, 1958; Venkataraman and Natarajan, 1959; Balakrishnan, 1961; Patel, 1961; Noor, 1965; Prasad *et al.*, 1973; Sarma and Agrawal, 1980 and Sarma and Bala, 1983), no attention has been given on any of its varieties. Present communication deals with the nuclear cytology of *R. hieroglyphicum* and its two varieties *viz.* var. *subterrestra* and var. *macromeres*. Also the karyotypes of investigated three taxa have been worked out which appears to be a maiden attempt of the author.

Materials and Methods

Three taxa under investigation were collected from different corners of Ranchi during the different parts of the year (Table 13.1).

Table 13.1: Details of Collection

Sl.No.	Name of the Taxa	Place of Collection	Month
1.	*R. hieroglyphicum var. macromeres*	Well, Nilratna Street, Ranchi	April
2.	*R. hieroglyphicum var. subterrestra*	Pond, Therpakhana, Rancni	August
3.	*R. hieroglyphicum* (Ag.) Kutz.	Hermu River, Kishoreganj, Ranchi	October

Fixation of the filaments for cytological preparations were made on the collection spots during morning hours between 8 to 10 a. m. Acetic Acid and Ethanol in equal proportion was used as the fixative. Iron-Alum-Acetocarmine method (Godward, 1948) was followed for cytological preparations.

Total chromatin length, coefficient of variations (C.V.) among the chromosome complements of a karyotype and the TF per cent were computed (Verma, 1980).

Observations

The interphase nuclei of three taxa under study are found similar being spherical and surrounded by a distinct nuclear membrane. Its number in a cell has been recorded from 2 to 8. In *R. hieroglyphicum* var. *macromeres*, the uninucleate cells have also been observed. Contrary to above similarities, the diameter of interphase nuclei varied appreciably among three taxa at hand (Table 13.2).

The presence of nucleolus as densely stained spherical structure and chromocenter like bodies as darkly stained thick structure in the interphase or early prophase nuclei has been marked as a common feature of all the three taxa under reference. Their numbers in a nucleus has been counted up to two (Table 13.2). The latter has invariably been observed either in association with the former or any where in the nuclear cavity. The single big nucleolus, which may be presumed to be the result of fusion of two smaller ones, have been measured as indicated in the Table 13.2.

Table 13.2: Details of Nuclear Measurements

Sl.No.	Name of the Taxon	Nucleus		Nucleolus		No. of Chromocenter like Bodies
		No/Cell	Diameter (μm)	No. of Nucleolus	Diameter (μm)	
1.	*R. hieroglyphicum*	2–8	4.5–5	1 – 2	2–3	1–2
2.	*R. hieroglyphicum* var. *subterrestra*	2–8	2.8–3.3	1	1–1.5	1
3.	*R. hieroglyphicum* var. *macromeres*	1–8	5–6	1 – 2	3	2

Events of nuclear divisions have been observed on typical mitotic pattern. At the onset of prophase, the nuclei become double of their original diameter in all the three taxa except *R. hieroglyphicum* var. *subterrestra* where the prophase nuclei have been measured only one and half of interphase nuclei.

At metaphase, the chromosomes appear thick rod shaped structures with distinctly localized centromere discernible as non-stained or constricted area anywhere along the length. In certain instances, the chromosomes have been observed with distinct chromatids marking the late metaphase and at this stage centromere becomes comparatively more distinct (Plate 13.1–13.-5). Since, these chromosomes are highly condensed and lie scattered in one plane, the chromosome counts for investigated taxa have been ascertained at this stage.

A count of 24-chromosomes has been ascertained for all the three investigated taxa (Plates 13.1–13.5). The length of each chromosome in a karyotype has been measured (Table 13.3) and the idiograms of each karyotype has been made out (Plates 13.6). The total chromatin length, C.V. and TF per cent of each of the three karyotypes has been computed (Table 13.3).

Table 13.3: Details of Chromosome and Karyotype

Sl.No.	*Name of the Talon*	*Chromosomes*		*Total Chromatin Length (μm)*	*C.V.*	*T.F.%*
		Number	*Length (μm)*			
1.	*R. hieroglyphicum*	n–24	1.6–3.5	57.4	24.9%	42.33
2.	*R. hieroglyphicum* var. *subterrestra*	n–24	0.5–1.6	23.4	28.9%	50.00
3.	*R. hieroglyphicum* var. *macromeres*	n–24	1.6–4.6	69.6	27.3%	44.97

Discussion

The number of nuclei per cell in three worked out taxa has been found varying up to 8 (Table 13.2). Brand (1889), Carter (1919) and Peterschilka (1923) have reported highly variable number of nuclei in the cells of *R. hieroglyphicum*. Balakrishnan (1961) observed 1-4 (rarely up to 8) nuclei in the sub aerial form and 4-8 (rarely up to 16) nuclei in the aquatic form of the same species. Prasad *et al.* (1973) reported the number of nuclei between 1-4 in their two collections of *R. hieroglyphicum*. Further, Sinha (1958) and Noor (1965) noted the increase in number of nuclei per cell when the filaments were transferred from natural habitat to the nutrient solution. Under the circumstances, the number of nuclei in the cells cannot be taken as a reliable taxonomic tool.

The number of nucleoli in a nucleus has been counted up to two in *R. hieroglyphicum* and *R. hieroglyphicum* var, *macromeres* and only one in *R. hieroglyphicum* var. *subterrestra*. In the first two cases, also a single big nucleolus has been observed, which may be presumed to be the result of the fusion of two smaller nucleoli (Geitler, 1936). Sinha (1958) has suggested an association of a single nucleolus either with a set of 12 or 18 chromosomes in the genus *Rhizoclonium*. In the present instances, single nucleolus seems to be associated with a set of 12 chrosomosomes. in *R. hieroglyphicum* var, *subterrestra* only one nucleolus has been recorded along with 24

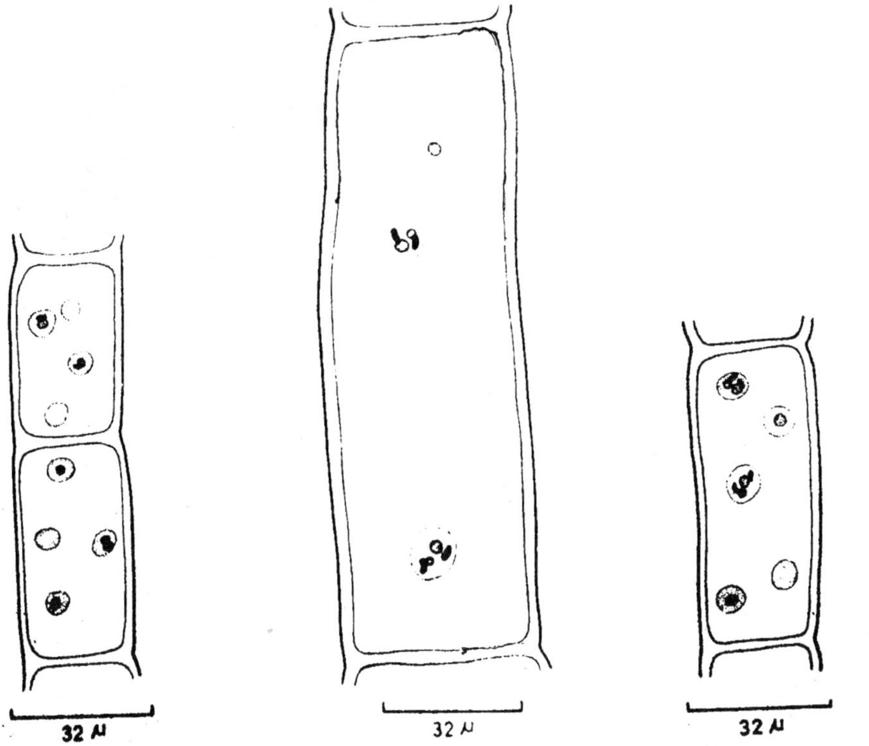

Fig. 1.1 Fig. 1.2 Fig. 1.3

Plate 13.1: Interphase Nuclei Showing Nucleoli and Chromocentre Like Bodies Figures 1.1: *R. hieroglyphicum* var. *subterrestra;* 1.2: *R. hieroglyphicum* var *macromeres*; 1.3: *R. hieroglyphicum.*

chromosomes. It is probable that the fusion of two nucleoli in this case might have taken place earlier.

The chromocenter like body has invariably been observed associated with the nucleolus. Sinha (1958) and Noor (1965) regard such associated structure as N.O. region of the chromosome.

The count of n = 24 has been ascertained for all the three taxa under study. From the available literature it is apparent, while the present count in *R. hieroglyphicum* confirms some of the previous reports, it differs from some others (Table 13.4). It is further apparent from the same table that there has been no previous report of chromosome number in *R. hieroglyphicum* var. *subterrestra*. and *R. hieroglyphicum* var. *macromeres*, except Verma (1969) who also reported the same number of chromosomes. Hence the chromosome numbers ascertained in them by the author are confirmation of the previous report.

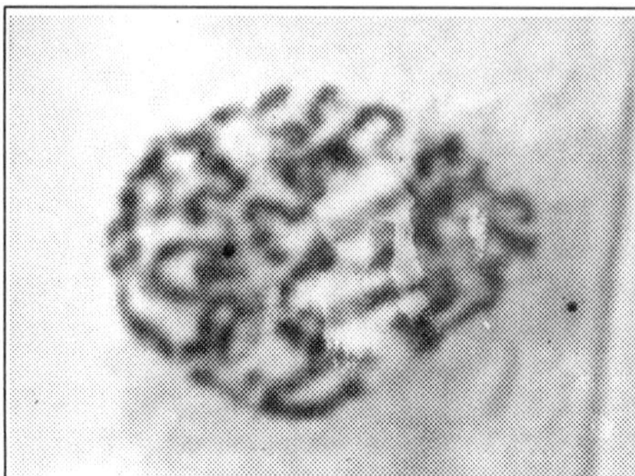

Fig. 2.1

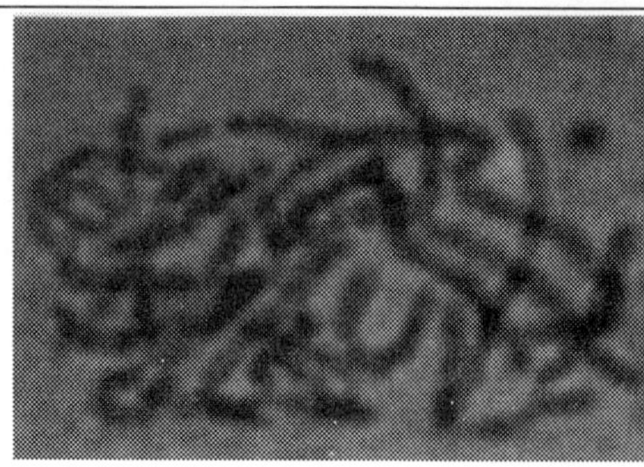

Fig. 2.2

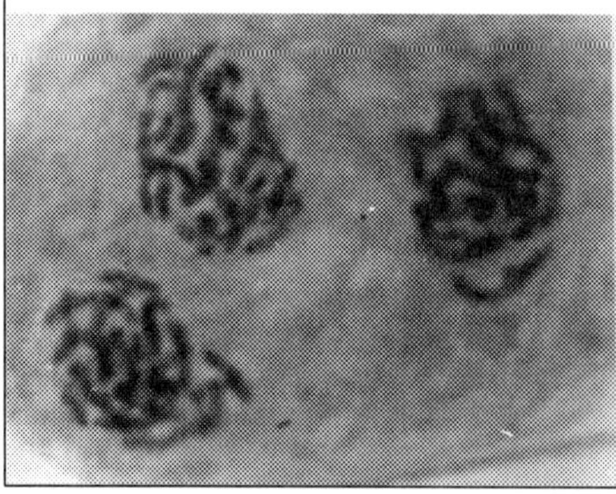

Fig. 2.3

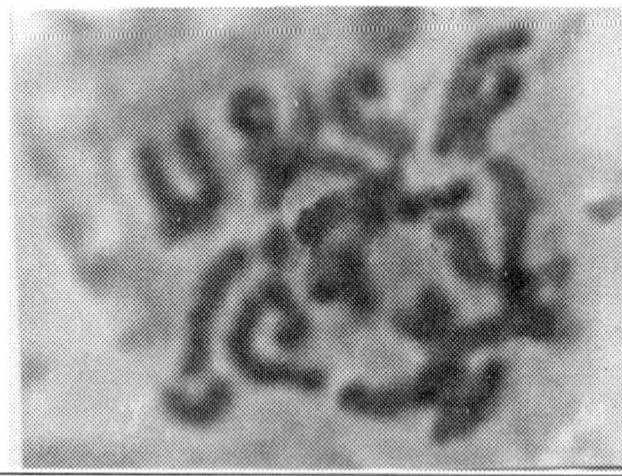

Fig. 2.4

Plate 13.2

Figures 2.1–2.4: *Rhizoclonium hieroglyphicum* var. *macromeres* showing different stages of prophase (magnification Figure 2.3: X 2400 and rest three x 3600).

From the table under reference it is also evident that *R. hieroglyphicum* exists in nature at different level and that three morphologically distinct taxa at hand share a common chromosome number. Therefore, the chromosome number seems to be of no taxonomic importance at least in the present context.

Polyploidy is prevalent in the order Cladophorales and it has been a successful means of speciation within the group (Godward, 1966). Additional means of asexual reporoduction and rapid vegetative propagation among the members of the order are well suited for the establishment of new polyploids. But, it does not seem applicable in the present instances where three distinct taxa are sharing a common count *i.e.* 2 = 24 (Plates 13.1–13.5). It may be possible that the structural manipulation within the chromosomes without involving any numerical changes might have taken place underlying the sub speciation of *R. hieroglyphicum*. Observed karyotypic differences (Table 13.3) among three karyotypes are consistent with the present assumption.

Taking all the karyotypic parameters together, no conclusion can be drawn as regards the relative primitiveness or advancement of the taxa under consideration.

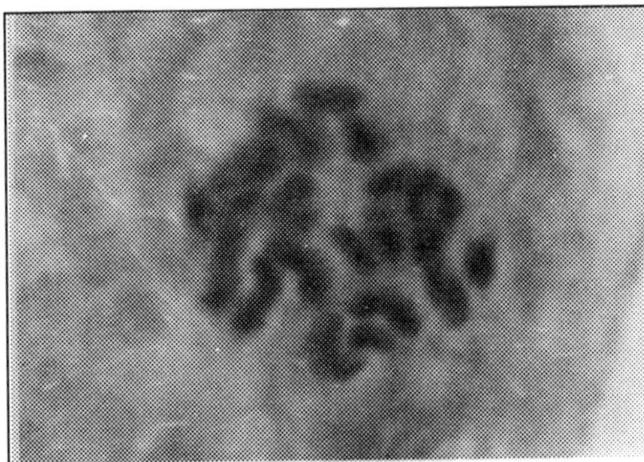

Fig. 3.1

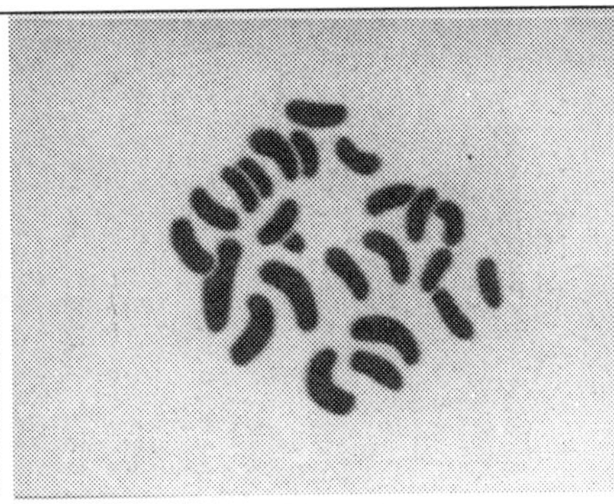

Fig. 3.2

Fig. 3.3

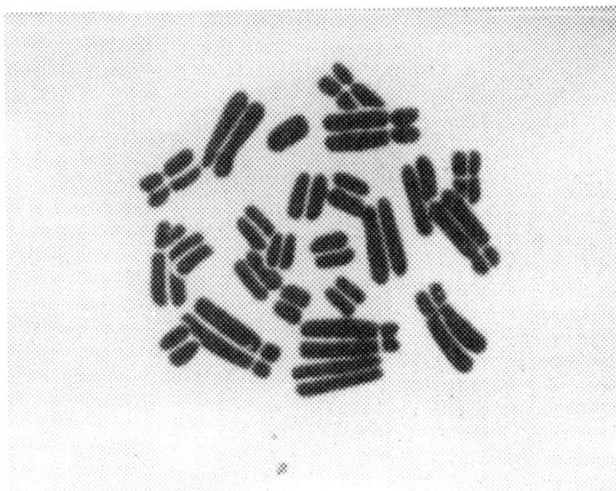

Fig. 3.4

Plate 13.3: Metaphase Showing Chromosome Number n = 24 in *Rhizoclonium hieroglyphicum*

Figures 3.1: Mitotic metaphase chromosomes (X 3600) and 3.2: Its outline drawing; 3.3: Mitotic chromosomes showing lengthwise splitting into chromatids (X 3600) and 3.4: Its outline drawing.

However, the karyotype of *R. hieroglyphicum* with the lowest C.V. seems most symmetrical and hence most primitive. Increased C.V. in rest two taxa entails involvement of chromosomal rearrangement. This assumption is based on the fact that the C.V. and TF per cent among the three karyotypes have increased parallel to each other (Table 13.3).

Distinction of three investigated taxa on the basis of morphology seems justified also on the basis of nuclear cytology. Possession of almost equal number of nuclei per cell and sharing of a common count apparently go against their existence as a distinct taxonomic unit but difference in their minute karyotypic details (Table 13.3), amply support it. As indicated in the idiograms (Plates 13.6), the chromosomes in each of the three karyotypes are graded suggesting their derivation from one another. Conservation of single count by all three taxa under reference further support the

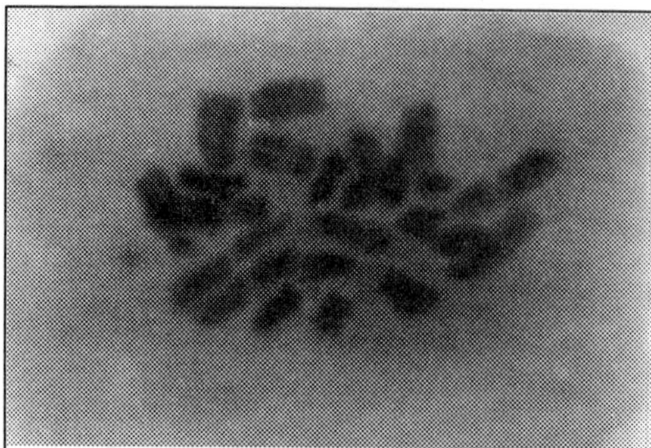

Fig. 4.1

Fig. 4.2

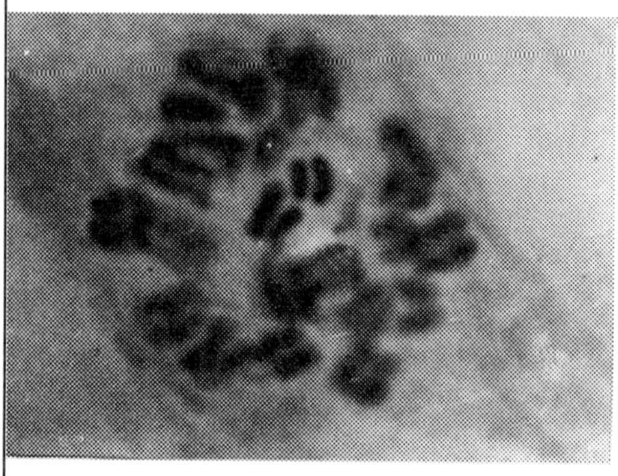

Fig. 4.3

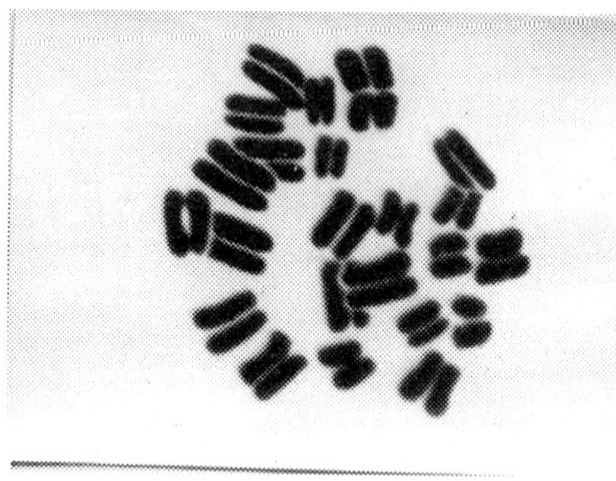

Fig. 4.4

Plate 13.4: Metaphase Showing Chromosome Number n= 24 in *Rhizoclonium hieroglyphicum* var. *macromeres*

Figures 4.1 and 4.3: Metaphase chromosomes with distinct lengthwise splitting into chromatids (X 3600); 4.2 and 4.4: Outline drawing of 4.1 and 4.3 Respectively.

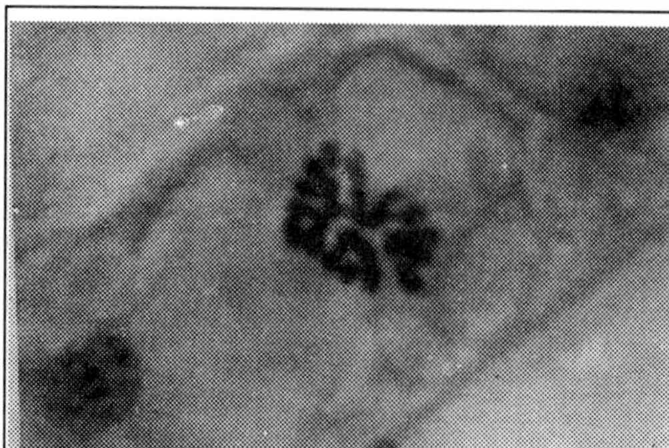

Fig. 5.1

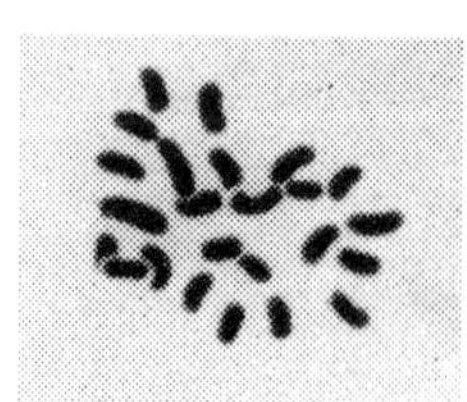

Fig. 5.2

Plate 13.5

Figure 5.1: Metaphase Showing Chromosome number n = 24 in *Rhizoclonium hieroglyphicum* var. *subterrestra* (X 3600) and 5.2: Its Outline Drawing

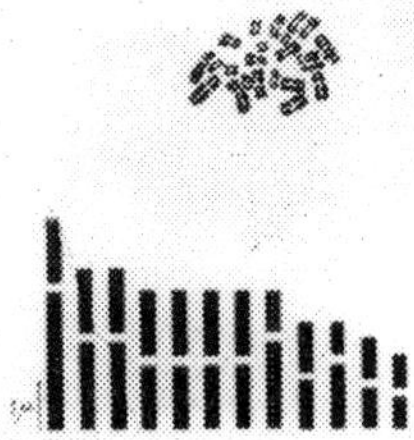

Fig. 6.1

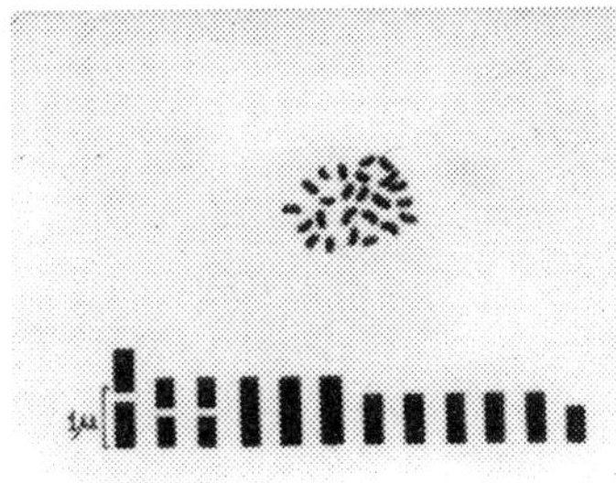

Fig. 6.2

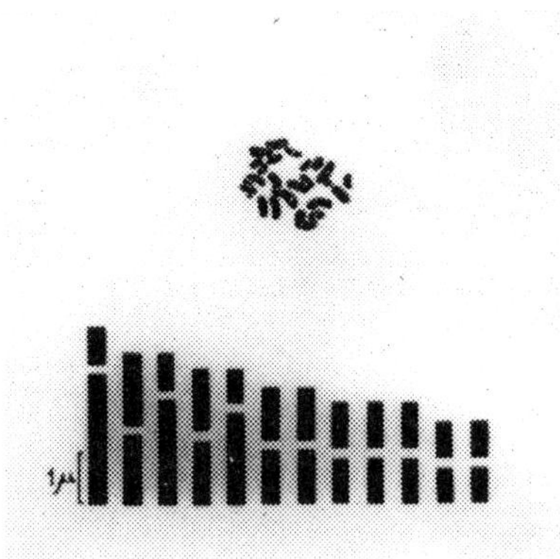

Fig. 6.3

Plate 13.6: Idiograms

Figures 6.1: *Rhizoclonium hieroglyphicum* var-*macromeres;* 6.2: *R. hieroglyphicum* var. *subterrestra*; 6.3: *R. hieroglyphicum.*

recognition of *R. hieroglyphicum* var. *subterrestra* and *R. hieroglyphicum* var. *macromeres* as the subtaxa of *R. hieroglyphicum*.

Summary and Conclusions

Three taxa *viz. R. hieroglyphicum., R. hieroglyphicum* var. *subterrestra* and *R. hieroglyphicum* var. *macromeres* have been collected from Ranchi (Bihar) and their karyomorphological studies have been made.

Number of nuclei per cell has been considered as no reliable taxonomic tool in the genus *Rhizoclonium*. Association of single nucleolus with a set of 12-chromosomes suggested by Sinha (1958) has been confirmed. Chromocenter link structure in association of nucleolus has been regarded as the N.O. region of the chromosome.

Table 13.4: Chromosome Numbers Reported by the Authors and Previous Workers in the Investigated Taxa

Sl.No.	Name of the Taxon	Present count	Previous Report		
			Chromosome No.	Author	Year
1	*R. hieroglyphicum* (Plate 13.1, Figures 1.1 and 1.2)	24	30	Peterschilka	1923
			24	Geitler	1936
			24	Sinha	1958
			24	Venkataraman and Natarajan	1959
			24	Verma	1969
			24 and 48	Patel	1961
			24 and 36	Balakrishnan	1961
			24	Noor	1965
			28 and 36	Prasad *et al.*	1973
2	*R. hieroglyphicum* var. *subterrestra* (Plate 13.3)	24	24	Verma	1969
3	*R. hieroglyphicum* var. *macromeres* (Plates 13.4 and 13.5)	24	24	Verma	1969

A count of N = 24 has been ascertained for all the three taxa under reference. Present count in the first named taxon confirms some of the pervious reports and differs from some others. The present counts in rest of the two taxa are new to the science.

On the basis of karyotype studies (Plate 13.6), it has been suggested that the chromosomal differentiation has played a significant role in sub speciation and that *R. hieroglyphicum* is primitive and rest two have been derived from it.

Taxonomic recognition of three taxa has been amply supported on the basis of the differences in their minute karyotypic details *i.e.* total chromatin length, coefficient of variations and T.F. per cent.

References

Balakrishnan, M.S., 1961. A cytotaxonomical investigation of Indian members of Cladophoraceae. *Prof. S. P. Agharkar Comm. Vol.*, pp. 34–40.

Brand, F., 1898. Culturvesuche mitzwi *Rhizoclonium. Arten. Botanischa. Centralbl.*, 73: 193–202.

Carter, N., 1919. The cytology of the Cladophoraceae. *Ann. Bot. Land.*, 33: 467–78.

De Toni, 1989. *Sylloge Algarum*, 1: 264–355.

Geitler, L., 1936. Vergleichende untersuchungen uberden feineren kern and chromosome bau der Cladophoraceen. *Plant*, 35: 530–78.

Godward, M.B.E., 1948. The iron-alum acetocarmine method for Algae. *Nature*, 16: 203.

Godward, M.B.E., 1966. *The Chromosomes of the Algae*. Edward Arnold (Publ.) Ltd., London.

Noor, M.N., 1965. *Ph.D. Thesis*, Ranchi Univ., Ranchi.

Patel, R.J., 1961. *Ph.D Thesis*, London Univ., London.

Peterschilka, F., 1923. Uber die Kernteilung and die vielkernigkeit und uber die Bezichungen zwischen Epithytisums und kern zahl bei *Rhizoclonium hideroglyphicum* Kutz. *Arch. Protistenk*, 47: 325– 49.

Prasad, B.N., Jain, Vijay and Dutta, S., 1973. Cytological studies on the genus *Rhizoclonium*. *Phykos*, 12(1 and 2): 106–113.

Sarma, T.A. and Bala, Vinod, 1983. Cytological studies on the genus *Rhizoclonium* Kutz. *J. Indian Bot. Soc.*, 62(2): 113–119.

Sarma, Y.S.R.K. and Agrawal, S.B., 1980. Effect of gamma-rays on the karyology of *Rhizoclonium hieroglyphicum* Ag. (Kutz.). *Phykos*, 19(2): 190–196.

Sinha, J.P., 1958. *Ph.D. Thesis*, London Univ., London.

Venkataraman, G.S. and Natarajan, K.V., 1959. Propiono-carmine quash technique and chromosome spreading in Algae. *Stain. Tech.*, 34: 233–234

Verma, B.N., 1969. *Ph.D. Thesis*, Ranchi University, Ranchi.

Verma, B.N., 1980. Karyotype analysis in three species of Rhizoclonium. Kutz. *Cytologia*, 45: 433–40.

Chapter 14

Environment Friendly, Nutrient Rich Food of the Millennium–*Spirulina*

Pushpa Srivastava

Department of Botany, University of Rajasthan, Jaipur – 302 004, Rajasthan

Algae in general are aquatic plants well adapted to fresh and alkaline habitats. These plants range from single celled organisms to giant kelps. Most of them are autotrophic plants which manufacture their food by way of photosynthesis. These algae take in CO_2 and liberate O_2 to the atmosphere. This has led to the belief that Oxygenic environment of the earth was created by algae as they date back to 3.5 billion years, perhaps then the only first life force on earth was born. These environment friendly life forms are being tried by NASA (National Aeronautics and Space Administration) to grow algae at space stations to convert CO_2 exhaust of the space station to O_2.

Of all the algae, the most primitive are procaryotic algae, otherwise known as blue green algae. To this group belongs *Spirulina,* (Figure 14.1) a popular food among the early men. Historical documents stand as a witness to the fact that *Spirulina* was in use as human food several centuries back. "Kanembu" people living near lake Chad in Central Africa used to eat "Dihe" which was a hardened cake of sun dried *Spirulina*. Aztecs, the native Mexicans living near the lake Texcoco called it "Sacred Power Plant" due to its high energy providing property. Aztecs were skilled workers, great warriors and proficient in arts who consumed *Spirulina* as food, specially during marathans for instant energy.

Figure 14.1

A: Commercial cultivation of Spirulina;
B: Drying of Spirulina slurry on plastic sheets;
C: Various preparations of Spirulina in the market.

In recent years *Spirulina* has become a popular health food in many developed and developing countries *i.e.* USA, Mexico, Peru, Vietnam, China, Malaysia, Thailand, Singapore and India. The extensive research work on nutritional and clinical aspects of *Spirulina* have proved it to be a most powerful and well balanced health food of the millennium on this planet. NASA found it to be an excellent compact space food for astronauts as 1Kg of *Spirulina* is equivalent to 1000 Kgs of assorted fruits and vegetables. WHO has hailed it as "Greatest super foods on earth".

The excellence of *Spirulina* is due to its protein quality (Table 14.1), which contains all the 8-essential amino acids and quantity 65-72 per cent, which no plant on this planet is known to provide. This highly digestible *Spirulina* yields higher energy than soyabean, corn, egg and milk. Similarly, vitamins and minerals are essential components of our diet. Vitamins in natural foods *i.e.* vegetables and *Spirulina* are biochelated with natural components of food *i.e.* proteins, carbohydrates and lipids and are completely assimilated in contrast to synthetic vitamins which are not conjugated to any biological molecule. *Spirulina* is a whole food source of B-complex vitamins and the only vegetarian source of vitamin B_{12}. It also contains vitamins B_1, B_2, B_3 and B_{12} and vitamin E in addition to folic acid, biotin, incitol and pantothenic acid. Limiting amino acids in *Spirulina* are sulphur containing methionin and cystine, but their level in this alga is higher than in grain seed, vegetable and legume proteins. *Spirulina* also contains sufficient quantity of calcium, magnesium, manganese, potassium, copper, zinc chromium and selenium. More than any other mineral, *Spirulina* is better known for its biochelated iron contents. It has been suggested to be 60 per cent better absorbed than ferrous sulphate which is mostly administered as iron-supplement to human body.

Table 14.1: Chemical Composition of Rajasthan Isolate of *Spirulina* at Burthal–Jaipur

Protein	65 per cent	Carbohydrates	14 per cent
Moisture	7 per cent	Ash	8 per cent
Lipids	6 per cent	Crude Fibre	0.9 per cent
Chla	1.15 per cent	Phycocyanin	14 per cent
Carotenoids	0.37 per cent		
	Nucleic Acids		
DNA	1.1 per cent	RNA	3.3 per cent
	Essential Amino Acids		
Isoleucine	Phenylalinin	Leucine	Thrionine
Lysine	Tryptophan	Methionine	Valine
	Minerals		
Calcium	10 per cent	Magnesium	10 per cent
Potassium	3 per cent	Phosphorus	9 per cent
Sodium	1 per cent	Iron	0.080 per cent

Spirulina has a beautiful spectrum of colourful natural pigments including chla and phycobilinins *i.e.* phycocyanin, phycoerythrin and allophycocyanin, above all yellow orange carotenoids. Besides, *Spirulina* is a rich source of beta-carotene, the precursor of vitamin A. Its carotenoid contents are 20 times higher than that of carrots, it contains the 9-cis isomeric form of beta-carotene which is more bioavailable than synthetic beta-carotene. Increase in cancer rate has been reported to be due to environmental pollution and change in dietary habits as well. Beta-carotene being the main source of vit.A prevents formation of carcinogens, increases detoxification and controls transformation of cell replication, expression of malignancy, cell differentiation and cell to cell communication. The US Research Council has recommended beta-carotene rich vegetables to reduce cancer risk and *Spirulina* is known to have rich store of β-carotene. Chlorophyll, commonly referred as "green blood" is as high as 1.1 per cent in *Spirulina*, higher than any other plant. The phycocyanin evolved a billion years before chlorophyll. It has both magnesium and iron in its molecular ring and might be parental molecule of both chlorophyll and haemoglobin. It is suggested that phycocyanin gives a cleansing effect to the kidneys.

Spirulina further contains 15-20 per cent carbohydrate which offer quick energy without taxing pancreas. The enzymes catalyse several important reactions in body. The dried *Spirulina* powder contains a number of enzymes including remarkable enzyme superoxide dismutase (SOD). 90 per cent of degenerative diseases such as diabetes, cataract, atherosclerosis, arthritis, aging and cancer are caused by free radicals (FR), which attack normal functioning of DNA and RNA and form harmful lipid peroxides. *Spirulina* has antioxidant property to scavenge free radicals and replenishes body's own antioxidant SOD inhibiting lipid peroxides to below minimum levels, thus controlling secondary oxidative damage. It further contains glycolipids and sulpholipids which are found to be remarkably active against AIDS virus.

The human body requires dietary source of polyunsaturated fatty acids. 10gm of *Spirulina* contains an average of 225 mg of essential fatty acids in the form of linolenic and gamma lenolenmic acid (GLA). GLA is the precursor of body's prostaglandins, the master hormone, which helps to regulate blood pressure, cholestrol synthesis, inflammation and cell growth. GLA deficiency figures in many diseases and health disorders. Hence, there can not be a better source of GLA than *Spirulina* to overcome health problems.

Thus after ruling the world as a nutrient supplement in dietary food of human beings over and above half the century, *Spirulina* has now entered the field of pharmaceuticals with the advent of new century. It is amply being used in therapeutic and chemopreventive application. The metabolic diseases like diabetes, hypercholestrolemia, obesity hepatitis, pancreatitis, arthritis and hypothyroidism. The chemopreventive treatment include cancer and HIV prevention, nutritional blindness and nutritional anaemias. Perhaps, this is the only known to the date, which has such enormous utilisation as nutrient source and is also of medicinal significance.

The multifacet uses led to the commercial cultivation of *Spirulina* in the world as a whole. Earthrise company of USA is in lead. However, there is hardly any developing or developed country, which has not resorted to commercial level production or family size farming. In India, the initiation was made at CFTRI (Mysore) under Indo German Collaboration Project. It was soon followed by MCRC (Chennai) and NBRI (Lucknow) the village level production was started at Sevagram (Wardha). The latest of the series is at Burthal, a village of Jaipur, Rajasthan, followed by at Halvad, a taluka of Surendra Nagar, Gujrat, Figure 14.1. Due to extremes of seasonal variation in temperature, light intensity and photoperiods in semi arid zone of Rajasthan and Gujarat States, cultivation of *Spirulina* was an enigma. This spiral blue green alga is the high priced plant in National and International market, but it is a very delicate one which grows under limited physical and chemical factors. A technology was developed at the University of Rajasthan, round the year cultivation of *Spirulina* in semiarid environ of Rajasthan and Gujarat. The *Spirulina* cultivation from the laboratory was transferred to land at Burthal village of Bassi under Jaipur district of Rajasthan where ST/SC women are cultivating *Spirulina* as an additional source of income, selling it at a rate of Rs.700/- Kg. Similarly, at Halvad taluka under Surendra Nagar district of Saurashtra–Gujarat, earth quake affected women are cultivating *Spirulina* following the same technology developed at Botany Department of Rajasthan University. The unit is spead over 4000m^2 land with a cultivation area of 600 m^2, where women are selling *Spirulina* flakes at a rate of Rs.800/Kg and powder as Rs.900/Kg. Production of *Spirulina* is cost effective. The productions of 3.6 Kg of *Spirulina* costs Rs.1075.00 and sale value at the rate of Rs.800/Kg comes out to be Rs.2880.00 including labour charges with a net profit of Rs.62.67 per cent. Both these projects at Rajasthan and Gujarat have been sponsored by Department of Biotechnology–Government of India, New Delhi. The best part of these units at Burthal and Halvad is that the women have started self entrepreneurship by making preparations out of *Spirulina i.e.* Biscuits Snacks, Papads and Capsules of 250 and 500 mg potency etc.

Spirulina has taken in its folds purification of atmosphere by release of O_2 and supplementation of nutrition to malnourished children and anaemic mothers and medication for number of diseases in human beings. Once, the compounds responsible for curing the various disease are identified, *Spirulina* will prove the best gift of the millennium to the ever increasing population of the world.

Chapter 15

Algal Biofertilizer: An Overview

Abhay Kumar[1] *and K.K. Sahu*[2]

[1]*Reader in Botany, Samastipur College, Samastipur, Bihar*
[2]*Reader in Botany, B.R.B. College, Samastipur, Bihar*

Sustainable agriculture can be defined as an environmentally safe agro-ecosystem ensuring economic viability on long term basis. It can be achieved by: (*i*) Reducing the cost of production, (*ii*) Delaying the process of resource depletion, (*iii*) Minimizing the negative environmental impact.

Unfortunately, agricultural practices adopted today defy all the values of sustainability. The introduction of high yielding varieties has made the use of heavy amount of inorganic fertilizer which is inevitable and also require heavy irrigation. The trend is increasing fastly and it is expected that existing consumption of inorganic fertilizer (130 lakhs tones/annum) will rise to 233-243 lakhs tones/annum by the end of current century alone in our country.

If we look total investment in the fertilizer industry at the global level, it comes to about US $ 15 in 600 fertilizer plants consuming 2 million barrels of crude oil per day. This trend is alarming for the reason below:

- ☆ Cost of fertilizer is increasing tremendously adding to the cost of cultivation.
- ☆ Heavy energy crisis due to mushroom growth of inorganic fertilizer industries.
- ☆ Rapid depletion of nonrenewable energy sources like oil, naptha, natural gas sulphur etc.

- ☆ Production of inorganic fertilizer invokes varieties of pollutants posing environmental threat.
- ☆ Impact of inorganic fertilizers poses a heavy foreign exchange burden on the country.
- ☆ Inorganic fertilizer based cultivation degrades the soil environment by depleting nutrient resources.
- ☆ Due to heavy cost of inorganic fertilizer, small and marginal farmers are bypassed.

Degradation of soil through continuous use of inorganic fertilizer can be prevented only by introducing self regenerating fertilizer of biological origin among which Biofertilizer deserves special mention. The idea behind this concept is based on the practices of transportation of soil from a legume crop field for the cultivation of legume by farmers in some parts of our country. The science has added credence to the concept by developing providing strains of these organism, worked out their compatibility with different crops and suggested various methods of increasing the number of these organisms in the soil.

The term biofertilizer is generally extended to the preparation of live or latent microbial population for field application to mobilize or to enrich the soil. The induced efficient management of the crop nutrients leads to long term sustainability in crop production.

Bacteria, blue-green algae and to a lesser extent green algae are generally used as biofertilizer. Amongst the two nitrogen fixing biofertilizer. Cyanobacteria has an edge over the bacteria because of trophic independence both at the fixation of 'C' and 'N' levels and in having the potency to fix atmospheric nitrogen even in water logged condition–a condition very much prevailing in the rice field. Saprophytic cyanobacteria has been reported to grow and fix in upland (moist and shaded soil surface).

Blue-green algae, also known as cyanobacteria are a group of photosynthetic prokaryotic organisms showing striking similarity both at morphological and physiological levels with bacteria. They are cosmopolitan in distribution and are able to withstand extremes of temperature and moisture stress because of the presence of a higher proportion of main valency bonds in their proteins and protection provided by the polysaccharide envelope around their trichomes. Their main mode of reproduction is vegetative but asexual mode of reproduction is also common brought about by formation of exospore, endospore and akinetes. Sexual reproduction and model stage in the life cycle has not been reported. The food reserve is glycogen. Members of Nostocales and Stigonematales possesses highly specialized cells called heterocysts which are modified regetative cells having thick wall and have no pigments system II. These cells have nitrogenase enzyme which help in N_2-fixation. Unicellular and non-hetecocystous forms have also been found to fix nitrogen under microaerophelic conditions. Many heterocystous forms like *Anabaene, Nostoc, Cylindrospermum, Tolypothrix, Calothrix, Scytonema, Hapalosiphon, Stigonema* and *Westiellopsis* belong to order Stigonematales and Nostocales and many non-heterocystous cyanobacteria are now known which fix atmospheric nitrogen under anaerobic or microaerophilic condition except *Gloeocapsa*.

Cyanobacteria have been reported to fix nitrogen in the range of 18-45 kg ha^{-1} under field condition. This benefit seems to be around 25 kg N ha^{-1} $season^{-1}$. In addition to N_2 fixation, cyanobacteria also synthesize and liberate growth promoting substances (Kumar 1985) like auxin, gibberllin and vitamins. They also add a organic matter to the soil due to being photoautotrophic, the process of photosynthesis provides oxygen to the submerged roots and also helps in reducing oxidisable matter content in the soil. The mucilaginous sheath around the trichomes binds the soil particles and improves the soil aggregation. Increased size of aggregates improves the water holding capacity and aeration of soil. Cyanobacteria have also been reported to solubilize the rock phosphate both *in vitro* and *in vivo*. Repeated application of cyanobacteria also maintains level of organic carbon in soil. It reclaim the acidic (saline) soil and also relieves the soil from the problem of iron toxicity due to prolonged water logging by converting the soluble form of Iron compound–Fe^{++} to the insoluble term Fe^{+++}. In addition to the aforesaid contribution, cyanobacteria also release quite good amount of amino acids, proteins, sugar, vitamins and growth promoting substance partly during active growth cycle and party after death and decay (Kumar 1985).

It is important to mention that the biofertilizer are not used by the crops but its product are utilized hence there is always a possibility of making of its population in the soil which may sustain the benefits after a few super imposed application. These biofertilizer add upto 25 kg N ha^{-1} $season^{-1}$. The blue green algae provide the nutrients in the soil in a gradual and linear manner thus reduces the loss and helps the crop to utilize more of the applied nutrients. Thus algal application has been shown to increase the crop yield by 25-30 per cent. The effect of algal application is more pronounced at lower level. In presence of higher levels of fertilizer nitrogen, the supplementation effect is up to 25 per cent.

Nutrient Availability by Algae

Nitrogen

- Total nitrogen fixed industrially is estimated only 40 metric tonns as against those fixed biologically as 140 metric tonns.
- Percentage of Nitrogen fixed by Cyanobacteria in paddy fields is dependent mainly on chemical, climatic and biotic factors. It has been reported that N2 fixed in different countries are as follows: 18-23kg/ha in Phillipines, 15-53 kg/ha in India and 11-28 kg/ha in Japan.
- About 18-45 (average 25) kg of N2/ha/season has been reported to be fixed by cyanobacteria under field condition.
- The N_2 which is fixed by cyanobacteria is released either by exudation or by microbial decomposition.
- Nitrogen uptake studies with *Aulosira* sp. indicates about 39 per cent of N fixed spread on field soil and 51 per cent is utilized by rice crop.

Phosphorus

- It is second most important material required to exploit full potential of high yielding variety crops. It is also important because it is constituent of number of vital enzymes.
- However a good share of phosphorus applied is the field is fixed and is unavailable to the crop.
- Solubilization of mussorrie rock phosphate (a source of P_2O_5 and raw material for fertelizer industry) has been reported by BGA (*Tolypothrix tennuis, Scytonema cincinnatum* and *Hapalosiphon forntinalis*) *in vitro* and is expected also *in vivo*.

Carbon

- Repeated application of BGA also enrich soil by contributing organic carbon content through accumulating its dead biomass.
- The increase in carbon content of saline soil in Andhra Pradesh has been calculated upto 22 per cent compounds like amino acids, proteins, sugar, complex polysaccharides, vitamins and growth promoting substances that also stimulate growth of rice crop.

Effect of Algalization on Rice Crop

The effect of algalization in the crop yield is both substituting and supplementing type. It does not differ with the crop variety and soil types. However in acidic and saline-alkali soils, suitable soil amendments ensure the expected effect. Several workers like De(1939), Watanabe (1960), Singh (1961), Venketaraman (1966, 1979), Goyal (1987, 1993) etc. have suggested spontaneous fertility of the rice fields soil particularly in tropical regions.

About 265 per cent and 195 per cent increase in crop yield after algalization in absence and presence of 45 kg ammonium sulphate ha^{-1} has been reported by Rao *et al.* (1963) Venkataraman and Goyal (1969) also reported that at 112 kg N/ha algalization increased the crop yield by 12 per cent. Various workers (Subramanayan *et al.*, 1964, Goyal and Venkataraman 1971, Venkataraman 1972, Roger and Kalasooriyal 1981, Goyal 1987, 1993) have reported increase in yield from 10-40 per cent due to algalization.

Algalization has also been found to be effective under upland condition (Rao *et al.*, 1972) and this effect can be further enhanced by the use of saprophytic cyanobacteria which grow luxuriantly on moist and shaded soil surface (Kolte and Goyal 1985, Goyal, 1989).

From the observation of Goyal (1987, 1993), it may be concluded that algalization is as a supplement to nitrogenous fertilizer and the effect is more pronounced at lower fertilizer nitrogen level.

Establishment of Algal Biomass in the Crop Field

Unlike the symbiotic Rhizobium, Cyanobacteria are exposed to natural and man made vagaries of environment outside the crop plants. However they thrive efficiently on account of their high endurance and adaptability.

pH–Cyanobacteria grow over a wide range of pH ranging from 3.8-10.5 but the optimum pH range from 6.5-8.5.

Temperature–The optimum temperature for the growth of cyanobacteria ranges from 30-400C but there are several member which colonize in hot springs and snow. The temperature of rice crop field generally lie in between 30-40°C.

The establishment of algae is greatly influenced by type, quantity and mode of application of nitrogenous fertilizer and pesticides in the field. Generally the Nitrogen level which limits the contribution to enhance the biomass production, most of cyanobacteria show concentration dependent increase in growth and nitrogen uptake upto about 75 g N ml^{-1} (Goyal and Marwaha 1985; Kolte and Goyal 1989). The metabolization of nitrogen helps in preventing the loss of a part of applied nitrogen. The reactivation of nitrogenase activity on reduction of the concentration of the combined nitrogen in the medium leads to increased nitrogen fixation by the increased algal biomass produced in presence of the fertilizer nitrogen. Thus we see that algalization in presence of fertilizer nitrogen not only augments but also increases the availability of this important crop nutrients.

Several types of herbicides, insecticides, fungicides, organophosphate group of pesticides are spread in the crop field. Some common herbicides used by the farmers are ordram, stam, trifularin, butachlor benthiocarb, pandimethalin, oxadizon and 2,4, D. These herbicides do not interfere with the biological nitrogen fixation activity of B.G.A.

Similarly insecticides like Gamma, BHC, Phorate, Carbofuran, Endosulphan, Malathion and Parathion are used. These insecticides also do not interact with the blue-green algae at the recommended dose (Venketaraman and Rajalakshmi 1971, 1972, Sardeshpandey and Goyal 1982, Goyal 1982, 1986, 1989, 1993). Fungicides like MBC, Diflotan and Captan also do not limit the growth of blue green algae.

Organophosphate group of pesticides within the prescribed limit of concentration have been found to promote algal growth.

A review of the various regulatory and promotory environmental stress exerted by various pesticides, insecticide and fungicides, it is necessary to choose only those pesticides which are compatible to blue green algae and step should be taken to develop algal strain with wider tolerance.

Cultivation of Algal Mass

Being photolithotrophs, blue green algae require light, water and inorganic nutrients for their growth. N_2 fixing cyanobacteria do not require any nitrogen source in the medium because they obtain it from atmosphere. However phosphorus is an essential and indispensable nutrients as it forms an important component of many vital enzymes. Various culture methods have been suggested for cultivation of algal mass.

Open air tank culture method is invariably practiced to produce soil based algal inoculum as it is easily adaptable and cost effective. But some of the elements related with open air tank culture methods are:

1. Time consuming
2. Climate dependent
3. Easily infested by parasites and predators
4. Presence of desired organism in appropriate quantity.
5. Bulk of inoculums required per unit area is another limitation which substantially increase handling cost of productions.

Rural based method for bulk production of cyanobacterial inoculums was developed at IARI; New Delhi. This method can be scaled up by farmers depending upon their requirements.

Some other methods are usually adopted for production of desired algal biomass.

1. Trough method
2. Pit method
3. Field scale method
4. Nursery cum Algal production method.

Efficacy of Production Unit

Adequate information is available on different aspect of large scale production of algae like Chlorella, Coelastrum, Micractinium, Dunaliella, Scenedesmus and *Spirulina*. The average yield potential for different algae reported by different workers ranges between 7-20 gm dry wtm-2 day-, (exception Dunaliella–60 gm dry wt m day) In case of *Spirulina* the average production in about 10 kg dry wt m^{-2} day ton a pond area of 1000 m^2.

From the different literature it has been established that mixing of fresh algal biomass with the carrier material in equal proportion yields very good quality of inoculums with longer period of life when stored in ambient condition. This also ensures better establishments of the inoculums in the field. These inoculums start growing within three days in enrichment culture.

References

Becker, E.W., 1994. *Microalgae: Biotechnology and Microbiology,* (Eds.) Sir Tames Baddiley, Carey, N.H., Higgins, I.J. and Potter, W.G., Pub. Cambridge Univ. Press, Cambridge.

De, P.K., 1939. The role of blue-green algae in nitrogen fixation in rice fields. In: *Proc. Roy. Soc.,* London, B127: 121.

Goyal, S.K., 1982. Blue-green algae and rice cultivation. In: *Proc. Natl. Symp. BNF,* Indian Agricultural Research Institute, New Delhi.

Goyal, S.K., 1986. Interaction between pesticides and cyanobacteria. In: *Proc. Symp. Curr. Status of BNV*, Haryana Agri. University, Hissar.

Goyal, S.K., 1987. Algal biofertilizer: An appraisal. *J. Maharashtra Agri. Univ.*, 12(1): 11.

Goyal, S.K., 1993. Algal biofertilizer for vital soil and free nitrogen. In: *Proc. Indian. Natl. Sci. Acad.*, B59(3–4): 295.

Goyal, S.K. and Marwaha, T.S., 1985. Effect of ammonium nitrogen on growth, absorption and fixation of nitrogen by *Anabaena fertilissima. Phykos*, 24: 152.

Goyal, S.K. and Venkataraman, G.S., 1971. Effect of algalization on high yielding rice varieties II. Response of soil types. *Phykos*, 10: 32.

Kolte, S.O. and Goyal, S.K., 1985. Distributional pattern of blue–green algae in rice field soils of Vidarbha region of Maharashatra State. *Phykos*, 24: 156.

Kolte, S.O. and Goyal, S.K., 1989. Effect of ammonium nitrogen on growth and uptake and fixation of nitrogen by blue-green algae. *Acta Botanica Indica*, 17: 172.

Kumar, A., 1985. Studies on extra and intracellular products of some common algae of Muzaffarpur, Bihar. *Ph.D. Thesis*, Bihar University, Muzaffarpur.

Roger, P.A., Sing,h D.N., Venkataraman, G.S. and Goyal, S.K., 1972. Effect of algal inoculation an Jaya rice variety under field conditions. *Lab. J.Sci. Technol.*, 10A(3–4): 107.

Reyanault, J., Sasson, A., Pearson, H.W. and Stewart, W.D.P., 1975. Nitrogen fixing algae in Moroco. In: *Nitrogen Fixation by free living Microorganisms*, (Ed.) W.D.P. Stewart. Cambridge Univ. Press, Cambridge, pp. 339–348.

Sahu, K.K., 1990. Screening of antibiotic induced morpho-physiological variation in certain blue-green algae. *Ph.D. Thesis*, Bihar Universty.

Sardespade, J.S. and Goyal, S.K., 1982. Effect of insecticides on growth and nitrogen fixation by blue-green algae. In: *Proc. Natl. Symp. BNV*, Indian Agricultural Research Institute, New Delhi.

Singh, R.N., 1961. *Role of Blue-Green Algae in the Nitrogen Economy of Indian Agriculture.* Indian Council of Agricultural Research, New Delhi.

Subramanyan, R., Relwani, L.L. and Manna, G.B., 1964. Observation on the role of blue-green algae on rice yield compared with that of conventional fertilizers. *Curr. Sci.*, 33(16): 485.

Sundara Rao, W.V.B., Goyal, S.K. and Venkataraman, G.S., 1963. Effect of inoculation of *Aulosira fertilissima* on rice plants. *Curr. Sci.*, 32: 366.

Venkataraman, G.S., 1966. Algalization. *Phykos*, 5: 164.

Venkataraman, G.S., 1972. *Algal Biofertilizer and Rice Cultivation.* Today and Tomorrows, New Delhi.

Ventakaraman, G.S., 1979. Algal inoculation of rice fields. In: *Nitrogen and Rice.* International Rice Research Institute, Manila, Philippines.

Venkataraman, G.S. and Goyal, S.K., 1969. Some recent observation on the effect of nitrogen fixing blue-green algae on crop plants. *Mikrobiologiya,* 38(4): 709.

Ventakataraman, G.S. and Rajalakshmi, B., 1971. Tolerance of blue-green algae to pesticides. *Indian J. Agri. Sci.,* 42(2): 119.

Watanabe, A., 1960. Collection and Cultivation of nitrogen fixing blue-green algae and their effect on growth and crop yield of the rice plant. In: *Proc. Symp. Algology,* ICAR, New Delhi, pp. 162–166.

Watanabe, I. and Cholitkul, W., 1979. Field studies on nitrogen fixation in paddy soils. In: *Nitrogen and Rice.* Interactional Rice Research Institute, Manila, Philippines.

Chapter 16

Algal Spectrum of the River Burhi Gandak at Muzaffarpur

Md Khatibullah Hamidi[1], *K.K. Ambasta*[2] *and R.R. Sinha*[3]

[1]*Department of Botany, Dr. R.M.L.S. College, Muazaffarpur, Bihar*
[2]*Sunshine Prep and High School, Muzaffarpur, Bihar*
[3]*Department of Botany, M.D.D.M. College, Muzaffarpur, Bihar*

Introduction

The river Burhi Gandak is an important perennial river of North Bihar and is a source of drinking and domestic water, irrigation and fish protein for millions of people living here. But due to intense human interference, the ecosystem of the river is being constantly disturbed. Besides municipal wastes from the villages and towns and effluents from the industries, debris as well as residual pesticides and fertilizers from fields with receding flood water, chiefly contribute to such disturbances. Algae are involved in water pollution in a number of ways and as such acquire immense importance in water pollution studies. In the present work, an attempt has been made to study the algal spectrum of the river Burhi Gandak at Muzaffarpur. Muzaffarpur is the most important city of North Bihar. The city wastes are directly or indirectly discharged into the river Burhi Gandak.

Material and Methods

Phytoplankton samples were collected by hauling water through a plankton net (bolting silk no. 22). Periphytic algae were collected by scraping submerged stones, sticks, pilings, macrophytes and other available substrates. The algal samples thus collected were preserved by adding Lugol's solution. The material for the study of the diatoms was, however, treated differently. The diatoms were cleaned according to Braun's method.

Identification of the algae was mainly based on the works of Prescott and Scott (1942), Desikachary (1959), Randhawa (1959), Prescott (1962), Philipose (1959, 1967), Suxena and Venkateswarlu (1968), Palmer (1980), Gonzalves and Joshi(1946), Sarode and Kamat (1984) and Jha (1985).

Observations

Table 16.1 enumerating the list of the algal species identified and their seasonal distribution in the river Burhi Gandak at Muzaffarpur.

Table 16.1: Algae Found in the River Burhi Gandak at Muzaffarpur

Algae	*Summers*	*Rains*	*Winters*
CHLOROPHYCEAE			
Ankistriodesmus braunii, (Naeg.) Brunnth.	+	+	+
A. convolutus, Cord,	+	+	+
A. falcatus, (Corda) Ralfs.	+	–	+
Aphanochaete repens, A.Braun	+	–	–
A. vermiculoides, Wolle	–	–	+
Chaetophora elegans, (Roth) Ag.	–	–	*
C. attenuata, Hazen	+	–	+
Chara canescens, Loiseleur–Deslongchamps	–	–	+
Chlamydomonas sp, Ehrenb.	+	+	+
Chlorella vulgaris, Beijerinck	+	+	–
C. conglomerata, (Artari) Oltmanns	+	–	–
Chlorococcum humicola, (Naeg.) Rabenh.	+	+	–
Cladophora fracta, (Dillw.) Kuetz.	+	+	–
C. glomerata, (L.) Kuetz.	*	–	*
Closterium rostratum, Ehrenb.	–	+	–
C. intermedium, Ralfs.	–	+	+
C. turgidum, Breb.	+	–	–
Cosmarium connatum, Breb.	+	–	–
C. contractum, Kirchn.	+	–	*
C. gonoides, W. and G.S.West	–	+	–
Crucigenia quadrata, Morren	–	–	*
Gonium pectorale, Mueller	–	–	+
Hydrodictyon reticulatum, (L.) Lagerh.	+	–	+
H. indicum, Iyenger	+	–	+
Kirchneriella lunaris, (Kirchn.) Moebius	–	–	+
Nitella tenuissima, (Desv.) Kuetz.	+	–	+
Oedocladium sp. Hirn.	+	–	*
Oedogonium boscii, (Le Cl.) Wittrock	–	+	–

Contd...

Table 16.1–Contd...

Algae	Summers	Rains	Winters
Oocystis pusilla, Hansgirg	+	–	–
Pediastrum duplex, Meyen	+	–	+
Protococcus viridis, Ag.	+	–	*
Rhizoclonium hieroglyphicum (Ag.) Kuetz.	–	–	+
Scenedesmus acuminatus, (Lagerh.) Chodat	–	+	+
S. arcuatus, (Lemm.) Lemm.	+	–	–
S. bijugatus, (Turpin) Kuetz.	*	+	+
S. denticulatus, Lagerh.	+	+	+
S. dimorphus, (Turpin) Kuetz.	*	*	–
S. longus, Meyen	+	–	–
S. obliquus, (Turpin) Kuetz.	+	+	+
S. quadricauda, (Turpin) Breb.	+	–	+
Selenastrum gracile, Riensch	–	–	+
Spirogyra porticalis, (Muell.) Cl.	+	+	*
S. majuscula, Kuetz.	–	–	+
S. communis, (Hass.) Kuetz.	*	–	+
Stigeoclonium lubricum, (Dillw.) Kuetz.	–	+	–
S. subsecundum, Kuetz.	+	–	+
S. tenue, (Ag.) Kuetz.	+	+	+
Tetraedron. muticum, (A.Braun) Hansgirg	+	+	*
Ulothrix subconstricta, G.S.West	+	+	+
U. zonata, (Weber & Mohr) Kuetz.	–	+	+
Zygnema sp., Ag.	–	–	+
CYANOPHYCEAE			
Anabaena circinalis, Rabenh.	+	+	+
A. constricta, Szafer	+	–	–
A. variabilis, Kuetz.	+	+	–
Aphanocapsa pulchra, (Kuetz.) Rabenh.	+	*	+
Chroococcus limneticus, Lemm.	–	*	–
C. minor, (Kuetz.) Naég.	*	–	–
Dactylococcopsis fascicularis, Lemm.	–	–	*
Gloeothece sp., Naeg.	–	+	–
Gloeotrichia natans, (Hedwig) Rabenh.	–	+	–
Gomphosphaeria lacustris, Chodat	+	–	–
Lyngbya calcifera, Bruhl et Biswas	+	+	–
Merismopedia glauca, (Ehr.) Naeg.	–	–	+

Contd...

Table 16.1–Contd...

Algae	Summers	Rains	Winters
M. elegans, A. Braun	*	*	+
M. minima, Beck	+	+	+
M. tenuissima, Lemm.	+	+	–
Microcystis elabens, (Breb.) Kuetz.	+	*	–
Nostoc calcicola, Breb.	+	+	–
N. linckia, (Roth) B. and T.	+	+	–
Oscillatoria curviceps, Ag.	–	+	+
O. formosa, Bory	+	–	+
O. limnetica, Lemm.	+	–	+
O. limosa, (Roth) Ag.	+	–	–
O. minima, Gicklhorn	+	+	+
O. princeps, Vaucher	–	*	+
O. tenuis, Ag.	+	+	–
Phormidium ambiguum, Gomont	–	+	–
Spirulina major, Kuetz.	–	*	+
S. subsalsa, Oersted	+	+	+
Synechococcus elongatus, Naeg.	+	–	–
Synechocystis pevalekii, Ercegovic	+	–	–
BACILLARIOPHYCEAE			
Achnanthes affinis, Grun.	+	+	+
A. microcephala, (Kuetz.) Grun.	+	+	+
Actinella punctata, Lewis	–	–	+
Amphora ovalis, Kuetz.	*	+	+
Caloneis amphisbaena, Cl.	+	*	*
C. silicula v. truncatula, (Ehr.) Cl. (Grun.) Mayer	–	+	+
Cocconeis placentula, Ehr.	+	+	–
Cyclotella glomerata, Bachman	–	+	–
C. meneghiniana, Kuetz.	+	+	+
Cylindrotheca gracilis, (Breb.) Grun.	+	–	–
Cymbella aspera, (Ehr.) Cl.	–	+	+
C. tumida, (Breb.) V.H.	+	+	*
C. turgida, (Greg.) Cl.	+	+	–
Diatoma anceps, (Ehr.) Kirchn.	*	+	+
Fragilaria capucina, Desm.	–	+	+
F. intermedia, Grun.	+	–	+
F. virescens, Ralfs.	+	+	+

Contd...

Table 16.1–Contd...

Algae	Summers	Rains	Winters
Gomphonema acuminatum, Ehr.	–	*	–
G. augur, Ehr.	+	–	–
G. constrictum v. capitata, Ehr. (Ehr.) Cl.	+	+	+
G. lanceolatum, Ehr.	+	+	–
G. parvulum, (Kuetz.) Grun.	*	+	+
G. sphaerophorum, Ehr.	+	*	–
Gomphoneis sp.	–	–	+
Gyrosigma acuminatum, (Kuetz.) Rabh.	+	+	*
G. scalproides, (Rabh.) Cl.	+	+	–
G. spencerii, (Qubk) Griff. & Henfr	+	–	+
Mastogloia smithii v. lacustris, (Thwaites) Grun.	–	–	+
Melosira granulata, (Ehr.) Ralfs	–	–	+
M. juergensii, Ag.	+	+	–
M. varians, Ag.	+	+	*
Navicula anglica, Grun.	+	–	–
N. cari, Ehr.	–	+	+
N. cocconeiformis, Greg.	–	–	+
N. cryptocephala, Kuetz.	+	*	*
N. cuspidata, Kuetz. (Ehr.) W. Smith, Cl.	+	*	–
N. graciloides, May	–	+	–
N. lucidula, Grun.	–	+	–
N. microcephala, Grun.	+	+	+
N. mutica, Kuetz.	+	+	+
N. pupula v. elliptica, (Kuetz.) Hustedt	–	+	+
N. pygmaea, Kuetz.	–	+	+
N. terrestris, Boy. Pet.	+	+	–
N. viriduloides v. lanceolata, (Gandhi) Gandhi	–	–	+
Nitzschia affinis, Grun.	+	+	–
N. apiculata, (Greg.) Grun.	+	+	*
N. capitellata, Hustedt	+	*	+
N. frustulum, (Kuetz.) Grun.	*	+	+
N. palea, (Kuetz.) W. Smith	+	*	+
N. tryblionella, Hantz.	+	+	+
Neidium longiceps v. undulatum, (Greg.) Cl.	–	–	+
Pinnularia divergens, W. Smith	+	+	+
P. molaris, Grun.	+	*	+

Contd...

Table 16.1–Contd...

Algae	*Summers*	*Rains*	*Winters*
Pleurosigma spencerii, S. Sm.	+	+	–
Surirella angusta, Kuetz.	–	+	+
S. ovalis, Breb.	–	+	+
Synedra acus, Kuetz.	+	*	–
S. pulchella, Ralfs ex Kuetz.	+	*	+
S. ulna, (Nitz.) Ehr.	*	*	*
EUGLENOPHYCEAE			
Euglena acus, Ehr.	+	–	–
E. ehrenbergii, Klebs	+	–	–
E. gracilis, Klebs	–	–	+
E. minuta, Prescott	–	+	–
E. spirogyra, Ehr.	+	+	–
Phacus pyrum, (Ehr.) Stein	*	+	+

+: Present, *: Present in abundance, –: Absent.

The observations made on the algal spectrum of the river at Muzaffarpur reveal that the algal population of the river was dominated by three classes namely, the Chlorophyceae, the Cyanophyceae and the Bacillariopyceae. During the period of observation 146 species of algae belonging to 68 genera were identified. The Chlorophyceae were represented by 51 species belonging to 28 genera, the Cyanophyceae by 30 species belonging to 16 genera and the Bacillariophyceae by 59 species belonging to 22 genera. Besides these, 6 species belonging to Euglenophyceae were also recorded. During the summers a total of 99 algal species were recorded in the river water; 34 of which belonged to the green algae, 21 to the blue-greens, 40 to the diatoms and 4 to the euglenoids. The river water during the rains was inhabited by 90 species of algae, out which 21 belonged to the green algae, 20 to the blue-greens, 40 to the diatoms and the rest 3 to the euglenoids. During the winters a total of 92 species of algae were recorded in the river water, 36 of which belonged to the green algae, 13 to the blue-greens, 41 to the diatoms and the rest 2 to the euglenoids.

Discussion

Algal spectrum of the river Burhi Gandak at Muzaffarpur reveals that the maximum number of species was recorded during the summers followed by the winters and the rains. It is also evident that the algal flora was dominated by the diatom species, an observation also made by Venkateswarlu and Sampath Kumar (1982), Sankaran (1984), Bhatt *et al.* (1985), Descy *et al.* (1987), Manikya Reddy and Venkateswarlu (1987), and Srivastava and Sen (1987). However, Govindan and Sundaresan (1979), Somashekar (1984), Bilgrami *et al.* (1985), Sengar *et al.* (1985) and Sengar and Sharma (1985) reported the dominance of the species of the green algae, and Rana and Palria (1988) of blue-green algae in their studies of the algal community of the river water. The diatoms were followed by the species of green algae and the blue green algae. Other groups were represented by only a few species.

Both the clean water and the pollution tolerant species of algae were found the river water. However, the distribution of these algae was not very specific. Algae have been used as indicators of water pollution by a number of workers including Palmer (1969), Govindan and Sundaresan (1979), Venakateswarlu and Sampath Kumar (1982), Somashekar (1984) Sahai *et al.* (1985) and Manikya Reddy and Venakateswarlu (1987). Palmer (1969) made outstanding contribution to this field by preparing list of 60 genera and 80 species of algae tolerant to organic pollution. As per Palmer's list, in the present work 37 genera of algae tolerant to organic pollution were identified which are given below in order of decreasing tolerance: *Euglena, Oscillatoria, Scenedesmus, Chlorella, Nitzschia, Navicula, Stigeoclonium, Synedra, Ankistrodesmus, Phacus, Phormidium, Melosira, Gomphonema, Cyclotella, Closterium, Spirogyra, Anabaena, Pediastrum, Fragilaria, Ulothrix, Surirella, Lyngbya, Oocystis, Merismopedia, Spirulina,Cymbella, Cladophora, Diatoma, Achnanthes, Pinnularia, Chlorococcum, Cocconeis, Cosmarium, Gonium, Selenastrum, Dictyosphaerium,* and *Crucigenia.* The presence and distribution of the above mentioned algal taxa in the river Burhi Gandak is suggestive of the fact that the river has mild organic pollution all sites but specific sites receiving the effluents are more polluted as indicated by the increase in number and kinds of the pollution tolerant forms.

Acknowledgement

We are thankful to Prof. B. N. Verma for providing valuable literature and guidance in identification of the algal taxa.

References

Bhatt, S.D., Bisht, Y. and Negi, U., 1985. Hydrology and phytoplankton populatons in river Kosi of the Western Himalaya (Uttar Pradesh). *Indian J. Ecol.*, 12(1): 141–146.

Bilgrami, K.S., Datta Munshi, J.S., Bhowmick, B.N., Yadava, R.N., Singh, A.K., Singh, D.K. and Pandit, B., 1985. Ecology of river Ganges: Impact of human activities and conservation of aquatic biota (Patna to Frakka). *Final Technical Report, MAB Project*, pp. 97.

Descy, J.P., Servais, P., Smitz, J.S., Billen, G. and Everbecq, E., 1987. Phytoplankton biomass and production in the river Meuse (Belgium).*Wat. Res.*, 21(12): 1557–1566.

Desikachary, T.V., 1959. *Cyanophyta.* ICAR, New Delhi, pp. 686.

Gonzalves, E.A. and Joshi, D.B., 1946. Fresh water algae near Bombay. *J. Bombay Nat. Hist. Soc.*, 46(1): 154–176.

Govindan, V.S. and Sundaresan, B.B,.1979. Seasonal succession of algal flora in polluted region of Adyar river. *Indian J. Environ. Hlth.*, 21(2): 131–142.

Jha, B.C., 1985. Comparative ecological studies of periphytic communities with particular reference to the members of Bacillariophyceae in different freshwater bodies like Getalsud Reservoir and Swarnrekha River of the tribal belt of Ranchi, Bihar. *Ph.D. Thesis*, Ranchi University, Ranchi.

Manikya Reddy, P. and Venkateswarlu, V., 1987. Assessment of water quality and pollution in the river Tungabhadra near Kurnool, Andhra Pradesh. *J. Environ. Biol.*, 8(2): 109–119.

Palmer, C.M., 1969. A composite rating of algae tolerating organic pollution. *J. Phycology*, 5: 78–82.

Palmer, C.M., 1980. *Algae and water pollution*. Castle House Publications Ltd., England, pp. 123.

Philipose, M.T., 1959. Freshwater phytoplankton of inland fisheries. *Proc. Symp. Algology*, 4: 275–291.

Philipose, M.T., 1967. *Chlorococcales*. ICAR, New Delhi, pp. 365.

Prescott, G.W., 1962. *Algae of the Western Great Lakes Area*. WmC Brown Company Publications, Dubuque, Iowa, pp. 977.

Prescott, G.W. and Scott, A.M., 1942. The freshwater algae of the South United States. I. Desmids from Mississippi with description of new species and varieties. *Trans. Amer. Microsc. Soc.*, 61: 1–29.

Rana, B.C. and Palria, S., 1988. Phycological and physico-chemical evaluation of the river Ayad, Udaipur. *Phykos*, 27: 211–217.

Sahai, R., Saxena, P.K. and Jabeen, S., 1985. Ecological survey of the algal flora of polluted habitats of Gorakhpur. *Phykos*, 24: 4–11.

Sankaran, V., 1984. Algae of the Aliyar river. *Phykos*, 23(1,2): 75–79.

Sarode, P.T. and Kamat, N.D., 1984. *Freshwater Diatoms of Maharashtra*. Saikripa Prakashan, Aurangabad, India, pp. 338.

Sengar, R.M.S. and Sharma, K.D., 1985. Investigation on quantitative determination of phytoplankton in Yamuna river. *Phykos*, 24: 117–124.

Sengar, R.M.S., Sharma, K.D. and Pathak, P.D., 1985. Studies on distribution of algal flora in polluted and non-polluted regions in Yamuna river at Agra (U.P). *J. Ind. Bot. Soc.*, 64: 365–376.

Somashekar, R.K., 1984a. Contribution to the algal flora of river Kapila, Karnataka. I. Cyanophyceae and Chlorophyceae. *Phykos*, 32(1,2) 116–124

Somashekar, R.K., 1984b. Contribution to the algal flora of river Kapila, Karnataka. II. Diatoms. *Phykos*, 23(1,2): 125–19.

Srivastava, V. and Sen, N., 1987. Freshwater algae of the river Narmada near Jabalpur. *Phykos*, 26: 78–81.

Suxena, M.R. and Venkateswarlu, V., 1968. Desmids of Andhra Pradesh from Dhramasagar lake, Warangal. II. *J. Osmania University (Science)*, pp. 179–201.

Venakateswarlu, V. and Sampath Kumar, P.T., 1982. Chemical and biological assessment of pollution in the river Moosi, Hyderabad (A.P.), India. *Biol. Bull. India*, 4(1): 23–30.

Index

C

D